The pollution of the earth's freshwater habitats is a topic of major concern. This timely synthesis considers the effects of pollutants on aquatic animals via a series of research and review articles which present experimental evidence of sublethal and lethal effects of a range of toxicants at the physiological, cellular and subcellular levels, and which explore new techniques for detection of pollution damage. Topics covered include routes of uptake of toxicants; the effects of acute and chronic exposure to toxic metal ions, particularly zinc, copper and aluminium, with emphasis on the mechanisms of toxicity and responses to chronic exposure to sublethal levels; the impact on fish biology of two chemicals of current concern, nitrites and polyaromatic hydrocarbons which may act as oestrogenic substances or potent mutagens; and *in vitro* studies of the mechanisms of toxicity at the cellular and subcellular level, including damage of DNA, using cultured fish cells.

T0291577

SOCIETY FOR EXPERIMENTAL BIOLOGY
SEMINAR SERIES: 57

TOXICOLOGY OF AQUATIC POLLUTION
PHYSIOLOGICAL, CELLULAR AND MOLECULAR
APPROACHES

SOCIETY FOR EXPERIMENTAL BIOLOGY SEMINAR SERIES

A series of multi-author volumes developed from seminars held by the Society for Experimental Biology. Each volume serves not only as an introductory review of a specific topic, but also introduces the reader to experimental evidence to support the theories and principles discussed, and points the way to new research.

TOXICOLOGY OF AQUATIC POLLUTION
PHYSIOLOGICAL, CELLULAR AND MOLECULAR APPROACHES

Edited by

E.W. Taylor

Department of Biological Sciences
University of Birmingham

CAMBRIDGE UNIVERSITY PRESS

CAMBRIDGE UNIVERSITY PRESS
Cambridge, New York, Melbourne, Madrid, Cape Town, Singapore, São Paulo, Delhi

Cambridge University Press
The Edinburgh Building, Cambridge CB2 8RU, UK

Published in the United States of America by Cambridge University Press, New York

www.cambridge.org
Information on this title: www.cambridge.org/9780521105774

First published 1996
This digitally printed version 2009

A catalogue record for this publication is available from the British Library

Library of Congress Cataloguing in Publication data

Toxicology of aquatic pollution: Physiological, cellular and molecular
approaches/edited by E.W. Taylor
    p.   cm. – (Society for experimental biology seminar series: 57)
Includes index.
ISBN (invalid) 0–521–45524–X (hc)
1. Water – Pollution – Toxicology. 2. Aquatic organisms – Effect of water
pollution on. 3. Fishes – Effect of water pollution on. I. Taylor, E.W. II.
Series: Seminar series (Society for Experimental Biology (Great Britain));
57.
QH90.8.T68A66 1995
574.5′263 – dc20   95–16556 CIP

ISBN 978-0-521-45524-4 hardback
ISBN 978-0-521-10577-4 paperback

Contents

viii   *Contents*

Contributors

BEAUMONT, M.W.
School of Biological Sciences, University of Birmingham, Edgbaston, Birmingham B15 2TT, UK.

BRAUNER, C.J.
Department of Zoology, University of British Columbia, V6T 1Z4, Canada.

BROWN, J.A.
Department of Biological Sciences, Hatherly Laboratories, University of Exeter, Exeter EX4 4PS, UK.

BUTLER, P.J.
School of Biological Sciences, University of Birmingham, Edgbaston, Birmingham B15 2TT, UK.

CHIPMAN, J.K.
School of Biochemistry, University of Birmingham, Birmingham, B15 2TT, UK.

GEORGE, S.G.
School of Biochemistry, University of Birmingham, Birmingham, B15 2TT, UK.

HANDY, R.D.
Department of Biological Sciences, The University, Dundee DD1 4HN, UK.

HOGSTRAND, C.
McMaster University, Department of Biology, Hamilton, On L8S 4L1, Canada.

JENSEN, F.B.
Institute of Biology, Odense University, Campusvej 55, DK-5230 Odense M, Denmark.

JOBLING, S.
Department of Biology and Biochemistry, Brunel University, Uxbridge, Middlesex UB8 3PH, UK.

LEAVER, M.J.
NERC Unit of Aquatic Biochemistry, Department of Biological and Molecular Sciences, University of Stirling, Stirling FK9 4LA, UK.

LIVINGSTONE, D.R.
NERC Plymouth Marine Laboratory, Citadell Hill, Plymouth PL1
2PB, UK.
MAIR, J.
School of Biological Sciences, University of Birmingham, Edgbaston,
Birmingham B15 2TT, UK.
MUJALLID, M.S.I.
School of Biological Sciences, University of Birmingham, Edgbaston,
Birmingham B15 2TT, UK.
NEUMAN, J.F.
Fisheries Bioassay Laboratory, Montana State University, Bozeman,
Montana 59717, USA.
NUNN, J.W.
School of Biochemistry, The University of Birmingham, Edgbaston,
Birmingham B15 2TT, UK.
OLSSON, P.-E.
Department of Cellular and Developmental Biology, Umea
University, S-901 87 Umea, Sweden.
RANDALL, D.J.
Department of Zoology, University of British Columbia, V6T 1Z4,
Canada.
SUMPTER, J.P.
Department of Biology and Biochemistry, Brunel University,
Uxbridge, Middlesex UB8 3PH, UK.
TAYLOR, E.W.
School of Biological Sciences, University of Birmingham, Edgbaston,
Birmingham, B15 2TT, UK.
THURSTON, R.V.
Fisheries Bioassay Laboratory, Montana State University, Bozeman,
Montana 59717, USA.
TYLER, C.R.
Department of Biology and Biochemistry, Brunel University,
Uxbridge, Middlesex UB8 3PH, UK.
WARING, C.P.
MAFF Fisheries Laboratory, Pakefield Road, Lowestoft, Suffolk
NR33 0HT, UK.
WILSON, R.W.
School of Biological Sciences, University of Manchester, G 38
Stopford Building, Oxford Road, Manchester M13 9PT, UK.
WOOD, C.M.
McMaster University, Department of Biology, Hamilton, On L8S
4L1, Canada.

Preface

This book originated from a symposium on Aquatic Toxicology at the Easter meeting of the Society for Experimental Biology at the University of Kent in Canterbury in 1993. It is in some respects a sequel to the volume entitled *Acid Toxicity and Aquatic Animals* (Seminar Series volume 34), as it is soft, acid waters into which toxic metal ions are leached from rocks and industrial waste. In this previous volume the role of aluminium in determining acid toxicity was discussed, and this account is extended into the current volume and amplified by consideration of the mechanisms of toxicity of other metal species and of the aromatic hydrocarbons, at low to sublethal levels. New approaches to the problem of pollution monitoring are considered, including the construction of large-scale models for the exchange of xenobiotics between fish and their environment, and the alternative use of sensitive cellular and molecular markers of pollution. A contemporary account of this general area is also provided by *Aquatic Toxicology: Molecular, Biochemical and Cellular Perspectives* edited by Donald C. Malins and Gary K. Ostrander, published by Lewis Publishers, 1994. Fortunately, there is little direct overlap between the two accounts, and the overall approach is differerent so that the volumes complement one another and both provide valuable background for students and practitioners of aquatic toxicology.

Pollution of the aquatic environment is a problem with its origins in the urbanization of the human population which, in Europe, accompanied the industrial revolution about 250 years ago. I am writing this preface in Birmingham, at the centre of the industrial West Midlands, a city with no major river, a population of over one million, an economy based on heavy industry and associated high traffic flows. The problems of aerial and aquatic pollution are intense, with most local waterways devoid of fish. Public awareness of the problem is more recent, and the history of pollution control and monitoring has been one of slowly evolving standards and techniques. Although some progress is being made in northern post-industrial countries (for example, the return of salmon has been reported in the Thames) the problem is worldwide and continuing to escalate in the Far East, Asia and Africa to the point where human health has suffered, as in the incidence of Minimato disease in Japan, caused by the ingestion of shellfish contaminated with methyl mercury.

Large numbers of new chemicals are manufactured, then released into the environment each year, and assessment of the potential impact of these xenobiotic (i.e. 'foreign to life') contaminants on aquatic organisms is being afforded a progressively higher priority by many nations and international agencies. Monitoring gross environmental pollution has always been the province of chemists. Biological means of assessment have included observation and quantification of faunal changes in polluted water, with the development of marker species or communities as indicators of of levels of pollution, as well as routine testing of the toxicity of effluents on aquatic animals, typically fish, using the LC_{50} test of lethality. However, it is now recognised that the effects of chronic exposure to sublethal levels of toxicants, with the risk of their accumulation over time or via a food chain, can have important limiting effects upon the survival of individual animals and on the survival or species diversity of communities of aquatic organisms. Accordingly, the assessment of pollution has more recently broadened to include the study of chronic exposure to sublethal levels of toxicants and their effects on vulnerable stages in the life-cycle of species.

For a full understanding of the nature of the problem of aquatic pollution, it is now deemed necessary to study the mechanisms of action of pollutants on living organisms at physiological, cellular and molecular levels. This new approach is the subject of this volume which combines the expertise of fish physiologists interested in the basic mechanisms of respiratory gas exchange, ion regulation and endocrine control with that of cell and molecular biologists similarly interested in the fundamental responses of aquatic organisms to toxic pollutants. It assesses the effects of two major classes of aquatic pollutants, toxic metal ions and in particular aluminium, copper and zinc and the polyaromatic hydrocarbons (PAHs) derived from pesticides and products of the petrochemical industry. Both classes are lethally toxic at high concentrations, but the emphasis is on the study of sublethal effects of chronic exposure to these chemicals and their bioaccumulation in a range of aquatic species, and on the development of new methods of early detection of these more subtle effects.

Differences in the rates of accumulation of organic xenobiotics by fish, based on species differences, activity levels and the effects of environmental variables such as temperature are reported by Randall and Thurston who, together with their coworkers, have developed a model for accumulation of these compounds based on their partition coefficients between water and lipid (representing the membrane components of tissue) which, together with a database for rates of oxygen uptake, will enable them to predict the likely outcome of a measured pollutant load on a fish population. Accumulation of a range of PAHs and related compounds at lower levels have been found to have oestrogenic effects on fish. Their bioaccumulation up the food chain may result in sterilization of males with consequent loss of fecundity. This cause of public anxiety is considered by Sumpter and colleagues. The same range of chemicals can act as genetic toxicants to aquatic organisms, particularly in their active forms following biotransformation to procarcinogens or free radicals. These effects include DNA damage with consequent mutagenic and reproductive effects and are described by Nunn, Livingstone, and Chipman who consider the utility of detection of DNA damage as a sensitive marker of environmental contamination by pesticides and other organic residues.

The use of *in vitro* techniques using primary cultures of fish cells, such as hepatocytes, to detect pollution damage by organic hydrocarbons is explored by George, and this cellular approach is extended to the molecular level by Leaver, who considers the measurement of the levels of cytochrome P450 messenger RNA, protein or enzymic activity in fish as a means of monitoring low levels of organic pollution. This enzyme is involved in biotransformation, by insertion of oxygen, of aromatic hydrocarbons, often rendering them available for excretion, and consequently countering their progressive accumulation in the lipid components of tissues; though the transformed products may be potent mutagens.

This consideration of major topics of current public concern in aquatic toxicology is continued in a chapter on nitrate accumulation by Jensen. Nitrite arises from the incomplete bacterial decomposition of organic waste or from agricultural run-off and is toxic to aquatic animals. It is actively accumulated, apparently over chloride channels, affects oxygen transport and ionoregulation and many be carcinogenic, following its conversion to N-nitroso compounds.

Toxic metal ions at high concentrations cause gross, non-specific gill damage to fish, resulting in loss of ionoregulatory ability and ultimately a breakdown in respiratory gas exchange which proves fatal. These changes resulting from exposure of trout to acutely toxic levels of copper in freshwater are described in detail by Taylor and colleagues, while the effects of zinc pollution are described by Hogstrand and Wood, who also consider the role of zinc in metabolism, routes and mechanisms for its uptake over the gills and following ingestion over the gut wall and regulation of its internal concentration. The role of ingestion in the accumulation of a wide range of metal ions is covered in detail by Handy. Taylor and colleagues progress to a consideration of the effects of exposure to sublethal levels of copper upon the swimming performance of trout. These effects are temperature dependent and apparently associated with problems of oxygen supply to aerobic tissues and accumulation of ammonia, though the mechanisms underlying this accumulation and its effects on aerobic exercise are as yet not understood.

The toxic effects of sublethal levels of zinc and copper seem to reside in their competition for calcium channels in the apical membrane of gill epithelial cells and inhibition of calcium–ATPase on the basolateral membranes. Calcium is important in the regulation of cell permeability and adhesion. The toxic effects of aluminium, including the hormonal control of observed physiological responses to aluminium exposure in trout, are described by Brown and Waring. Aluminium, which is solubilized in soft, acid water, may precipitate on to the surface of the gills due to alkalinization of the microenvironment between the lamellae. Aspects of the complex nature of this microenvironment, which are influenced by active ionoregulatory activity, including an electrogenic proton pump, are considered by several authors, as there are common features of the toxic effects of a range of metal ions on the functioning of the gills in ionoregulation, acid–base regulation and ultimately in respiratory gas exchange.

Wilson considers the effects of chronic exposure of trout to low levels of aluminium, a common feature of acidified upland rivers. They can undergo a process of acclimation which increases their resistance to any subsequent exposure, this is accompanied by cell proliferation on the gills, including the production of mucocytes and subsequent increased mucous production, which will bind metal ions removing them from solution. However successful these processes of acclimation, chronic exposure has costs for the fish which may manifest themselves in reduced appetite, swimming performance, growth rate and fecundity. Another protective response to exposure to toxic metal ions is the production of metallothioneins, proteins which bind toxic metal ions intracellularly. Their occurrence, structure and the endocrinology and molecular biology of their induction in fish are considered by Olsson. He describes techniques for their measurement in the tissues of animals exposed to toxic metals and discusses their use as early indicators of environmental pollution.

Each chapter is written by experts in the field and they have provided an up-to-date list of references to support their texts. The index will enable many of the major themes in the book to be traced within and between chapters. I am grateful to all authors for their hard work on their manuscripts and to Maria Murphy of Cambridge University Press and Linda Kachur of the University of British Columbia for their cheerful patience during the final stages of the preparation of this book.

D.J. RANDALL, C.J. BRAUNER,
R.V. THURSTON and J.F. NEUMAN

Water chemistry at the gill surfaces of fish and the uptake of xenobiotics

Introduction

There is a massive production of chemicals synthesized to meet the demands of industry. Most of these chemicals are foreign to the body (xenobiotics). They enter the environment where they are available for uptake by organisms. Most animals possess enzymes (multi-function oxidases) which reduce the toxicity of the xenobiotics and facilitate their excretion. Many xenobiotics, however, are resistant to metabolic degradation, are accumulated in the body and can be extremely toxic.

Most organic chemicals must enter the body before they can exert their toxic effect(s). Uptake can occur through the food chain or by direct uptake from the environment across the respiratory surface and skin. The uptake of xenobiotics from water by fish is determined by numerous factors, the most important of which are the transfer capacity of the gills and the physio-chemical properties of the compound.

Exchange at the gills

In adult fish, the gills comprise the major surface area of the body, and constitute a thin but continuous barrier between the environment and the blood. Water flows over the gills counter to blood flow ensuring a constant partial pressure gradient of O_2 and CO_2 over the duration of blood transit through the gills. Conditions for diffusion of O_2 are further maximized by the presence of haemoglobin which binds molecular oxygen keeping the PO_2 in the blood low. The gills in fish are hyperventilated with water relative to blood to ensure adequate rates of oxygen uptake from the environment. The content of oxygen per unit volume in water is less than 10% of that in blood and consequently fish at rest maintain a gill ventilation : blood perfusion ratio between 10 and 15 (Cameron & Davis, 1970; Kiceniuk & Jones, 1977; Jones & Randall, 1978). This ensures that oxygen delivery to the gills by water is matched by the oxygen transport capacity of the blood. In general, conditions at the gills in fish are adjusted to meet the metabolic demand for oxygen (Randall, 1990). During exercise ventilation rate, diffusing capacity and cardiac output all increase to maximize gas exchange across the gills (Kiceniuk & Jones, 1977; Randall, 1990). These characteristics which make the gills so efficient for gas exchange, also augment xenobiotic uptake. Xenobiotic uptake from the water is predominantly across the gills (Neely, 1979; Gobas, Opperhuizen & Hutzinger, 1986; Randall & Brauner, 1993) and this constitutes the major exchange site with the water.

Toxicant uptake across the gills can be limited at three sites: delivery to the gills (VENTILATION LIMITATIONS), diffusion across the gill epithelium (DIFFUSION LIMITATIONS) and removal by the blood from the gills (PERFUSION LIMITATIONS) (Fig. 1). Toxicant delivery to the gills is determined by the product of ventilation volume and the water solubility of the toxicant. Diffusion of a xenobiotic across

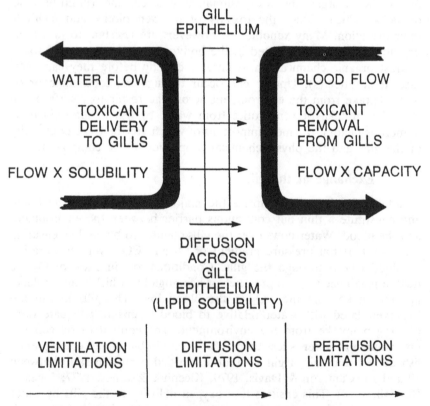

Fig. 1. Sites at which limitations to xenobiotic uptake exist at the gills.

the gill epithelium will depend largely on lipid solubility. Finally, removal of the toxicant by the blood will be determined by the product of blood flow and toxicant solubility in the blood. The amount of the toxicant that can be held in the blood will depend on the extent of binding of the chemical and the lipid solubility of the chemical and lipid content of the blood.

Octanol : water partition coefficient

Many organic substances are much more soluble in lipids than in water. These compounds enter animals because they are lipid soluble and then accumulate in the body fat of the animal. This BIOACCUMUL-ATION of substances in animals due to the high fat solubility of compounds has long been recognized. The relative solubility of chemicals in water and lipids is often measured as the octanol : water partition coefficient (K_{ow}). Octanol was chosen to represent 'biological lipids' and K_{ow} became the standard for quantifying the partitioning characteristics of compounds (Connell, 1990; Niimi, 1991).

In resting fish exposed to numerous non-dissociated chemicals, log uptake rate constant increases approximately linearly over a range of log K_{ow} between 1 and 4, remains constant between log K_{ow} 4 and 6 (Fig. 2, McKim, Schmieder & Veith, 1985) and may decrease above log K_{ow} 6. Assuming the lipid content in fish blood is 5%, during initial exposure to a xenobiotic with log K_{ow} of 2 or higher almost all of the xenobiotic in the blood will be bound to lipids and proteins (Schmieder & Henry, 1988) and the ability of the blood to remove the toxicant will far exceed the ability to deliver it and thus will never be limiting. Thus, uptake of xenobiotics with log $K_{ow} > 2$ will either be ventilation or diffusion limited (Randall & Brauner, 1993). An increase in K_{ow} could result from an increase in lipid solubility or a decrease in water solubility. In general, for chemicals with a log K_{ow} of up to 6, it is a decrease in water solubility which is responsible for an increase in log K_{ow} (Dobbs & Williams, 1983; Verschueren, 1983; Chessells, Hawker & Connell, 1992). Thus, for high K_{ow} chemicals (above log K_{ow} 4) water solubility is very low and uptake may be ventilation limited while at lower K_{ow} (log K_{ow} between 1 and 4) water solubility is relatively high and uptake is likely diffusion limited (Randall & Brauner, 1993).

Gill micro-environment

Uncharged molecules diffuse across non-polar lipid membranes much more easily than charged molecules. This is of significance in the uptake of compounds which dissociate into weak acids because the membrane

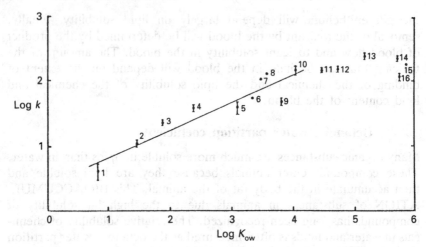

Fig. 2. Relationship between log uptake rate (where water pH<
pK_a) and log K_{ow} in guppies. The numbers refer to compounds:
1=butyric acid; 2=phenol; 3=benzoic acid; 4=4-phenylbutyric acid;
5=2,4-dichloro-phenol; 6=2-sec-butyl-4,6-dinitrophenol; 7=3,4-dichlo-
robenzoic acid; 8=2,6-dibromo-4-nitrophenol; 9=2,4,5-trichloro-
phenol; 10=2,4,6-trichlorophenol; 11=2,3,4,6-tetrachlorophenol; 12=
tetrachloroverathrol; 13=pentachlorophenol; 14=pentachloroanisol;
15=2,4,6-trichloro-5-phenylphenol; 16=DDT; 17=2,3,6-trichloro-4-
nitrophenol. (From Saarikoski *et al.*, 1986.)

diffusivity of the dissociated form of the compound is very low relative
to the non-dissociated form. The proportion of a compound which is
in the dissociated form is dependent upon the pH. The pH of the bulk
water in which a fish respires is easy to measure; however, it is the
pH in the gill micro-environment which determines the concentration
of the undissociated form in water next to the gill surface. The distance
between the secondary lamellae is very small relative to their height
and length and subsequently there is a large boundary effect on water
flow through the channel. Fish excrete protons into this boundary layer
(Lin & Randall, 1991) which acidifies expired water when bulk water
pH is above 5.5 (Lin & Randall, 1990). The best estimate of gill
micro-environment water pH is probably a mean of inspired and expired
values.

Saarikoski *et al.* (1986) measured the uptake rate of a variety of
weak acids over a range of water pH (Fig. 3). As bulk water pH
approached and exceeded the pK_a of the xenobiotic, the concentration
of the dissociated form increased and uptake rate decreased. Measured

uptake rates generally somewhat exceeded predictions based on the hypothesis that only the non-dissociated form was available for uptake. The concentration of the non-dissociated form over a range of water pH, however, was calculated from bulk water pH and pK_a. This difference between estimated and measured weak acid uptake may be due to the difference between bulk water and gill micro-environment pH. It may also indicate that there is some uptake of the dissociated form of a weak acid.

Modelling xenobiotic uptake across fish gills

There has been considerable effort to model the uptake of xenobiotics using a variety of models which range from simple, one compartment models to complex pharmaco-kinetic models (Barron, Stehly & Hayton,

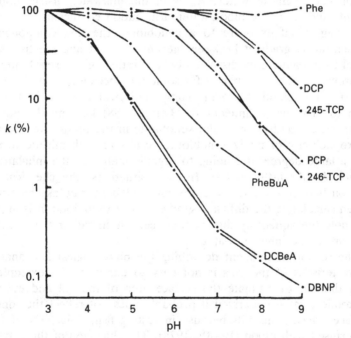

Fig. 3. Relationship between percentage rate of absorption ($k\%$) of weak acids by guppies and water pH. Phe=phenol (pK_a=10.05), DCP=2,4-dichlorophenol (pK_a=7.85), 245-TCP=2,4,5-trichlorophenol (pK_a=7.07), 246-TCP=2,4,6-trichlorophenol (pK_a=6.22), PCP=pentachlorophenol (pK_a=4.71), PheBuA=4-phenylbutyric acid (pK_a=4.76), DCBeA=3,4-dichlorobenzoic acid (pK_a=3.7), DBNP=2,6-dibromo-4-nitrophenol (pK_a=3.7). (From Saarikoski *et al.*, 1986.)

1990). Hayton and Barron (1990) have recently proposed a physiological model to describe xenobiotic uptake across the gills in fish:

$$\text{Uptake rate} = (C_w - C_f)[d(D_m^{-1})\ A\ K_m)$$
$$+ (h(D_a^{-1})A) + (V_b K_b^{-1}) + V_w^{-1}]^{-1} \tag{1}$$

This model consists of three components: 1) the xenobiotic concentration gradient between the environment and the fish, where C_w and C_f are the concentration of the xenobiotic in the external water and the plasma water, respectively, 2) physiological and anatomical characteristics of the gills, where d is epithelial thickness, h is aqueous stagnant layer thickness, A is gill surface area, V_b is effective gill blood flow, V_w effective water flow, and 3) physical constants specific to the compound, where D_m is diffusion coefficient in epithelium, D_a is diffusion coefficient in water, K_m is epithelium/water distribution coefficient, and K_b is the blood/water distribution coefficient.

During initial exposure to a xenobiotic, the first component, the concentration gradient between the environment and the fish, will be equal to the aqueously dissolved concentration of the xenobiotic in the environment. This will persist for some time because, immediately after entering the blood, lipophilic compounds will dissolve in lipids and bind to proteins (Schmieder & Henry, 1988) keeping the aqueously dissolved concentration of the xenobiotic in the plasma very low. The hydrophobicity of many xenobiotics results in both micelle formation and a large degree of binding to organic matter in the inhalant water (Black & McCarthy, 1988). In both situations, the chemical is not aqueously dissolved and therefore not available for uptake by diffusion. When calculating the diffusion gradient for a xenobiotic, it is important to know the aqueously dissolved concentration rather than xenobiotic content in the inhalant water.

The second component describing the physiological and anatomical characteristics of the gills is not easy to quantify. For example, it is very difficult to estimate the surface area of the gill and even if an anatomical value is derived it probably does not reflect the functional surface area of the gills because in resting fish, only 60% of the gill is perfused with blood (Booth, 1978). The thickness of the respiratory epithelium ranges from 0.5 to 11 μm depending upon the species (Hughes, 1984) and the thickness is probably not constant from the proximal to distal portion of the secondary lamellae. To complicate matters further many model parameters are not constant and are influenced by the metabolic rate of the fish. For instance, during exercise, there is an increase in cardiac output and ventral aortic blood pressure (Kiceniuk & Jones, 1977) which increases the perfused gill surface area and gill blood flow, and reduces the thickness of the respiratory epithelium (Randall, 1990). The thickness of the aqueous stagnant layer or boundary layer will also be greatly influenced by the velocity of water flow over the gills. Thus, all of these model parameters are difficult to measure, they are species specific but most importantly they are all adjusted to meet the metabolic demand of the animal to ensure adequate oxygen uptake.

Oxygen consumption rate can be measured easily and accurately. Thurston and Gehrke (1993) have compiled an oxygen data bank (OXYREF) which contains about 5000 measurements of oxygen consumption rate from over 300 species of fish exposed to different temperatures at rest and during exercise. Analyses of these data indicate that the main determinant of oxygen consumption rate over a broad range of fish species and temperatures is body weight. Oxygen is a lipid-soluble molecule which diffuses across the gill lamellae transcellularly as do most xenobiotics. Assuming conditions for oxygen transfer are indicative of those for toxicant transfer the development of a general coefficient (λ) describing xenobiotic uptake as a function of oxygen consumption rate can be used to replace the gill model constants in equation (1).

The third component in equation (1) has to do with characteristics specific to the xenobiotic to which the fish is exposed. As mentioned above, one of the most important physio-chemical characteristics which determines xenobiotic uptake rate is K_{ow}. Both K_m and K_b are dependent upon the K_{ow} of the xenobiotic and therefore a general K_{ow} term can be used in place of K_b and K_m. As described above, for a variety of compounds the logarithm of chemical uptake rate increases linearly with log K_{ow} between 1 and 4.3 and is constant between log K_{ow} 4.3 and 6 (McKim *et al.*, 1985). Thus, during initial exposure to a compound, equation (1) can be simplified to:

$$\text{Uptake} = (C_w)\ \lambda\ (K_{ow}\ s) \qquad (2)$$

where λ is a coefficient describing xenobiotic uptake as a function of oxygen consumption rate, and $K_{ow}\ s$ is the coefficient used to account for uptake of chemicals of various log $K_{ow}\ s$. For xenobiotics with log K_{ow} between 4.3 and 6, this coefficient is 1, for those between 1 and 4.3 this coefficient is $10^{0.429\log K_{ow}-1.842}$ (Randall & Brauner, 1993) calculated from Saarikoski *et al.* (1986).

The diffusion coefficient of a chemical in water is inversely proportional to the square root of the molecular weight (Opperhuizen *et al.*, 1985) and so, for small molecules, diffusion coefficients are relatively insensitive to differences in molecular weight. Saito *et al.* (1990) demonstrated that compounds greater than 2000 in molecular weight are not absorbed across the gills in carp while Zitko and Hutzinger (1976) have proposed that the upper molecular weight limit for xenobiotic uptake is approximately 600. Thus only small molecules cross the gills and, as a result, diffusivity parameters have been excluded from equation (2).

To predict λ, the uptake of xenobiotics in fish based upon MO_2 values retrieved from OXYREF, we used 1,2,4,5-tetrachlorobenzene (TCB) because it does not dissociate, is not easily metabolized, has a log K_{ow} of 4.97 and there is no influence of TCB exposure on MO_2 (Brauner *et al.*, 1994). In rainbow trout exposed to a large range of TCB concentrations there is no significant effect of toxicant exposure on oxygen consumption rate in either resting or active fish (Fig. 4(*a*) and 4(*b*)). This, and oxygen consumption rate in fish can be used to predict toxicant uptake.

The relationship between MO_2 and TCB uptake rate was examined in five species of fish similar in size, and for several different weight classes of rainbow trout. Forced exercise was chosen as the means to influence oxygen consumption rate because during exercise, fish can increase MO_2 relative to resting rates by up to 10 fold (Brett, 1964). In all tests fish were introduced into a 130 l Brett-type respirometer (described by Gehrke *et al.*, 1990) for two hours at a water velocity of 18 cm s^{-1}. Tetrachlorobenzene was added and a water sample was taken and fish were forced to swim for 2 h at either a low or a high water velocity (less than 80% of maximal swimming velocity) while MO_2 was measured. At the end of the test, a water sample was taken and the fish were removed from the respirometer, killed and immediately frozen. The fish and water samples were kept frozen at $-80\,°C$ and then transported to Montana State University in dry ice for TCB analyses. The entire fish was homogenized, TCB from the tissue was concentrated in hexane by Soxhlet extraction and total body and water TCB concentration was measured by gas chromatography.

Five species of fish: goldfish (*Carrasius auratus*, 11.43 ± 0.99 g), large mouth bass (*Micropterus salmoides*, 4.8 ± 0.18 g), channel catfish (*Ictalurus punctatus*, 3.7 ± 0.06 g), fathead minnow (*Pimephales promelas*, 3.92 ± 0.34 g) and rainbow trout (*Oncorhynchus mykiss*, 5.02 ± 0.07 g) were exposed to water TCB concentrations of 260 or 780 μg l^{-1} at low or high swimming velocity with exception of the large mouth bass which were only capable of swimming at the low velocity. There is a significant correlation between TCB uptake rate and MO_2 at both the low TCB concentration ($r^2 = 0.29$) and the high TCB

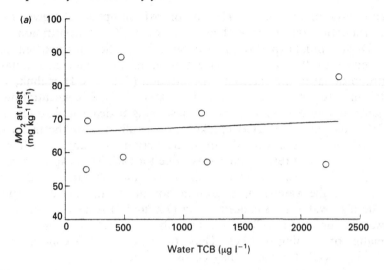

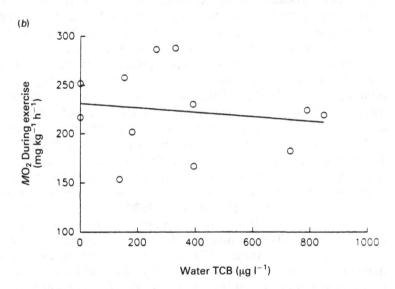

Fig. 4(*a*) The effect of water 1,2,4,5-tetrachlorobenzene (TCB) concentration on the oxygen consumption rate of adult rainbow trout at rest ($r^2=0.01$) or (*b*) swimming at 1.25 body lengths per second (Bl s^{-1}) ($r^2=0.02$).

concentration ($r^2 = 0.79$). The rate of toxicant uptake at a given oxygen consumption rate is dependent upon water TCB concentration.

During initial exposure to a toxicant, the gradient for toxicant uptake is equal to the toxicant concentration in the water and uptake is proportional to the external concentration (Spacie & Hamelink, 1982). In an attempt to standardize for external TCB concentration and generate one regression equation describing toxicant uptake relative to $\dot{M}O_2$, the measured TCB uptake rate was divided by the mean water TCB concentration over the exposure period. It is apparent, however, that two distinct relationships exist, one for each TCB exposure concentration. One explanation for this is that the measured TCB concentration in the water is an overestimation of the aqueous concentration. The high water TCB concentration (760 µg l^{-1}) is near to the maximal water solubility for TCB and it is possible that TCB is binding to organic matter or forming micelles. Thus, the aqueous TCB concentration in the high TCB treatment may be overestimated.

Lipid content in fish is important during long-term exposure to xenobiotics particularly as fish approach equilibrium with the environment (Connell, 1990; Geyer, Scheunert & Korte 1985), but the importance of lipid content to toxicant uptake during initial exposure is not known. The lipid content of the five species was measured and significant differences were found between goldfish and bass relative to channel catfish, fathead minnow and rainbow trout (Fig. 5). The inclusion of lipid in the relationship describing toxicant uptake as a function of $\dot{M}O_2$ significantly improved the coefficient of determination at the low TCB concentration ($r^2=0.85$) but marginally reduced the coefficient in the high TCB concentration (from 0.79 to 0.68) (Fig. 6). These results indicate that the body lipid content in fish may be an important determinant of toxicant uptake during initial as well as prolonged exposure to TCB; however this requires further investigation.

Medium (38.6 ± 0.96 g) and large (412.5 ± 6.9 g) rainbow trout were forced to swim at one of two water velocities during exposure to an aqueous TCB concentration of 260 µg l^{-1}. Toxicant uptake rate was divided by the mean TCB concentration during the exposure duration and when regressed against $\dot{M}O_2$ the coefficient of determination is significant ($r^2=0.59$). However, the slope of this regression line differs from that for the five species of small fish exposed to the low TCB concentration. The low coefficient of determination ($r^2=0.32$) for the five species of small fish exposed to the low TCB concentration is predominantly due to the low TCB uptake rate measured in goldfish and large mouth bass. The goldfish are more than twice the size of the other fish in this group and when the goldfish data are combined

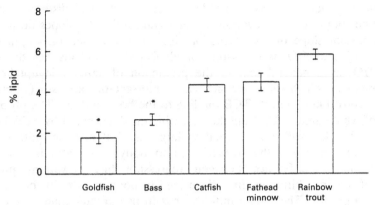

Fig. 5. The percentage body lipid determined for five species of fish. * indicates statistically different from those without symbol (*n*=8).

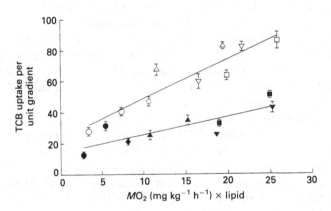

Fig. 6. The relationship between TCB uptake per unit gradient and the product of oxygen consumption rate and proportion of body lipid, in five species of fish exposed to one of two external TCB concentrations (260, open symbols, or 780 μg l⁻¹ TCB, closed symbols) at a water velocity of 2.25 or 3.75 Bl s⁻¹ for 2 h (circle, goldfish; diamond, large mouth bass; upright triangle, channel catfish; inverted triangle, fathead minnow; and square, rainbow trout) *n*=8, vertical lines represent 1 standard error, r^2=0.69 at the high TCB concentration and r^2=0.85 at the low concentration).

with the large rainbow trout data, there are still two distinct regression lines but only the elevations differ significantly, the slopes do not. The constant slope of the regression lines indicates that TCB uptake rate relative to $\dot{M}O_2$ is the same for all fish sizes and justifies the use of $\dot{M}O_2$ as a general tool in the prediction of xenobiotic uptake. The point at which the regression lines intersect the abscissa (when $\dot{M}O_2$ is zero) likely reflects TCB binding to the body surface. The proportion of whole animal TCB uptake rate which is determined by TCB binding to the skin will not be great in large fish but may be quite large in small fish where the surface area to body mass ratio is large. The significant difference between regression line elevations probably reflects this difference in surface area to body mass ratios between the two groups. These data indicate that in fish greater than 11 g there is very little effect of fish size on the uptake rate of TCB due to TCB binding to the skin, but there is a large effect in fish less than 11 g. Therefore, in fish greater than 11 g, TCB uptake during initial exposure in fish can be predicted for a variety of fish species using equation (2) where $\lambda = 0.0820 - O_2 + 15.63$ and $\dot{M}O_2$ is in mg kg^{-1} h^{-1}.

The uptake of TCB during initial exposure can also be expressed as total toxicant accumulated per hour (standardized for water TCB concentration) as a function of total oxygen consumed in mg h^{-1}. This results in one highly significant relationship ($r^2 = 0.95$) which describes the data for all fish combined. Equation (2) can again be used to predict initial uptake of TCB where $\lambda = 113.36 - O_2 + 252.34$. $\dot{M}O_2$ in this equation is in mg h^{-1}. The derived uptake rate in this case will not be standardized for body weight. In the OXYREF oxygen data bank (Thurston & Gehrke, 1993), the main determinant of oxygen consumption rate over a broad range of species, activity levels and temperatures, is body weight. Thus, if the fish weight is known, an approximate value for total oxygen consumed (mg h^{-1}) can be obtained from OXYREF and the uptake rate for TCB can be estimated for different fish species and sizes.

As discussed above, there is very little effect of log K_{ow} on log uptake rate for xenobiotics with log K_{ow} between 4.3 and 6 (Saarikoski *et al.*, 1986). Tetrachloroguaiacol (TCG) has a log K_{ow} of 4.41 and molecular weight of approximately 262 (Xie & Dryssen, 1984) and was chosen to determine whether the relationship between TCB uptake rate and $\dot{M}O_2$ can be applied to other xenobiotics. Large (410 ± 2.0 g) and medium (44 ± 2.2 g) rainbow trout were exposed to 260 µg l^{-1} TCG at low or high swimming speed for 2 h and large-scale suckers (*Catastomus macroheilus*) (706 ± 5.0 g), which could not be forced to swim at a high velocity, were exposed to 260 µg l^{-1} TCG at two

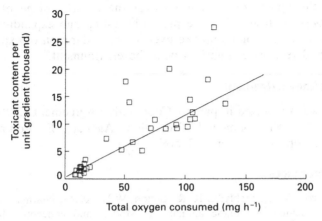

Fig. 7. The effect of total oxygen consumed (mg h^{-1}) on total TCG content accumulated (mgTCG h^{-1}) per unit gradient TCG (mg l^{-1}). Solid regression line is that for TCB uptake in similar fish under similar conditions.

different temperatures (9 or 18 °C). Tetrachloroguaiacol dissociates and the concentration of the non-dissociated form was calculated from water pH and the pK_a for TCG of 6.19 (Xie & Dryssen, 1984). The water pH during these experiments was approximately 5.5 and thus, there will be very minor changes in expired water pH relative to bulk water pH (Lin & Randall, 1990). The uptake rate of TCG was divided by the concentration of non-dissociated TCG and is significantly correlated with $\dot{M}O_2$ ($r_2 = 0.97$).

When the data for TCG (Fig. 7) is superimposed upon the regression line of total TCB accumulated per hour (standardized for water TCB concentration) as a function of total oxygen consumed in mg h^{-1}, it is apparent that the relationship for TCB describes that for TCG. Thus, equation (2) can be used to predict the uptake rate of both TCB and TCG, and likely other xenobiotics similar in log K_{ow}, as a function of $\dot{M}O_2$. Although an estimate of the diffusion coefficient for TCG in water based upon differences in molecular weight is approximately 9% below that for TCB, the effect on uptake rate appears to be minimal.

The model proposed above at present is restricted to conditions of normoxia, during initial exposure to xenobiotics. It is simple, and requires limited information about the xenobiotic to which a fish is exposed (log K_{ow} and pK_a if it is a weak acid) and the oxygen uptake of the fish. The rate of oxygen uptake can be measured directly or estimated from the oxygen database OXYREF. The use of this model to predict the uptake of xenobiotics over the log K_{ow} range of 1 to 4.3 remains to be tested. We are presently testing and expanding this model to predict xenobiotic uptake over a longer duration of exposure as the animal reaches equilibrium with the environment.

Acknowledgement

This work was supported in part by Cooperative Agreement CR816369 from the US Environmental Protection Agency, Environmental Research Laboratory, Athens, Georgia.

References

Barron, M.G., Stehly, G.R. & Hayton, W.L. (1990). Pharmacokinetic modelling in aquatic animals. I. Models and concepts. *Aquatic Toxicology*, **18**, 61–86.

Black, M.C. & McCarthy, J.F. (1988). Dissolved organic macromolecules reduce the uptake of hydrophobic organic contaminants by the gills of rainbow trout (*Salmo gairdneri*). *Environmental Toxicology and Chemistry*, **7**, 593–600.

Booth, J.H. (1978). The distribution of blood flow in the gills of fish: application of a new technique to rainbow trout (*Salmo gairdneri*). *Journal of Experimental Biology*, **73**, 119–29.

Brauner, C.J., Randall, D.J., Neuman, J.F. & Thurston, R.V. (1994). The effect of exposure to 1,2,4,5 tetrachlorobenzene and the relationship between toxicant and oxygen uptake in rainbow trout (*Oncorhynchus mykiss*) and cutthroat trout (*Oncorhynchus clarki*) during exercise. *Environmental Toxicology and Chemistry*, **13**, 1813–20.

Brett, J.R. (1964). The respiratory metabolism and swimming performance of young sockeye salmon. *Journal of the Fisheries Research Board of Canada*, **21**(5), 1183–226.

Cameron, J.N. & Davis, J.C. (1970). Gas exchange in rainbow trout (*Salmo gairdneri*) with varying blood oxygen capacity. *Journal of the Fisheries Research Board of Canada*, **27**, 1069–85.

Chessells, M., Hawker, D.W. & Connell, D.W. (1992). Influence of solubility in lipid on bioconcentration of hydrophobic compounds. *Ecotoxicology and Environmental Safety*, **23**, 260–73.

Connell, D.W. (1990). *Bioaccumulation of Xenobiotic Compounds*. 213 pp. CRC Press, Boca Raton, Florida.

Dobbs, A.J. & Williams, N. (1983). Fat solubility – a property of environmental relevance? *Chemosphere*, **12**(1), 97–104.

Gehrke, P.C., Fidler, L.E., Mense, D.C. & Randall, D.J. (1990). A respirometer with controlled water quality and computerized data acquisition for experiments with swimming fish. *Fish Physiology and Biochemistry*, **8**(1), 61–7.

Geyer, H., Scheunert, I. & Korte, F. (1985). Relationship between the lipid content of fish and their bioconcentration potential of 1,2,4-trichlorobenzene. *Chemosphere*, **14**(5), 545–55.

Gobas, F.A.P.C., Opperhuizen, A. & Hutzinger, O. (1986). Bioconcentration of hydrophobic chemicals in fish: relationship with membrane permeation. *Environmental Toxicology and Chemistry*, **5**, 637–46.

Hayton, W.L. & Barron, M.G. (1990). Rate-limiting barriers to xenobiotic uptake by the gill. *Environmental Toxicology and Chemistry*, **9**, 151–7.

Hughes, G.M. (1984). *General anatomy of the gills*. In *Fish Physiology*. Hoar, W.S. and Randall, D.J. (eds.), Vol. 10 A pp. 1–63. Academic Press, New York.

Jones, D.R. & Randall, D.J. (1978). *The respiratory and circulatory systems during exercise*. In *Fish Physiology*. Hoar, W.S. and Randall, D.J. (eds.), Vol. 7 pp. 425–501. Academic Press, New York.

Kiceniuk, J.W. & Jones, D.R. (1977). The oxygen transport system in trout (*Salmo gairdneri*) during sustained exercise. *Journal of Experimental Biology*, **69**, 247–60.

Lin, H. & Randall, D.J. (1990). The effect of varying water pH on the acidification of expired water in rainbow trout. *Journal of Experimental Biology*, **149**, 149–60.

Lin, H. & Randall, D.J. (1991). Evidence for the presence of an electrogenic proton pump on the trout gill epithelium. *Journal of Experimental Biology*, **161**, 119–34.

McKim, J., Schmieder, P. & Veith, G. (1985). Absorption dynamics of organic chemical transport across trout gills as related to octanol–water partition coefficient. *Toxicology and Applied Pharmacology*, **77**, 1–10.

Neely, W.B. (1979). Estimating rate constants for the uptake and clearance of chemicals by fish. *Environmental Science and Technology*, **13**(12), 1506–10.

Niimi, W.B. (1991). Solubility of organic chemicals in octanol, triolein, and cod liver oil and relationships between solubility and partition coefficients. *Water Research*, **25**(12), 1515–21.

Opperhuizen, A., Velde, E.W.v.d., Gobas, F.A.P.C., Liem, D.A.K. & Steen, J.M.D.v.d. (1985). Relationship between bioconcentration in fish and steric factors of hydrophobic chemicals. *Chemosphere*, **114**(11–12), 1871–96.

Randall, D.J. (1990). *Control and co-ordination of gas exchange in water breathers*. In *Advances in Comparative and Environmental Physiology*. Boutilier, R.G. (ed.), Vol. 6, pp. 253–78. Springer-Verlag, Berlin, Heidelberg.

Randall, D.J. & Brauner, C.J. (1993). Toxicant uptake across fish gills. *Proceedings of an International Symposium, Sacramento, California,*

September 18–20, 1990. United States Environmental Protection Agency Environmental Research Laboratory, Athens, Georgia, USA. EPA/600/R-93/157 (ed. Russo, R.R. and Thurston, R.V.) pp. 109–116. Fish Physiology, Fish Toxicology, and Water Quality Management (in press).

Saarikoski, J., Lindstrom, R., Tynela, M. & Viluksela, M. (1986). Factors affecting the absorption of phenolics and carboxylic acids in the guppy (*Poecilia reticulata*). *Ecotoxicology and Environmental Safety*, **11**, 158–73.

Saito, S., Tateno, C., Tanoue, A. & Matsuda, T. (1990). Electron microscope autoradiographic examination of uptake behaviour of lipophilic chemicals into fish gill. *Ecotoxicology and Environmental Safety*, **19**, 184–91.

Schmieder, P.K. & Henry, T.A. (1988). Plasma binding of 1-butanol, phenol, nitrobenzene and pentachlorophenol in the rainbow trout and rat: a comparative study. *Comparative Biochemistry and Physiology*, **91**C(2), 413–18.

Spacie, A. & Hamelink, J.L. (1982). Alternative models for describing the bioconcentration of organics in fish. *Environmental Toxicology and Chemistry*, **1**, 309–20.

Thurston, R.V. & Gehrke, P.C. (1993). Respiratory oxygen requirements of fishes: Description of OXYREF, a datafile based on test results reported in the published literature. *Proceedings of an International Symposium, Sacramento, California, September 18–20, 1990.* United States Environmental Protection Agency Environmental Research Laboratory, Athens, Georgia, USA. EPA/600/R-93/157 (ed. Russo, R.R. and Thurston, R.V.) pp. 95–108. Fish Physiology, Fish Toxicology, and Water Quality Management (in press).

Verschueren, K. (1983). *Handbook of Environmental Data on Organic Chemicals.* pp. 24–5. Van Nostrand Reinhold Co., New York.

Xie, T.M. & Dryssen, D. (1984). Simultaneous determination of partition coefficients and acidity constants of chlorinated phenols and guaiacols by gas chromotaography. *Analytica Chimica Acta*, **160**, 21–30.

Zitko, V. & Hutzinger, O. (1976). Uptake of chloro- and bromobiphenyls, hexachloro- and hexabromobenzene by fish. *Bulletin of Environmental Contamination and Toxicology*, **16**, 665–73.

R.V. THURSTON, J.F. NEUMAN,
C.J. BRAUNER, and D.J. RANDALL

Bioaccumulation of waterborne 1,2,4,5-tetrachlorobenzene in tissues of rainbow trout

Introduction

During the past three years we have been studying the rates of uptake by fishes of selected synthetic organic chemicals, foreign to natural systems, known as xenobiotics. We have conducted several tests using an automated respirometer in which we have exposed fish to 1,2,4,5-tetrachlorobenzene (TCB). Our principal test animal has been the rainbow trout (*Oncorhynchus mykiss*), although we have also tested other fish species to compare results. One of our long-range objectives is to determine to what extent uptake and depuration rates of xenobiotics by fishes may be related to their respiration rates. Using that information, we are working toward developing a simple predictive model to describe the rates of uptake and depuration of xenobiotics as a function of the oxygen uptake rate and chemical properties of the chemical in question. Some of our laboratory research, and our progress to date on developing the predictive model, are reported in Chapter 1, Randall *et al.* (1995). In the present paper, we will provide details of the methods used while conducting our respirometry experiments, and also provide results of some of the other experiments conducted during early stages of our study.

Methods

Test conditions

Many of our preliminary tests were conducted at Fisheries Bioassay Laboratory, Montana State University (MSU), and most of our respirometry experiments were conducted in a modified Brett-type respirometer at the Zoology Department, University of British Columbia (UBC). All fish tissue and water analyses were conducted at MSU. The UBC respirometer was designed and built specifically for our studies, and has been described in some detail by Gehrke *et al.* (1990). The respirometer can test fishes up to 2 kg in size and at swimming speeds as great as 2.5 m/s. Experiments can be conducted over several days with water velocity, temperature, pH, dissolved oxygen, and carbon dioxide all computer controlled at predetermined levels, and data monitored continuously by the computer as well. For the tests at UBC described here, fish were sacrificed immediately after removal from the respirometer, and whole fish or fish tissues, and test water samples, were frozen and stored at $-20\,°C$ until shipment to MSU for chemical analysis; shipment was by overnight air carrier with samples packed in dry ice. Fish tested at MSU were prepared for analysis the same as those tested at UBC. All samples were stored at $-20\,°C$ at MSU prior to analysis.

TCB analysis

After receipt at MSU, small whole fish (<10 g), sub-samples of homogenates of larger fish, and individual organ tissues, were blended to a fine powder with the aid of dry ice and anhydrous Na_2SO_4 while still frozen. At least two separate samples were prepared from homogenates of larger fish. Only single samples were prepared from smaller fish and fish tissues because of limitations in sample quantity. TCB was soxhlet extracted ($\geqslant 8$ hours) from each sample using hexane. Surrogate (1,2,3,4-TCB) in amount comparable to the amount anticipated for TCB results (usually 100 µg) was added at the initiation of the extraction step. Lipids were removed from portions of this extract using florisil column chromatography; TCB and surrogate were eluted with 5% methyl-*t*-butyl ether/hexane. Pentachlorobenzene (PCB) internal standard (0.25 µg) was added to the purified extract and component concentrations were determined by electron capture gas chromatography (ECD–GC). Packed GC columns (1.8 m × 2 mm i.d.) containing 3% SE-30 or 3% Carbowax on 100/120 Supelcoport™ were used. A multi-point calibration curve (0.5–10 µg/l) was employed during quantitation. Both standards and sample extracts were injected in duplicate. When plasma samples were required, the caudal peduncle was severed immediately after the fish was sacrificed and blood was collected from the caudal artery into heparinized capillary tubes. Plasma was separated by centrifugation and 5.0 µl of plasma was removed by means of a 10 µl glass syringe and transferred to a 10 ml volumetric flask containing 0.05 µg internal standard (PCB). This solution was made to volume with hexane, and the contents mixed for approximately 1 minute using a vortex mixer. TCB was extracted from test water samples using hexane, and internal standard was added to the diluted extracts. Both plasma and water samples were analysed for TCB as described above.

Lipid analysis

Total lipid content in whole small fish (<10 g) or samples from homogenized larger fish (>30 g) was determined by a modification of a method described by Folch, Lees & Stanley (1957). While still frozen, small whole fish or sub-samples of tissue homogenates from larger fish were ground to a powder using dry ice and anhydrous Na_2SO_4. Lipids were extracted by blending this mixture twice for 3 minutes with 2 : 1 chloroform:methanol. Both extracts and rinsings were vacuum filtered and combined into a 500 ml separatory funnel. Methanol was partitioned from the chloroform into aqueous KCl solution. The chloroform layer was vacuum filtered using a 0.2 µ Nylon-66 filter and the volume adjusted to 250 ml. A measured aliquot was transferred to a tared beaker and chloroform was removed using low heat and a gentle stream of dry N_2. The residue was dried (≥12 hours) in a vacuum desiccator and weighed.

Tests conducted, results, and discussion

In several of our initial experiments rainbow trout were exposed to TCB for different time periods up to 6 hours under either static test conditions at MSU or forced swimming conditions in the respirometer at UBC. After each test, fish were sacrificed and different tissues analysed to determine any apparent sequence among them in increase of toxicant concentration. Tissues analysed were blood plasma, gills, liver, kidney, brain, spleen, heart, upper gut, lower gut, white muscle, pink muscle, and adipose. We also reviewed the data to determine if the rate of toxicant concentration increase in any one tissue might be representative of the fish as a whole. Because TCB is lipophilic, this pre-supposes one tissue might have a fat content representative of the fish as a whole, or at least with a fairly constant ratio to that of the fish as a whole. Standard deviations from the mean concentrations of TCB among the tissues analysed were so great that we were unable to detect any sequential build-up among the tissues. Indeed, concentrations were not appreciably different between 1 and 6 hours except in the case of adipose tissue and possibly muscle (Table 1). We also looked at the tissue/blood plasma ratio of the test fish, and once again did not see any consistent ratio (Table 2).

It was apparent we would need to measure TCB concentrations in whole fish, and our next tests were designed to compare blood plasma TCB concentration against that of the whole body over time. In our first experiment the concentrations of TCB in blood plasma and whole

Table 1. *Fish tissue concentrations after exposure to TCB*

Exposure (hours)	n	Plasma	Gills	Liver	Kidney	Spleen	U. gut	L. gut	W. muscl	P. muscl	Heart	Brain	Adipose
					Test 118	500–25 µg/l TCB		0 BL/s					
0.5	2	5.73	4.68	8.70	8.10	<1.5	2.50	3.44	4.03	11.85	10.40	10.60	2.99
1.0	2	10.34	8.65	36.35	13.15	9.84	4.23	10.41	6.82	31.30	15.45	25.90	1.17
3.0	3	10.09	8.58	39.73	17.85	10.55	5.86	16.18	9.49	83.00	19.97	42.00*	20.97
6.0	3	8.39	5.20	14.87	14.20	7.64	7.28	6.19	7.26	63.33	16.00	15.43	23.83
					Test 105	400–50 µg/l TCB		1.25 BL/s					
1.0	2	3.74	4.13	7.85	5.38	2.78	1.93	7.68	3.54	18.50	5.33	11.15	20.59
2.0	2	6.85	4.24	11.25	6.43	2.59	3.32	7.34**	5.20	18.00	6.13	11.29	8.22
6.0	2	3.17	2.96	10.78	9.29	2.21	4.22	no data	7.09	38.55	5.19	8.99	59.75

Plasma µg/ml, tissue µg/g * = two values, ** = one value.

Table 2. *Ratio of fish tissue concentrations to blood plasma after exposure to TCB*

Exposure (hours)	n	Plasma µg/ml	Gills/ plasma	Liver/ plasma	Kidney/ plasma	Spleen/ plasma	U.gut/ plasma	L.gut/ plasma	W. muscl/ plasma	P. muscl/ plasma	Heart/ plasma	Brain/ plasma	Adipose/ plasma
					Test 118		500–25 µg/l TCB		0 BL/s				
0.5	2	5.73	0.816	1.52	1.41	0.0421	0.437	0.600	0.703	2.07	1.82	1.85	0.523
1.0	2	10.3	0.839	3.53	1.28	0.956	0.411	1.01	0.663	3.04	1.50	2.51	0.114
3.0	3	10.1	0.849	3.93	1.77	1.04	0.580	1.60	0.940	8.22	1.98	3.40	2.08
6.0	3	8.39	0.619	1.77	1.69	0.911	0.860	0.737	0.866	7.55	1.91	1.84	2.84
					Test 105		400–50 µg/l TCB		1.25 BL/s				
1.0	2	3.87	1.07	2.03	1.39	0.720	0.499	1.98	0.915	4.78	1.38	2.88	5.32
2.0	2	6.85	0.619	1.64	0.939	0.378	0.484	1.05	0.759	2.63	0.896	1.65	1.20
6.0	2	3.17	0.934	3.40	2.93	0.699	1.33	no data	2.24	12.2	1.64	2.83	18.8

body were very similar after 2 hours' exposure, but blood plasma concentration at 6 hours had not increased, whereas whole body concentration had more than doubled (Fig. 1). The obvious conclusion was that TCB equilibrium between exposure water and blood plasma was achieved within 2 hours or less, and after reaching saturation the blood was acting as a conduit to pass TCB from gills to body tissues. Thus, blood plasma could not be used to indicate either total body content or uptake rate of the toxicant.

In our next experiment 5 g rainbow trout were exposed to nominal TCB concentrations of 10, 50, and 100 µg/litre in holding tanks for several days, during which they were transferred to a fresh solution every 24 hours. Four fish were removed from each tank at 1, 2, 4, and 8 days, sacrificed, and blood plasma and whole body TCB concentrations measured. A tight pattern of TCB increase over time emerged for both blood plasma and whole body, and these patterns were also correlated with exposure concentrations (Fig. 2), but once again equilibrium between TCB in exposure water and whole body lagged behind TCB equilibrium reached in blood plasma.

Earlier studies at MSU had shown the rate of uptake of some of the chemicals with which we might be working can be extremely rapid, with equilibrium between whole body and water sometimes reached in

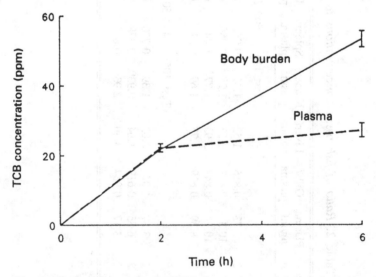

Fig. 1. Short-term uptake of TCB in plasma and in whole body of rainbow trout.

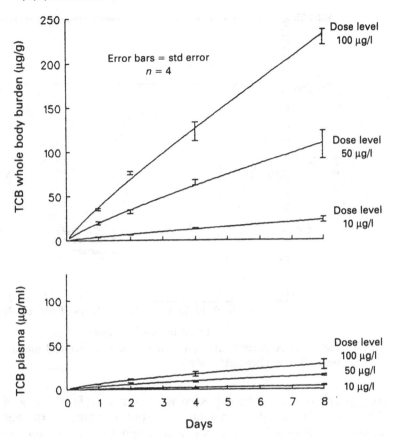

Fig. 2. TCB bioconcentration in plasma and in whole body of rainbow trout vs. time.

just a few hours. Tischmak (1984) demonstrated that equilibrium for 2,4,6-trichloroaniline (log P = 3.80) between fathead minnows (*Pimephales promelas*) and their water environment was achieved in less than a day (Fig. 3), and although equilibrium for TCB (log P = 4.67) required 6–8 days, it was also apparent that the rate of uptake of TCB decreased considerably between initial exposure and 24 hours into test (Fig. 4).

Our next step was to conduct a series of short-term tests in which we exposed rainbow trout to TCB in the UBC respirometer to measure TCB uptake vs. oxygen consumed. We chose a 2-hour test period for several reasons, but principally because this was long enough to measure

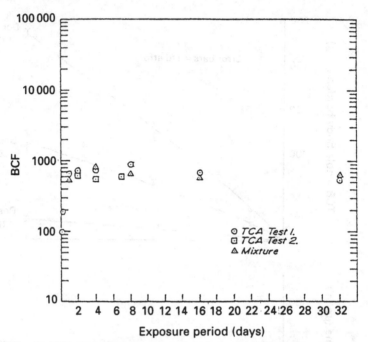

Fig. 3. TCA bioconcentration in whole body of fathead minnows vs. time. (From Tischmak, 1984.)

oxygen differences in the exposure water for as little as 100 g of test fish, and short enough so that rate of toxicant uptake by the test fish would not measurably vary between start and finish of the test period. The uptake of TCB by rainbow trout per gram of fish for three different weight classes is shown in Fig. 5. Because of the lipophilicity of TCB, we also measured lipid content in whole fish samples from each weight class, and the data were plotted as uptake of TCB against total lipid (Fig. 6). Finally, we looked at uptake of TCB in relation to oxygen consumption rate for each of the three different weight classes of rainbow trout, each tested at two different swimming speeds (Fig. 7). This latter curve demonstrates that there was a striking correlation between oxygen uptake and TCB uptake by these rainbow trout under our test conditions.

Separate from these chemical uptake studies, we have compiled a file of data from the published literature on respiratory oxygen require- ments of fishes (Thurston & Gehrke, 1992). This file, called OXYREF, contains data from over 6800 individual laboratory tests in which oxygen consumption was measured. The data in our file include fish species,

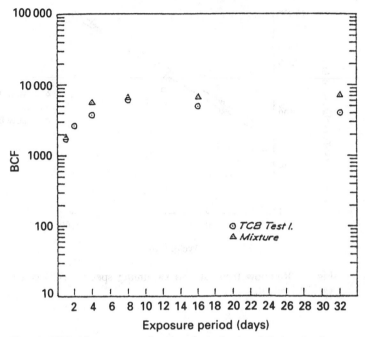

Fig. 4. TCB bioconcentration in whole body of fathead minnows vs. time. (From Tischmak, 1984.)

fish weight, certain test water conditions, measured respiratory oxygen requirements, and mode at which each fish was tested: 'standard' (resting), 'routine' (moving about), or 'active' (measured swimming rate). OXYREF contains data from 1514 tests in which fish were swum under 'active swimming' conditions, the mode under which we conducted our TCB tests at UBC. We plotted oxygen consumption vs. weight for these 'active swimming' tests and we superimposed the data from our rainbow trout TCB experiments on the resultant curve (Fig. 8). One sees an excellent fit. The bulk of the data points from the tests we conducted at the higher swimming speed, in which the fish respired at an elevated rate, are, not surprisingly, above the curve which represents the mean of the 'active swimming' tests from OXYREF. Correspondingly, the data for most of the tests conducted at slower swimming speeds fall below the curve.

There is a clear correlation between the rate of oxygen uptake and the rate of uptake of TCB among the freshwater fish species we tested. If this correlation holds for other representative fish species, and if the oxygen uptake rate and the water concentration of a chemical to which

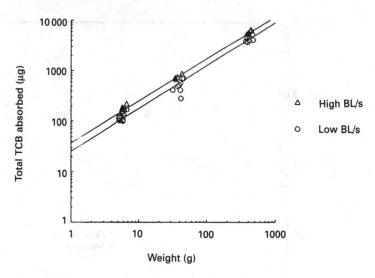

Fig. 5. Rainbow trout at two swimming speeds, TCB concentration vs. fish weight.

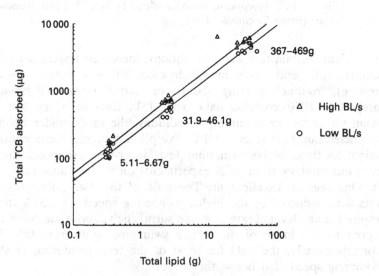

Fig. 6. Rainbow trout of three weight classes, TCB concentration vs. lipid concentrations.

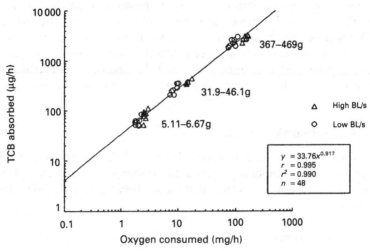

Fig. 7. Rainbow trout of three weight classes, TCB concentration vs. oxygen consumption.

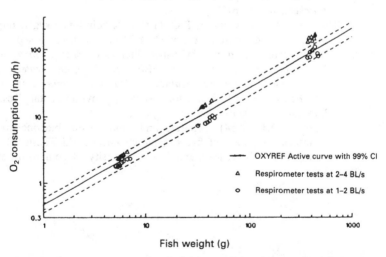

Fig. 8. Oxygen consumption by three weight classes of rainbow trout compared with 'active swimming' regression curve from OXYREF.

a fish is exposed are known, it may be possible to predict uptake rates of that chemical during exposure. We are continuing to expand OXYREF as more data become available in the literature, but in the present laboratory study if we can establish correlations between oxygen consumption and toxicant uptake and depuration, OXYREF can provide a powerful tool for the prediction of bioaccumulation of xenobiotics by fishes.

Acknowledgement

This research was funded in part by the US Environmental Protection Agency, Environmental Research Laboratory, Athens, Georgia, through Cooperative Agreements CR 813424, CR 816369, and CR 995189.

References

Folch, J., Lees, M. & Stanley, G.H.S. (1957). A simple method for the isolation and purification of total lipids from animal tissues. *Journal of Biological Chemistry*, **226**, 497–509.

Gehrke, P.C., Fidler, L.E., Mense, D.C. & Randall, D.J. (1990). A respirometer with controlled water quality and computerized data acquisition for experiments with swimming fish. *Fish Physiology and Biochemistry*, **8**, 61–7.

Thurston, R.V. & Gehrke, P.C. (1992). Respiratory oxygen requirements of fishes: description of OXYREF, a data file based on test results reported in the published literature. *Second International Symposium on Fish Physiology, Fish Toxicology, and Water Quality Management. Sacramento, California (USA), September 1990*. USA: US Environmental Protection Agency, Environmental Research Laboratory, Athens, Georgia.

Tischmak, D.J. (1984). Separate and simultaneous bioconcentration in fathead minnows of five organic chemicals. MS Thesis, Department of Chemistry, Montana State University, Bozeman, Montana, USA.

R.D. HANDY

Dietary exposure to toxic metals in fish

Introduction

The ingestion of contaminated food is considered a primary route of metal intoxication in terrestrial organisms. Consequently, the mammalian literature contains a wealth of information on the ingestion, absorption and oral toxicity of metals. In contrast, the effects of oral doses of metals in fish are mostly unknown, except in the context of nutrition (Ogino & Yang, 1978, 1980; Ketola, 1979; Knox, Cowey & Adron, 1982, 1984; Hardy & Shearer, 1985; Poston, 1991). The general belief that the gills are the main route of toxicant absorption in fish has also diverted attention from oral toxicity studies. Early laboratory investigations confirmed that fish can absorb metal poisons across the gut (Aoyama, Inoue & Inoue, 1978a,b; Patrick & Loutit, 1978), but the toxicological significance of these observations remains unclear despite field data on contamination levels in food species (Krantzberg & Stokes, 1989; Yevtushenko, Bren & Sytnik, 1990; Wren & Stephenson, 1991; Miller, Munkittrick & Dixon, 1992). However, at least one research team has confirmed the diet as an important route of contamination in wild fish (Dallinger & Kautzky, 1985; Dallinger et al., 1987). Aquatic microcosms have also been used successfully in trophic studies (Rodgers et al., 1987), while more recently radiotracer studies in whole ecosystems have yielded information on the trophic transfer of toxic metals (Cope, Wiener & Rada, 1990; Harrison, Klaverkamp & Hesslien, 1990). The available information on the occurrence, intestinal absorption, excretion, accumulation, and oral toxicity of Hg, Cd, Al, Cu and Zn are summarized here. References to the mammalian literature are made for comparison and where data on fish are lacking.

Mercury

The trophic transfer of mercury compounds have been given particular attention because of poisoning in man from the consumption of contaminated fish (Nakanishi et al., 1989; Kudo & Miyahara, 1991). Mercury occurs at picogram levels in open seawater and is typically 1–7 ng l^{-1} in freshwater (Craig, 1986; Bernhard & George, 1986), but concentrations may be much higher in waters receiving industrial effluents or the atmospheric deposition of mercury (Craig, 1986; Wiener et al., 1990). Anthropogenic mercury sources are mostly inorganic species which may be methylated to methylmercury at the water/sediment interface or by bacteria associated with the gut of fish (Rudd, Furutani & Turner, 1980; Rudd et al., 1983; Craig, 1986; Winfrey & Rudd, 1990). The total mercury concentration in contaminated invertebrate food species range from <0.1–10 μg g^{-1} ww (wet weight), with between 1 and 17% of the metal stored in the methylmercury form (Huckabee, Elwood & Hildebrand, 1979; Wren & Stephenson, 1991). Unlike invertebrates, fish accumulate the poison almost exclusively as methylmercury, regardless of their position in the food chain, with total mercury concentrations in prey species around 0.1 μg g^{-1} ww or greater. Fish store most of the methylmercury in the muscle and therefore the pollutant is rapidly biomagnified through the aquatic food chain (Boudou & Ribeyre, 1985; Riisgård & Hansen, 1990; Jackson, 1991; Bloom, 1992).

Several authors have studied the uptake of mercury compounds across the gut of fish (Hannerz, 1968; Backström, 1969; Wobeser, 1975; Pentreath, 1976; Huckabee et al., 1978; Philips & Buhler, 1978; Phil-

Table 1. The distribution of mercury between internal organs of rainbow trout after 42 days exposure to 10 g kg^{-1} dw mercuric chloride in the food

Organ	Mean mercury concentration (μg g^{-1} ww)		Mercury content of whole organ (μg)		Proportion of body burden in each organ (%)	
	Control	Exposed	Control	Exposed	Control	Exposed
Muscle	0.08	0.41	4.77	24.48	89.07	17.58
Gill	0.026	8.17	0.08	26.96	1.49	19.36
Liver	0.045	2.07	0.05	2.40	0.93	1.72
Kidney	0.08	5.30	0.11	7.47	2.03	5.36
Whole body	0.045	1.17	5.355	139.23	–	–

The mean mercury concentrations taken from Fig. 1 are multiplied by whole organ wet weights for a 119 g fish. The proportion of the body burden in each organ are whole organ metal contents divided by that of whole fish minus the gut.

ips & Gregory, 1979; Turner & Swick, 1983; Boudou & Ribeyre, 1985; Rodgers *et al.*, 1987; Riisgård & Hansen, 1990; Skak & Baatrup, 1993; see also reviews Huckabee *et al.*, 1979; Spry & Wiener, 1991). These investigations reveal that methylmercury is absorbed five times faster than the inorganic species and is assimilated from the gut with an efficiency of about 20%. Methylmercury has a relatively high lipid solubility which suggests the methyl form crosses the intestinal barrier by diffusion (Olson, Bergman & Fromm, 1973; Boudou & Ribeyre, 1985). The inorganic species is also taken up from the gut lumen, but most of the metal is bound to mucus and other anionic ligands on the gut epithelium and is unavailable for uptake (Pärt & Lock, 1983). Furthermore, the active transport of Hg^{2+} is unlikely given its inhibitory effects on ATPases and ion channels (Baatrup, Doving & Winberg, 1990; Gill, Tewari & Pande, 1990). Only about 6% of the inorganic mercury present in food may be absorbed by fish (Boudou & Ribeyre, 1985).

Information on the excretion of oral doses of mercury compounds in fish is very sparse. Mammals excrete mercury compounds in the bile and by intestinal sloughing. Mercury is also excreted in the urine (Mehra & Choi, 1980; Refsvik, 1983). The excretion routes in fish have not been studied for dietary inputs, but high intestinal mercury concentrations in fish dosed via injection suggest biliary excretion as the principal mode of detoxification (Weisbart, 1973). Renal excretion is probably limited owing to pathology in the kidneys of exposed animals (Fig. 2; Wobeser, 1975). The gill has been suggested as an excretory route for mercury in the goldfish, *Carassius auratus* (Weisbart, 1973). This might also occur in the rainbow trout, *Oncorhynchus mykiss*, since mercury accumulation in the gills in the absence of gill pathology and dissolved mercury suggest branchial detoxification (Fig. 1; Boudou & Ribeyre, 1985). Detoxification by the secretion of contaminated mucus also occurs since mercury appears in the external mucus of fish held in clean water (Fig. 1).

Once inorganic mercury has crossed the intestinal barrier, it is accumulated in most internal tissues including the liver, kidney, muscle, brain, spleen, gill and blood (Boudou & Ribeyre, 1985; Riisgård & Hansen, 1990; Fig. 1). The general contamination of all the internal organs with the inorganic metal (Fig. 1) probably reflects the bulk transport of inorganic mercury in the blood plasma and extracellular fluid, whereas methylmercury is usually assimilated by red blood cells (Olson *et al.*, 1973; Riisgård & Hansen, 1990). Orally administered methylmercury accumulates in the muscle, blood, liver, gills, kidneys, spleen, brain and olfactory system of fish (Wobeser, 1975; Boudou &

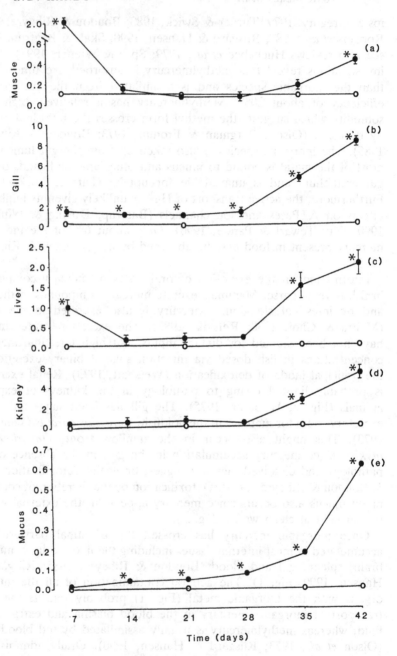

Ribeyre, 1985; Baatrup & Doving, 1990). The muscle tends to support a substantial proportion of the body burden whether the metal is presented as inorganic or organic mercury, and regardless of the route of uptake (for review see Huckabee *et al.*, 1979; McKim *et al.*, 1976; Boudou & Ribeyre, 1985; Harrison *et al.*, 1990; Riisgård & Hansen, 1990; Table 1). Mercury accumulation from the diet can be complex when toxicity occurs. Food regurgitation may reduce the exposure dose (Wobeser, 1975), but disintegration of the gut may enhance diffusive entry of the toxicant. In Fig. 1 an initially high mercury uptake was reduced with progressive gastrointestinal damage and food regurgitation, on day 28 the gut finally disintegrated, allowing the diffusive entry of mercury into the blood and an apparently high uptake rate (Fig. 2).

There are no acute or chronic oral toxicity data for mercury compounds from which median lethal concentrations may be derived for fish. Mortalities have been notably absent in previous trophic studies, even when the food is heavily contaminated (Wobeser, 1975; Boudou & Ribeyre, 1985; Fig. 1). Toxic effects include regurgitation and extensive gastrointestinal damage (Fig. 2), swelling of the epithelium around the Bowman's capsule and increased melanomacrophage activity in the kidney (Wobeser, 1975; Fig. 2). However, extensive gill damage has been absent (Wobeser, 1975). Hypertrophy, hyperplasia, and increased 'chloride cell' numbers and the fragmentation of pavement cell microridge patterns were not observed in trout fed 10 g kg^{-1} mercury dry weight of food. Mucus hypersecretion at the gills was also absent (Fig. 2). The severe gastrointestinal pathology and the inhibition of enzymes in the gut (Fig. 2; Gill *et al.*, 1990) indicate that digestion and nutrient absorption may be impaired, although no data are available for fish.

Aluminium

Aluminium enters the aquatic environment by natural processes such as the weathering of mineral deposits and atmospheric deposition of

---

Fig. 1. Total mercury concentrations (μg g^{-1} ww) in selected tissues from rainbow trout, *Oncorhynchus mykiss*, fed normal pellets (open circles) or pellets enriched with 10 g kg^{-1} dw mercuric chloride (closed circles) over 42 days. * significant difference from control by ANOVA and multiple range test. Mucus values in μg ml^{-1}. Food regurgitation commenced on day 21, but no mortalities occurred during the study.

34 R.D. HANDY

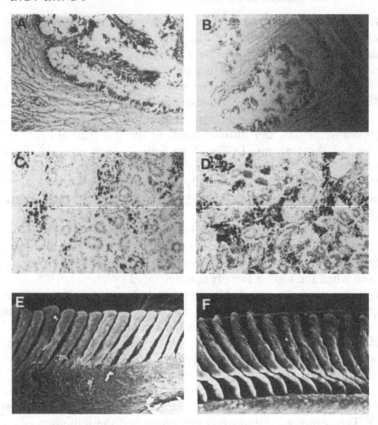

Fig. 2. Histopathology in rainbow trout exposed to 10 g kg^{-1} dw mercuric chloride via the diet for 42 days. Transverse sections of gut in A control, B exposed fish, and TS of kidney in control C and exposed D trout. Sections (8 μm, mag. × 250) were stained with haematoxylin and eosin. Electron micrographs of the gills of control E and exposed F animals show no pathology.

dust particles (Turnpenny *et al.*, 1988; Galvin, 1991). Anthropogenic sources of the metal are mainly derived from water treatment processes and pulp mill effluent (e.g. Benschoten & van Edzwald, 1990). There is no evidence of Al biomagnification through the aquatic food chain. Nevertheless, invertebrates typically contain 37–2130 μg g^{-1} ww of Al (Wren & Stephenson, 1991), while whole body concentrations in fish are about 1–20 μg g^{-1} ww (Brumbaugh & Kane, 1985; Cleveland, Buckler & Brumbaugh, 1991).

There have been numerous studies on waterborne Al exposure in fish which demonstrate toxicity, accumulation responses, and physiological effects (for review see Spry & Wiener, 1991). In contrast, there is very little information on the effects of oral doses of the metal and only a small amount of data are published (Park & Shimizu, 1989; Poston, 1991; Handy, 1993).

The exact mechanism of Al uptake across the gut of animals is unclear at present, mainly because the pH-dependent luminal speciation of the metal has not been determined. Evidence from mammalian studies suggest Al is absorbed across the intestine by an energy-dependent carrier-mediated process, which apparently interacts with both Na and Ca transport. The metal might also cross the small intestine via a paracellular route as a diffusible citrate complex, but this complex also shows saturable uptake kinetics, perhaps indicating transcellular absorption. In addition, Al^{3+} might diffuse across the stomach wall of mammals (Stewart, 1989; Van der Voet, 1992).

The mechanism of absorption in fish is completely unknown. The fact that mammals can absorb Al in the acidic, proximal duodenum (Ondreicka, Kortus & Ginter, 1971; Kaehny *et al.*, 1977; Van der Voet, 1992) implies that fish might absorb the metal in a correspondingly acidic region, perhaps the stomach and pyloric appendices. However, the protective mucus barrier covering the gut maintains circumneutral pH values adjacent to the epithelial surface, even when the lumen is strongly acidic. Thus much of the soluble Al from the lumen will be precipitated and bound to mucus before it can be absorbed. Consequently the bioavailability of oral doses of the metal is low in mammals (1–2% Kaehny *et al.*, 1977; Recker *et al.*, 1977; Yokel & McNamara, 1985) and rainbow trout (Handy, 1993).

The excretion of Al is also controversial. Aluminium is excreted in the urine of mammals, but the renal mechanisms by which this is achieved are not understood (Kaehny *et al.*, 1977; Recker *et al.*, 1977; Stewart, 1989). Excretion in the bile might occur given the high Al content of mammalian livers compared to other organs (Fairweather-Tait *et al.*, 1991). On the other hand, patients with renal failure accumulate the metal rather than excrete it by an alternative route (Stewart, 1989). Excretion mechanisms have not been specifically investigated in fish, but the appearance of Al in the external mucus of rainbow trout fed a metal-enriched diet indicates mucus secretion as a means of detoxification. Elevation in gill Al levels relative to the liver or kidney might also indicate branchial excretion (Table 2). Urinary Al levels have not been measured in freshwater species, and renal excretion is presumably negligible in marine fish because of their low glomerular filtration rates.

Once Al has crossed the intestinal barrier it is carried in the blood to most internal organs of mammals, but is particularly accumulated in the bone and nervous tissues (Recker et al., 1977; Sideman & Manor, 1982). Dietary sources of Al also appear in the liver, kidney, gills and muscle of rainbow trout (Table 2), but do not cause increased Al deposition in the vertebrae of Atlantic salmon, Salmo salar (Poston, 1991) or the bones of the Japanese eel, Anguilla japonica (Park & Shimizu, 1989).

Trace levels of Al may have some nutritional benefit in fish (Poston, 1991), but the transition between nutritional value and toxicity cannot be identified owing to the paucity of sublethal effects data. Dietary supplements which increased the Al content of food to about 80 mg kg^{-1} ww enhanced the growth and condition factor of Japanese eels (Park & Shimizu, 1989), while 2 g Al kg^{-1} dw had no beneficial or adverse effect on the growth or survival of Atlantic salmon (Poston, 1991). The slightly higher Al concentration of 10 g kg^{-1} dw of food produced signs of toxicity in rainbow trout (Table 3). Food regurgitation occurred in exposed fish after 30 days, usually 1–2 hours after the daily meal, resulting in a reduced oral dose and consequent reductions in tissue Al levels (Table 2). Melanomacrophage activity in the kidney also increased, but damage to the gills and gastrointestinal tract were notably absent, except for marginal changes in branchial mucus and chloride cell numbers (Table 3). Moreover, the maintenance of body weight and the occurrence of only one mortality in the rainbow trout experiment suggest that acute lethal toxicity would not be reached with 10 g Al kg^{-1} dw of food. If a ration size of 2% body weight per day is assumed, a sublethal dose of 20 µg 100 g^{-1} fish day^{-1} is derived for the trout. A lethal oral dose for salmonids would presumably be greater. There are no data on the sublethal physiological effects of oral doses of Al. Influences on nutrient absorption and digestion are also unknown in fish, although disturbances to phosphate metabolism and iron absorption are possible given their occurrence in mammalian systems (Ondreicka et al., 1971; Fairweather-Tait et al., 1991).

Cadmium

Cadmium is a rare metal found primarily as a sulphide in Zn-containing ores and can therefore be found at low levels (about 0.4 µg l^{-1}) in natural waters. Contamination of the aquatic environment mainly arises as leachate from coal ash and fertilizers, the dumping of sewage sludge

Table 2. *Aluminium concentrations in organs of rainbow trout fed 10 g Al kg^{-1} dw of food for 42 days*

Time (days)	Al concentration in tissues (μg g^{-1} ww)					
	Muscle	Gill	Liver	Kidney	Mucous	Whole body
Control	0.17±0.05(6)	0.29±0.05(5)	0.22±0.06(7)	0.60±0.07(7)	0.04±0.005(5)	0.12 0.17
7	0.16±0.06(5)	*0.52±0.10(7)	*0.74±0.32(7)	0.85±0.08(7)	0.07±0.02(9)	0.10 0.23
14	0.13±0.03(7)	*0.84±0.24(7)	*0.63±0.17(7)	0.69±0.03(7)	0.06±0.02(9)	0.75 0.04
21	*0.39±0.09(5)	*3.74±1.00(7)	*3.0±0.70(7)	*4.6±1.71(7)	*0.36±0.07(9)	1.32
28	—	*2.80±0.97(5)	*4.29±0.97(5)	*1.46±0.30(7)	*0.42±0.09(9)	1,32
35	*1.15±0.13(7)	*2.76±0.830(7)	*4.36±1.12(7)	*5.07±0.55(7)	*0.66±0.12(9)	8.53
42	*0.57±0.09(7)	*6.10±2.39(7)	*3.12±0.54(7)	*3.73±0.29(6)	*0.58±0.12(5)	7.63

Data are means ± S.E., except for whole body Al concentrations which are for individual animals. Values in parenthesis indicate the number of observations for each mean. * significant difference from controls by analysis of variance and multiple range test. Significance level, $P = 0.05$ Control data are pooled values for control animals on days 7 and 14. Food regurgitation commenced on day 30 of exposure.

Table 3. *Toxicity and pathology in rainbow trout exposed to 10 g metal kg^{-1} dw of food*

Metal	Exposure duration	Mortality	Food regurgitation	Foregut lesions	Hindgut lesions	Gills	Kidney
Hg	42 days	none	yes (day 21+)	yes (severe)	yes (slightly)	normal	increased melanomacrophages
Al	42 days	one fish	yes (day 30+)	yes (slight)	none	normal with slight increase in Cl and mucus cell numbers	increased melanomacrophages
Cd	28 days	42%	yes (day 7+)	no data	no data	no data	no data
Cu	28 days	one fish	yes (day 20+)	yes (slight)	none	normal, rare swelling at base of lamellae	increased melanomacrophages

at sea, and the precipitation of flue dust (Craig, 1986). Many industrial effluents also contain Cd, but the disposal of such wastes in freshwater are strictly regulated; receiving waters should not contain more than 5 µg l^{-1} Cd in the UK (ENDS, 1992). Invertebrates may become heavily polluted with bioconcentration factors of up to 4000 (Coombs, 1979; Dallinger & Kautzky, 1985; Krantzberg & Stokes, 1989; Yevtushenko *et al.*, 1990; Wren & Stephenson, 1991). Early estimates of Cd levels are between 0.03 and 295 µg g^{-1} dw for invertebrates (Coombs, 1979), recent values are more conservative (0.13–56.6 µg g^{-1} dw, Wren & Stephenson, 1991) with a mean of about 3 µg g^{-1} dw. These high levels of contamination are not biomagnified through the aquatic food chain to fish (Taylor, 1983; Dallinger & Kautzky, 1985). Consequently fish contain lower Cd levels of around 0.02–0.09 µg g^{-1} dw (Coombs, 1979; Spry & Wiener, 1991).

The mechanism of gastrointestinal Cd uptake has not been specifically investigated in fish, but circumstantial evidence from accumulation studies suggest that the mechanism is similar to that in mammals. Cadmium is primarily absorbed across the bush border of the duodenum in mice. The metal is initially bound to the luminal surface of mucosal cells by electrostatic attraction and is then 'internalized' by a process related to membrane fluidity. This process apparently requires no energy and is therefore not pinocytosis or active transport. Once inside the mucosal cells, Cd is rapidly complexed by cytosolic metallothionein (MT) to form a Cd–MT complex. Cadmium ions derived from the MT complex eventually cross the basolateral membrane into the blood, perhaps via a Ca^{2+}-dependent ATPase (Andersen, 1989).

The situation in fish is probably similar. Cadmium speciation is not particularly affected by pH and thus the metal will probably exist in the Cd^{2+} form or as cationic chloride complexes in the gut. These ions will probably irreversibly bind to the fixed anionic matrix of gut mucus, since binding efficiencies in fish mucus are around 95% (Pärt & Lock, 1983). Much of the bound toxicant will be sloughed into the lumen and thus the oral bioavailability of Cd is only 1–2% of the ingested dose (Haesloop & Schirmer, 1985; Harrison & Klaverkamp, 1989; Handy, 1992; Weber *et al.*, 1992). The mechanism of uptake across the mucosal cell surface is unknown, although active uptake by a Ca/Mg ATPase has been suggested in fish (Pratap *et al.*, 1989). Once the metal is inside mucosal cells it is bound to a low molecular weight material in the cytosol, which is presumably MT, before entering the blood (Weber *et al.*, 1992). Cadmium probably crosses the basolateral membrane of mucosal cells by a Na$^+$/Ca^{2+} exchanger, but a Ca^{2+}-ATPase may also be involved (Schoenmakers *et al.*, 1992).

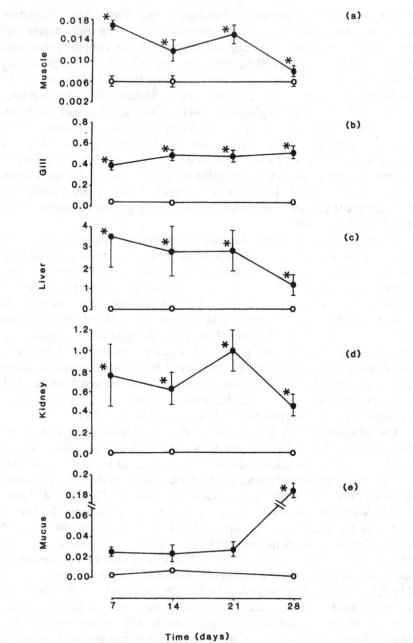

Accumulated Cd is mainly excreted in the bile of mammals, but may also be lost with the urine. Large quantities of the metal are also lost in the faeces when intestinally adsorbed Cd is sloughed with mucosal cells (Andersen, 1989). Similarly in fish, intestinal sloughing may account for about 66% of the apparent excretion during depuration, secretion into the bile might also occur (Haesloop & Schirmer, 1985; Harrison & Klaverkamp, 1989). Unlike mammals, fish do not excrete accumulated Cd in the urine (Giles, 1988). Instead, Cd is excreted by the gills (Harrison & Klaverkamp, 1989; Pratap & Wendelaar Bonga, 1993). The mechanism of branchial Cd excretion is unknown, but might not arise from diffusion since gill pavement cell tight junctions remain intact (Pratap & Wendelaar Bonga, 1993).

Dietary Cd is usually accumulated in most internal organs of fish. The muscle may show minimal or negligible contamination (Haesloop & Schirmer, 1985; Handy, 1992), except when high toxic doses are used (Fig. 3). The gut supports most of the body burden (60–90% Harrison & Klaverkamp, 1989; 76% Handy, 1992), reflecting the adsorption of Cd by the intestinal mucosa. Cadmium is carried in the blood (Koyama & Itazawa, 1977a; Harrison & Klaverkamp, 1989; Handy, 1992), probably as a Cd–MT complex (Weber *et al.*, 1992). The metal also accumulates in the liver, kidney, spleen, gall bladder and the bone to a lesser extent (Koyama & Itazawa, 1977b, 1977c; Haesloop & Schirmer, 1985; Harrison & Klaverkamp, 1989; Handy, 1992). Surprisingly, the gills of fish may support up to 10% of the body burden and can even contain more Cd than the liver (e.g. rainbow trout Harrison & Klaverkamp, 1989; Handy, 1992). Branchial contamination persists well into the depuration phase of exposure implicating the gills as an excretory organ, a notion supported by histological observations (Pratap & Wendelaar Bonga, 1993). On the other hand, high levels of contamination in the gill might also be associated with the regurgitation of toxic food (Fig. 3).

Oral toxicity data for Cd are limited and median lethal concentrations have not been determined. Nevertheless, the information is not as

---

Fig. 3. Cadmium concentration ($\mu g\ g^{-1}$ ww) in selected tissues of the rainbow trout fed normal pellets (open circles) and pellets enriched with 10 g kg^{-1} Cd dw (closed circles) for 28 days. * significant difference from control by ANOVA and multiple range test. Mucus values in $\mu g\ ml^{-1}$. Food regurgitation commenced on day 7 and 42% mortality occurred during the study.

sparse as that for Hg or Al. Exposure to 10 g Cd kg^{-1} dw of food over 28 days in rainbow trout, equivalent to 68 mg kg^{-1} fish day^{-1}, caused 42% mortality and should therefore be close to the LC$_{50}$ dose (Fig. 3). Interestingly, this is within the range of acutely lethal doses reported for rodents (50–400 mg kg^{-1}, Andersen, 1989). In a previous study with trout (Handy, 1992) 1.4 mg Cd kg^{-1} fish day^{-1} caused only 3% mortality and therefore represents a sublethal dose during acute exposure. This compares to 1.14 mg kg^{-1} fish day^{-1} which is the average dose of the waterborne LC$_{50}$ for trout ventilating at 37 ml min^{-1} in soft water (Cusimano, Brakke & Chapman, 1986). Guppies, *Poecilia reticulata*, experienced 17% compared to >50% mortality when exposed via the food and water respectively for 30 days (Hatakeyama & Yasuno, 1982). Clearly, orally administered Cd is much less toxic than water-borne exposure.

Sublethal effects include hypocalcemia (Pratap *et al.*, 1989), which may be partially ameliorated by the mobilization of bone Ca deposits in freshwater fish, or by Ca uptake from seawater in marine species (Koyama & Itazawa, 1977a,b, c). Blood osmolality and NaCl levels are unaffected by dietary Cd, even though branchial Na$^+$K$^+$ATPase activity may be inhibited (Pratap *et al.*, 1989; Pratap & Wendelaar Bonga, 1993). Interestingly, Cd exposure increases the Cu content of the liver and serum of fish by the induction of Cu-binding proteins (Weber *et al.*, 1992). Histopathology includes the appearance of granular deposits in the liver, atrophy of the proximal renal tubules, and increases in chloride cell turnover at the gills (Koyama & Itazawa, 1977b; Pratap & Wendelaar Bonga, 1993). There is no information on metabolic effects of dietary Cd, although inhibition of liver oxidase and ATPase enzymes might occur, along with the induction of acetylcholine esterase synthesis to produce hyperexcitability (see Coombs, 1979). Effects on digestion are also unknown.

Copper

Copper is found in a variety of mineral deposits and therefore occurs naturally in seawater (1.5–58.2 µg l^{-1} Bernhard & George, 1986) and freshwater (< µg l^{-1} EIFAC, 1976). Pollution of the aquatic environment is mainly anthropogenic since Cu is used in many industrial processes and occurs in effluents at concentrations between 40 000 and 90 000 µg l^{-1} (Craig, 1986). The speciation of copper is pH-dependent with Cu^{2+} being the predominant dissolved form in circumneutral waters (Sylva, 1976; Cusimano *et al.*, 1986). However, the divalent ion is rapidly complexed in the presence of organic matter, and for this reason sediments and invertebrates may accumulate copper. Levels of contamination in invertebrates is often around 10–100 µg g^{-1} dw, but may be >1 mg g^{-1} dw in animals found near effluent discharges (Dallinger & Kautzky, 1985; Yevtushenko *et al.*, 1990; Van Hattum, Timmermans & Govers, 1991; Miller *et al.*, 1992). Some of this contamination is passed on to fish through the aquatic food chain (Patrick & Loutit, 1978; Dallinger & Kautzky, 1985), but is not biomagnified since fish are able to regulate copper accumulation. The average Cu concentration in fish from the USA is 0.65 µg g^{-1} ww (Schmitt & Brumbaugh, 1990), but contamination may range between <1 and 22 µg g^{-1} dw (Bohn & Fallis, 1978; Dallinger & Kautzky, 1985; Laurén & McDonald, 1987).

The gastrointestinal absorption of Cu by fish has not been specifically investigated, but the uptake mechanism may be derived by comparing absorption in mammals with that of fish. Copper is absorbed almost exclusively in the small intestine of mammals and probably enters the gut mucosal cells by diffusion as the Cu^{2+} ion (Linder, 1991). Small amounts of Cu may also be pinocytosed as Cu–protein complexes (Mills, 1986). Some of the luminal Cu will be precipitated in the gut mucus layer, and in the case of ruminants, form complexes with sulphide or molybdenum which are either insoluble or unavailable for uptake (Mills, 1986). The bioavailability of Cu is relatively high compared to Cd or Al, with absorption ranging between 28 and 75% in man (Linder, 1991), but as low as 10% in cattle (Buckley, Huckin & Eigendorf, 1985). The mechanism by which Cu crosses the basolateral surface of the mucosal cells is unclear, but involves complexation with MT for cytosolic storage, then transport by a carrier-mediated process (Linder, 1991). The metal is temporarily bound to plasma albumins en route to the liver, where it is incorporated into ceruloplasmin before bulk transport in the blood to other internal organs (Linder, Weiss & Wirth, 1985; Goode, Dinh & Linder, 1989).

In fish, Cu^{2+} will be complexed by mucus present on the gut wall (Miller & Mackay, 1982). This might explain the low the bioavailability of dietary Cu in rainbow trout (3.2% Handy, 1992). The mechanism by which the metal crosses the gut epithelium of fish is unknown, although circumstantial evidence suggests that MT in the mucosal cells sequesters Cu (Lanno, Singer & Hilton, 1985*a*). Once in the blood, the metal is incorporated into a plasma protein which is probably ceruloplasmin, and as with mammals, a small percentage of the toxicant is associated with erythrocyte membranes (Bettger *et al.*, 1987).

Copper excretion in mammals occurs primarily in the bile (10–75% of the dose Linder, 1991), but also in the urine (2–7% of the dose

Johnson, 1989; Linder, 1991). Renal excretion in marine species of fish is presumably negligible because of their low glomerular filtration rates. Urinary Cu levels have not been measured in freshwater fish, although dietary sources of the metal are accumulated in the kidney (Lanno et al., 1985a; Lanno, Singer & Hilton, 1985b; Handy, 1992). However, Cu deposits in the kidney of fish cause tubular necrosis and such renal dysfunction is incompatible with urinary excretion (Benedeczky et al., 1991). Lanno, Hicks and Hilton (1987) suggests biliary excretion as an important mode of detoxification since Cu-rich granules are deposited in the liver. On the other hand, contamination of the gut does not persist in the depuration phase of exposure (Handy, 1992) suggesting that excretion via the bile might be limited. Branchial excretion of dietary Cu has not been demonstrated, but the gills of fish are involved in copper homeostasis (Laurén & McDonald, 1986). Copper deposition in the gills might indicate branchial Cu excretion (Dallinger & Kautzky, 1985; Handy, 1992), but might also reflect food regurgitation into the buccal cavity during acute toxicity (see Fig. 4). Small amounts of dietary Cu are also secreted in the mucus of trout (Handy 1992; Fig. 4).

Dietary Cu accumulates in the intestine, liver, kidney, heart, pancreas, thymus, spleen and to a lesser extent in muscle of mammals (Koh et al., 1989; Failla, Babu & Seidel, 1988; Linder, 1991). This general pattern of bioaccumulation is repeated in fish with uptake by the liver, kidney, blood, gills and skin (Murai, Andrews & Smith, 1981; Lanno et al., 1985a,b, 1987; Handy 1992; Fig. 4). Similar to mammals, the muscle of fish does not become heavily contaminated (Murai et al., 1981; Handy, 1992), except perhaps with very high oral doses (Fig. 4). The distribution of the body burden also reflects the route of uptake in fish with 53%, 11.9% and 1% of the contamination associated with the gut, liver and gills respectively (Handy, 1992).

There are few toxicity data for oral doses of the metal even though copper is an essential trace mineral for fish (Ogino & Yang, 1980). The transition between nutritional benefit and oral toxicity has, how-

Fig. 4. Copper concentration ($\mu g \, g^{-1}$ ww) in selected tissues of the rainbow trout fed normal pellets (open circles) and pellets enriched with 10 g kg^{-1} Cu dw (closed circles) for 28 days. * significant difference from control by ANOVA and multiple range test. Mucus values in $\mu g \, ml^{-1}$. Food regurgitation commenced on day 20 and only one mortality occurred during the study.

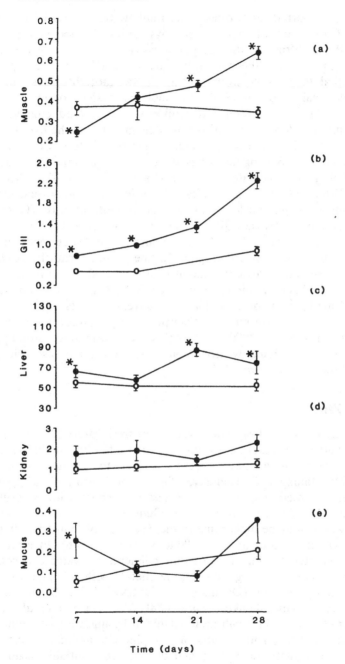

Time (days)

ever, been reported for two fish. The rainbow trout survived exposure to 10 g Cu kg^{-1} dw of food for 28 days, although food regurgitation and epithelial lifting in the foregut were observed (see also Fig. 4). Dietary exposure to 3088 mg Cu kg^{-1} dw of food for 8 weeks proved more lethal to rainbow trout causing 28.8% mortality, even though gastrointestinal legions were absent. Sublethal toxicity occurs at about 730 mg Cu kg^{-1} dw of food and manifests as decreased growth, food refusal, and the deposition of Cu-containing granules in the liver (Lanno et al., 1985b, 1987). The maximum tolerable dietary Cu level for rainbow trout is 665 mg kg^{-1}, based on normal plasma Cu, glucose, haemoglobin and haematocrit values (Lanno et al., 1985b). Knox et al. (1982) found that 150 mg Cu kg^{-1} dw of food had no adverse effects on plasma ions, plasma levels of aspartic aminotransferase (GOT) or alanine aminotransferase (GPT) which are normally lost from the liver during toxicity. Levels of the Cu-containing enzyme superoxide dismutase (SOD) were also unchanged. In the channel catfish, Ictalurus punctatus, sublethal toxicity manifests as reduced growth rates and anemia with diets containing more than 8 mg Cu kg^{-1} dw, but levels of 1.5–4 mg kg^{-1} dw of food have no adverse effects (Murai et al., 1981). Toxic effects on nutrient absorption and digestion are unknown for most fish. However, 800 mg Cu kg^{-1} dw of food reduces iron uptake from the gut of rainbow trout, but does not affect ascorbic acid metabolism (Lanno et al., 1985a).

Zinc

This metal occurs at low concentrations in freshwaters (20 µg l^{-1}) due to the natural weathering of mineral deposits, but pollution from anthropogenic sources may increase Zn levels to 150 µg l^{-1} (Miller et al., 1992; Dallinger & Kautzky, 1985; Van Hattum et al., 1991). Seawater collected from the open ocean may contain lower levels of total zinc (10–600 ng l^{-1} Bernhard & George, 1986), but pollution in coastal waters may be higher due to the discharge of industrial effluents with Zn concentrations of 20–30 000 µg l^{-1} (Craig, 1986). Invertebrates become polluted by Zn uptake from sediments or water and contain between 8 and 1290 µg g^{-1} dw of the metal, with 150 µg g^{-1} dw an average body burden (Dallinger & Kautzky, 1985; Krantzberg & Stokes, 1989; Yevtushenko et al., 1990; Van Hattum et al., 1991; Miller et al., 1992). Fish will accumulate Zn by ingesting contaminated invertebrates, but do not biomagnify trophic contamination since fish are able to regulate Zn uptake and do not preferentially accumulate the toxicant in the muscle. Consequently, Zn levels in fish are generally lower than that of invertebrates from the same watercourse (Patrick &

Loutit, 1978; Dallinger & Kautzky, 1985). Body burdens of Zn in US freshwater fish are an average of 21.7 μg g^{-1} dw but may reach 118 μg g^{-1} ww (Schmitt & Brumbaugh, 1990).

The intestinal uptake of Zn has been relatively well studied in mammals and evidence suggests that the mechanism of uptake is similar in fish. In mammals, Zn is complexed in the intestinal mucus before it enters the mucosal cells by a process which is probably facilitated diffusion. The metal may also diffuse across the gut epithelium as low molecular weight complexes, particularly as Zn–citrate. Once inside the mucosal cell, Zn is complexed by MT, but is eventually transported into the blood by a carrier mediated process located on the basolateral surface of the mucosal cells. Interestingly, the Zn uptake pathway is also used for Cu and thus copper status will have a considerable effect on Zn absorption and vice versa (Mills, 1986; Linder, 1991; Wiseman & Aggett, 1985; Parry *et al.*, 1985). The bioavailability in mammals is usually about 20%, but can be much higher (70%) in Zn deficient animals (Sandström *et al.*, 1985; Matsusaka, Berg & Kollmer, 1985). Once Zn has entered the blood plasma, roughly 30% is irreversibly bound to alpha-2-macroglobulin, the remainder is reversibly complexed with mainly albumin but also transferrin (Favier, Faure & Arnaud, 1985).

Dietary Zn uptake in fish occurs primarily in the upper region of the small intestine and is initiated by ionic adsorption on to the soluble proteins on the intestinal mucosa (Shears & Fletcher, 1979, 1983; Hardy, Sullivan & Koziol, 1987). The metal then enters the mucosal cells by an unknown process where it is incorporated into a low molecular weight protein, which is presumably MT (Shears & Fletcher, 1979; Weber *et al.*, 1992). A basolateral carrier has not been identified in the mucosal cells of fish, and unlike mammals, the transfer of Zn into the blood might be a passive process (Shears & Fletcher, 1983). However, once in the blood, Zn is complexed by plasma proteins which show similar binding characteristics to those of mammals, indicating complexation with albumins (Fletcher & Fletcher, 1978; Bettger *et al.*, 1987). Similar to mammals, the Zn uptake pathway in fish is also probably used by Cu since the absorption of the metals appears antagonistic at high doses (Shears & Fletcher, 1979, 1983; Knox *et al.*, 1984). The bioavailability of dietary Zn covers a broad range in fish (35–72.8%), but values of about 70% are considered typical for animals with normal homeostasis (Pentreath, 1973; Milner, 1982; Hardy *et al.*, 1987).

Zinc excretion in mammals, as with Cu, is principally via the bile and intestinal sloughing, with only a small urinary loss. Evidence from radiotracer studies of Zn homeostasis in rainbow trout suggests that the situation in fish is similar, with about 27% of the dose lost with the faeces and <1% excreted in the urine (Hardy et al., 1987; Nakatani, 1966). This is hardly surprising given the extremely low urinary excretion rate for Zn in trout (15 nequiv kg^{-1} h^{-1} Spry & Wood, 1985). Fish also excrete dietary Zn across the gills; trout excreted 13.7% of the ingested dose in the study of Hardy et al. (1987).

The accumulation of dietary Zn in mammals reflects the route of uptake with the greatest contamination in the gut>carcass>kidney> liver. The heart, blood, brain and bones may also become contaminated (Matsusaka et al., 1985; Tjioe & Van den Hamer, 1985). The situation in rainbow trout is similar with the gut supporting 43% of the body burden and remainder distributed in the liver>gill>kidney>spleen> skin>gonad>blood>bile>eye>bone>muscle (Hardy et al., 1987). Fish are able to retain a third to a half of the absorbed Zn. The percentage retained in the rainbow trout is 35–58.8% (Hardy & Shearer, 1985; Hardy et al., 1987) and 17–47% in the plaice, Pleuronectes platessa (Milner, 1982).

Zinc is an essential trace element and there have been several studies on the nutritional requirements of fish, but there is very little information on the sublethal toxic effects of dietary Zn and no data on acute oral toxicity. Rainbow trout require 15–30 μg g^{-1} Zn in purified diets, but may need 150 μg g^{-1} Zn in foods containing high levels of Ca or P to avoid deficiency (Ogino & Yang, 1978; Ketola, 1979; Hardy & Shearer, 1985). Rainbow trout can tolerate 1700 μg g^{-1} Zn in the food without lethal toxicity, even though the metal accumulated in the gills, liver and blood (Wekell, Shearer & Houle, 1983). Metabolic effects include reduced Mn-superoxide dismutase activity in the liver of trout fed 1000 μg g^{-1} Zn (Knox et al., 1984). Surprisingly, dietary Zn does not induce MT in the liver or kidney of turbot, Scophthalmus maximus (Overnell, Fletcher & McIntosh, 1988). Elevating dietary Zn from 1 to 590 μg g^{-1} dw in the presence of 7 μg l^{-1} dissolved Zn reduced the Na, Cl, K, Ca, Mn, V and Br whole body concentrations in trout, but increased Fe uptake without any obvious toxic effects (Spry, Hodson & Wood, 1988). Toxic effects of high dietary Zn on digestion are unknown in fish.

Relative accumulation from food and water

Metals are usually toxic to fish at μg l^{-1} (ng ml^{-1}) levels when exposure is via the water (see Spry & Weiner, 1991). In contrast, it seems that dietary sources of metals have sublethal toxicity at about 0.1– 1 mg g^{-1} dw of food and that acutely lethal oral doses probably exceed

Table 4. Relative contributions of dietary and waterborne sources to metal accumulation

Species	Metal	[Water] (ng ml^{-1})	[Food] (µg g^{-1} dw)	% of body burden from the diet	Reference
Rainbow trout (*Oncorhynchus mykiss*)	Zn	7	590	43	Spry *et al.* (1988)
Plaice (*Pleuronectes platessa*)	Zn	100 600	2.47–3.12 2.47–3.12	>90 50	Milner (1982)
Mosquito fish (*Gambusia affinis*)	Zn	0.0015	61	78	Willis & Sunda (1984)
Juvenile spot (*Leiostomus xanthurus*)	Zn	0.0015	61	82	Willis & Sunda (1984)
Rainbow trout (*Oncorhynchus mykiss*)	Pb	3–120	3.8–117.9	Negligible	Hodson *et al.* (1978)
Rainbow trout (*Oncorhynchus mykiss*)	Cd	0.00125	0.0307	*68	Harrison & Klaverkamp (1989)
Lake Whitefish (*Coregonus clupeaformis*)	Cd	0.00125	0.0674	*75	Harrison & Klaverkamp (1989)
Rainbow trout (*Oncorhynchus mykiss*)	CH$_3$–Hg	0.33	3.08	71	Philips & Buhler (1978)

*estimated from accumulation data assuming dietary and waterborne sources are additive.

1 g g^{-1} dw of food. This huge difference in toxicity might suggest that uptake from the water greatly exceeds that from the gut.

There have only been a few studies which compare uptake from the water and food simultaneously, or within a series of experiments (Hodson, Blunt & Spry, 1978; Philips & Buhler, 1978; Milner, 1982; Willis & Sunda, 1984; Spry et al., 1988; Harrison & Klaverkamp, 1989). All of these experiments have used sub-lethal concentrations of metals in the water to allow comparison with gastric uptake in the absence of mortality. Consequently, metal concentrations in the water have been much lower than in the food, and perhaps not surprisingly, the toxicant body burden appears to be derived principally from the food except in the case of lead (Table 4). A comparison based on exposure dose would be more reliable, but most authors have either failed to record branchial ventilation rates and/or ration size and thus dose is impossible to calculate. However, Harrison and Klaverkamp (1989) calculated the exposure doses for Cd in the food and water for two salmonids. In their experiments rainbow trout and lake whitefish, *Coregonus clupeaformis*, received only 23% and 32% of the total dose in the food, respectively. However, both fish accumulated at least twice as much Cd from the food compared to the water. Thus it is possible that fish absorb metals primarily from the food when waterborne levels are 'normal'. This situation is reversed for Zn when dissolved concentrations become toxic, since 100% of the body burden can be explained by uptake from the water (Spry et al., 1988). Branchial uptake might therefore only become important in cases of waterborne toxicity.

Acknowledgement

The experimental work on trout was completed while the author was in receipt of a NERC fellowship at Dundee University. William Penrice prepared all the histology in the experiments.

References

Andersen, O. (1989). Oral cadmium exposure in mice: toxicokinetics and efficiency of chelating agents. *Critical Reviews in Toxicology*, **20**, 83–112.

Aoyama, I., Inoue, Y. & Inoue, Y. (1978a). Experimental study on the concentration process of trace elements through a food chain from the view point of nutritional ecology. *Water Research*, **12**, 831–6.

Aoyama, I., Inoue, Y. & Inoue, Y. (1978b). Simulation analysis of the concentration process of trace metals by aquatic organisms from the view point of nutritional ecology. *Water Research*, **12**, 837–42.

Baatrup, E. & Doving, K.B. (1990). Histochemical demonstration of mercury in the olfactory system of salmon (*Salmo salar* L.) following treatments with dietary methylmercuric chloride and dissolved mercuric chloride. *Ecotoxicology and Environmental Safety*, **20**, 277–89.

Baatrup, E., Doving, K.B. & Winberg, S. (1990). Differential effects of mercurial compounds on the electroolfactogram (EOG) of salmon (*Salmo salar* L.). *Ecotoxicology and Environmental Safety*, **20**, 269–76.

Backström, J. (1969). Distribution studies of mercuric pesticides in quail and some freshwater fishes. *Acta Pharmacologica et Toxicologica*, **27**, 5–103.

Benedeczky, I., Nemcsok, J., Albers, C. & Götz, K.M. (1991). Effect of hypoxia and copper sulphate on the structure of liver and kidney of carp. In *Bioindicators and Environmental Management*. D.W. Jeffrey & B. Madden (eds.) pp. 379–87. Academic Press, London.

Benschoten, J.E. & van Edzwald, J.K. (1990). Chemical aspects of coagulation using aluminium salts I. Hydrolytic reactions of alum and polyaluminium chloride. *Water Research*, **24**, 1519–26.

Bernhard, M. & George, S.G. (1986). Importance of chemical species in uptake, loss, and toxicity of elements for marine organisms. In *The Importance of Chemical 'Speciation' in Environmental Processes*. Bernhard, M., Brinchman, F.E. & Sadler, P.J. (eds.). pp. 385–422. Springer-Verlag, Berlin, Heidelberg.

Bettger, W.J., Spry, D.J., Cockell, K.A., Cho, C.Y. & Hilton, J.W. (1987). The distribution of zinc and copper in plasma, erythrocytes and erythrocyte membranes of rainbow trout (*Salmo gairdneri*). *Comparative Biochemistry and Physiology*, **87C**, 445–51.

Bloom, N.S. (1992). On the chemical form of mercury in edible fish and marine invertebrate tissue. *Canadian Journal of Fisheries and Aquatic Sciences*, **49**, 1010–17.

Bohn, A. & Fallis, B.W. (1978). Metal concentrations (As, Cd, Cu, Pb and Zn) in shorthorn sculpins, *Myoxocephalus scorpius* (Linnaeus) and arctic char, *Salvelinus fontinalis* (Linnaeus), from the vicinity of Strathcona Sound, Northwest Territories. *Water Research*, **12**, 659–63.

Boudou, A. & Ribeyre, F. (1985). Experimental study of trophic contamination of *Salmo gairdneri* by two mercury compounds–$HgCl_2$ and CH_3HgCl–analysis at the organism and organ level. *Water, Air, and Soil Pollution*, **26**, 137–48.

Brumbaugh, W.G. & Kane, D.A. (1985). Variability of aluminum concentrations in organs and whole bodies of smallmouth bass (*Micropterus dolomieui*). *Environmental Science and Technology*, **19**, 828–31.

Buckley, W.T., Huckin, S.N. & Eigendorf, G.K. (1985). Stable isotope tracer methods for determining absorption of dietary copper in dairy cattle. In *Trace Elements in Man and Animals*. Mills, C.F., Bremner, I. & Chesters, J.K. (eds.). pp. 678–82. Commonwealth Agricultural Bureau, Slough.

Cleveland, L., Buckler, D.R. & Brumbaugh, W.G. (1991). Residue dynamics and effects of aluminum on growth and mortality in brook trout. *Environmental Toxicology and Chemistry*, **10**, 243–8.

Coombs, T.L. (1979). Cadmium in aquatic organisms. In *The Chemistry, Biochemistry and Biology of Cadmium*. M. Webb (ed.) pp. 93–139. Elsevier/North-Holland Biomedical Press, Amsterdam.

Cope, W.G., Wiener, J.G. & Rada, R.G. (1990). Mercury accumulation in yellow perch in Wisconsin seepage lakes: relation to lake characteristics. *Environmental Toxicology and Chemistry*, **9**, 931–40.

Craig, P.J. (1986). Chemical species in industrial discharges and effluents. In *The Importance of Chemical 'Speciation' in Environmental Processes*. Bernhard, M., Brinckman, F.E. & Sadler, P.J. (eds.) pp. 443–64. Springer-Verlag, Berlin, Heidelberg.

Cusimano, R.F., Brakke, D.F. & Chapman, G.A. (1986). Effect of pH on the toxicities of cadmium, copper, and zinc to steelhead trout (*Salmo gairdneri*). *Canadian Journal of Fisheries and Aquatic Sciences*, **43**, 1497–503.

Dallinger, R. & Kautzky, H. (1985). The importance of contaminated food for the uptake of heavy metals by rainbow trout (*Salmo gairdneri*): a field study. *Oecologia* (Berlin), **67**, 82–9.

Dallinger, R., Prosi, F., Segner, H. & Back, H. (1987). Contaminated food and uptake of heavy metals by fish: a review and proposal for further research. *Oecologia*, (Berlin), **73**, 91–8.

EIFAC (1976). Report on copper and freshwater fish. *European Inland Fisheries Advisory Commission Technical Paper 27*. pp. 1–21. FAO, Rome.

ENDS (1992). *Dangerous Substances in Water: A Practical Guide*. Environmental Data Services Ltd, London.

Failla, M.L., Babu, U. & Seidel, K.E. (1988). Use of immunoresponsiveness to demonstrate that the dietary requirement for copper in young rats is greater with dietary fructose than dietary starch. *Journal of Nutrition*, **118**, 487–96.

Fairweather-Tait, S.J., Piper, Z., Fatemi, S.J.A. & Moore, G.R. (1991). The effect of tea on iron and aluminium metabolism in the rat. *British Journal of Nutrition*, **65**, 61–8.

Favier, A.A., Faure, H. & Arnaud, J. (1985). Determination of ultrafilterable zinc, transferrin bound and albumin bound zinc using ultrafiltration and flameless A.A.S. In *Trace Elements in Man and Animals*. Mills, C.F., Bremner, I. & Chesters, J.K. (eds.) pp. 666–70. Commonwealth Agricultural Bureau, Slough.

Toxic metals in the diet 53

Fletcher, P.E. & Fletcher, G.L. (1978). The binding of zinc to the plasma of winter flounder (*Pseudopleuronectes americanus*): affinity and specificity. *Canadian Journal of Zoology*, **56**, 114–20.

Galvin, R.M. (1991). Study of evolution of aluminium in reservoirs and lakes. *Water Research*, **25**, 1465–70.

Giles, M.A. (1988). Accumulation of cadmium by rainbow trout, *Salmo gairdneri*, during extended exposure. *Canadian Journal of Fisheries and Aquatic Sciences*, **45**, 1045–53.

Gill, T.S., Tewari, H. & Pande, J. (1990). Use of the fish enzyme system in monitoring water quality: effects of mercury on tissue enzymes. *Comparative Biochemistry and Physiology*, **97C**, 287–92.

Goode, C.A., Dinh, C.T. & Linder, M.C. (1989). Mechanisms of copper transport and delivery in mammals: review and recent findings. In *Copper Bioavailability and Metabolism*. Kies, C. (ed.) pp. 131–44. Plenum Press, New York.

Haesloop, U. & Schirmer, M. (1985). Accumulation of orally administered cadmium by the eel (*Anguilla anguilla*). *Chemosphere*, **14**, 1627–34.

Handy, R.D. (1992). The assessment of episodic pollution II. The effects of cadmium and copper enriched diets on tissue contaminant analysis in rainbow trout (*Oncorhynchus mykiss*). *Archives of Environmental Contamination and Toxicology*, **22**, 82–7.

Handy, R.D. (1993). The accumulation of dietary aluminium by rainbow trout, *Oncorhynchus mykiss*, at high exposure concentrations. *Journal of Fish Biology*, **42**, 603–6.

Hannerz, L. (1968). Experimental investigations on the accumulation of mercury in water organisms. *Report of the Institute of Freshwater Research of Sweden*, **48**, 120–76.

Hardy, R.W. & Shearer, K.D. (1985). Effect of dietary calcium phosphate and zinc supplementation on whole body zinc concentrations of rainbow trout (*Salmo gairdneri*). *Canadian Journal of Fisheries and Aquatic Sciences*, **42**, 181–4.

Hardy, R.W., Sullivan, C.V. & Koziol, A.M. (1987). Absorption, body distribution, and excretion of dietary zinc by rainbow trout (*Salmo gairdneri*). *Fish Physiology and Biochemistry*, **3**, 133–43.

Harrison, S.E. & Klaverkamp, J.F. (1989). Uptake, elimination and tissue distribution of dietary and aqueous cadmium by rainbow trout (*Salmo gairdneri* Richardson) and lake whitefish (*Coregonus clupeaformis* Mitchill). *Environmental Toxicology and Chemistry*, **8**, 87–97.

Harrison, S.E., Klaverkamp, J.F. & Hesslien, R.H. (1990). Fates of metal radiotracers added to a whole lake: accumulation in fathead minnow (*Pimephales promelas*) and lake trout (*Salvelinus namaycush*). *Water, Air and Soil Pollution*, **52**, 277–93.

Hatakeyama, S. & Yasuno, M. (1982). Accumulation and effects of cadmium on guppy (*Poecilia reticulata*) fed cadmium-dosed *Cladocera*

(*Moina macrocopa*). *Bulletin of Environmental Contamination and Toxicology*, **29**, 159–66.

Hodson, P.V., Blunt, B.R. & Spry, D.J. (1978). Chronic toxicity of waterborne and dietary lead to rainbow trout (*Salmo gairdneri*) in Lake Ontario water. *Water Research*, **12**, 869–78.

Huckabee, J.W., Janez, S.A., Blaylock, B.G., Talmi, Y. & Beauchamp, J.J. (1978). Methylated mercury in brook trout (*Salvelinus fontinalis*): absence of an *in vivo* methylating process. *Transactions of the American Fisheries Society*, **107**, 848–52.

Huckabee, J.W., Elwood, J.W. & Hildebrand, S.G. (1979). Accumulation of mercury in freshwater biota. In *The Biogeochemistry of Mercury in the Environment*. Nriagu, J.O. (ed.) pp. 278–302. Elsevier/North-Holland Biomedical Press, Amsterdam.

Jackson, T.A. (1991). Biological and environmental control of mercury accumulation by fish in lakes and reservoirs of Northern Manitoba, Canada. *Canadian Journal of Fisheries and Aquatic Sciences*, **48**, 2449–70.

Johnson, P.E. (1989). Factors affecting copper absorption in humans and animals. In *Copper Bioavailability and Metabolism*. Kies, C. (ed.) pp. 71–9. Plenum Press, New York.

Kaehny, W.D., Arlene, P., Hegg, B.S. & Alfrey, A.C. (1977). Gastrointestinal absorption of aluminium from aluminium-containing antacids. *The New England Journal of Medicine*, **296**, 1389–90.

Ketola, H.G. (1979). Influence of dietary zinc on cataracts in rainbow trout (*Salmo gairdneri*). *Journal of Nutrition*, **109**, 965–9.

Knox, A., Cowey, C.B. & Adron, J.W. (1982). Effects of dietary copper and copper:zinc ratio on rainbow trout *Salmo gairdneri*. *Aquaculture*, **27**, 111–19.

Knox, A., Cowey, C.B. & Adron, J.W. (1984). Effects of dietary zinc intake upon copper metabolism in rainbow trout (*Salmo gairdneri*). *Aquaculture*, **40**, 199–207.

Koh, E.T., Reiser, S., Fields, M. & Scholfield, D.J. (1989). Copper status in the rat is affected by modes of copper delivery. *Journal of Nutrition*, **119**, 453–7.

Koyama, J. & Itazawa, Y. (1977a). Effects of oral administration of cadmium on fish – I. Analytical results of the blood and bones. *Bulletin of the Japanese Society of Scientific Fisheries*, **43**, 523–6.

Koyama, J. & Itazawa, Y. (1977b). Effects of oral administration of cadmium on fish –II. Results of morphological examination. *Bulletin of the Japanese Society of Scientific Fisheries*, **43**, 527–33.

Koyama, J. & Itazawa, Y. (1977c). Effects of oral administration of cadmium on fish – III. Comparison of the effects on the Porgy and the Carp. *Bulletin of the Japanese Society of Scientific Fisheries*, **43**, 891–5.

Krantzberg, G. & Stokes, P.M. (1989). Metal regulation, tolerance, and body burdens in the larvae of the genus *Chironomous*. *Canadian Journal of Fisheries and Aquatic Sciences*, **46**, 389–98.

Kudo, A. & Miyahara, S. (1991). A case history; Minamata mercury pollution in Japan – from loss of human lives to decontamination. *Water Science and Technology*, **23**, 283–90.

Lanno, R.P., Singer, S.J. & Hilton, J.W. (1985*a*). Effect of ascorbic acid on dietary copper toxicity in rainbow trout (*Salmo gairdneri* Richardson). *Aquaculture*, **49**, 269–87.

Lanno, R.P., Singer, S.J. & Hilton, J.W. (1985*b*). Maximum tolerable and toxicity levels of dietary copper in rainbow trout (*Salmo gairdneri* Richardson). *Aquaculture*, **49**, 257–68.

Lanno, R.P., Hicks, B. & Hilton, J.W. (1987). Histological observations on intrahepatocytic copper-containing granules in rainbow trout reared on diets containing elevated levels of copper. *Aquatic Toxicology*, **10**, 251–63.

Laurén, D.J. & McDonald, D.G. (1986). Influence of water hardness, pH, and alkalinity on the mechanisms of copper toxicity in juvenile rainbow trout, *Salmo gairdneri*. *Canadian Journal of Fisheries and Aquatic Sciences*, **43**, 1488–96.

Laurén, D.J. & McDonald, D.G. (1987). Acclimation to copper by rainbow trout, *Salmo gairdneri*: biochemistry. *Canadian Journal of Fisheries and Aquatic Sciences*, **44**, 105–11.

Linder, M.C. (1991). *Biochemistry of Copper*. pp. 15–161. Plenum Press, New York.

Linder, M.C., Weiss, K.C. & Wirth, P.L. (1985). Copper transport within the mammalian organism. In *Trace Elements in Man and Animals*. Mills, C.F., Bremner, I. & Chesters, J.K. (eds.) pp. 323–8. Commonwealth Agricultural Bureau, Slough.

Matsusaka, N., Berg, D. & Kollmer, W.E. (1985). Influence of changing Zn supply on ^{65}Zn absorption and retention in rats. In *Trace Elements in Man and Animals*. Mills, C.F., Bremner, I. & Chesters, J.K. (eds.) pp. 394–7. Commonwealth Agricultural Bureau, Slough.

McKim, J.M., Olson, G.F., Holcombe, G.W. & Hunt, E.P. (1976). Long-term effects of methyl-mercuric chloride on three generations of brook trout (*Salvelinus fontinalis*): toxicity, accumulation, distribution and elimination. *Journal of the Fisheries Research Board of Canada.*, **33**, 2726–39.

Mehra, M. & Choi, B.H. (1980). Distribution and biotransformation of methylmercuric chloride in different tissues of mice. *Acta Pharmacologica et Toxicologica*, **49**, 28–37.

Miller, T.G. & Mackay, W.C. (1982). Relationship of secreted mucus to copper and acid toxicity in rainbow trout. *Bulletin of Environmental Contamination and Toxicology*, **28**, 68–74.

Miller, P.A., Munkittrick, K.R. & Dixon, D.G. (1992). Relationship between concentrations of copper and zinc in water, sediments, benthic invertebrates, and tissues of white sucker (*Catostomus commersoni*) at metal-contaminated sites. *Canadian Journal of Fisheries and Aquatic Sciences*, **49**, 978–84.

Mills, C.F. (1986). The influence of chemical species on the adsorption and physiological utilization of trace elements from the diet or environment. In *The Importance of Chemical 'Speciation' in Environmental Processes*. Bernhard, M., Brinckman, F.E. & Sadler, P.J. (eds.) pp. 71–83. Springer-Verlag, Berlin, Heidelberg.

Milner, N.J. (1982). The accumulation of zinc by O-group plaice, *Pleuronectes platessa* (L.), from high concentrations in sea water and food. *Journal of Fish Biology*, **21**, 325–36.

Murai, T., Andrews, J.W. & Smith, R.G. (1981). Effects of dietary copper on channel catfish. *Aquaculture*, **22**, 353–7.

Nakanishi, H., Ukita, M., Sekine, M. & Murakami, S. (1989). Mercury pollution in Tokuyama Bay. *Hydrobiologia*, **176/177**, 197–211.

Nakatani, R.E. (1966). Biological responses of rainbow trout (*Salmo gairdneri*) ingesting ^{65}Zn. In *Proceeding on the Disposal of Radioactive Wastes in Seas, Oceans, and Surface Waters*. Guillon, A. (ed.), pp. 809–23. International Atomic Energy Agency, Vienna.

Ogino, C. & Yang, G.Y. (1978). Requirement of rainbow trout for dietary zinc. *Bulletin of the Japanese Society of Scientific Fisheries*, **44**, 1015–18.

Ogino, C. & Yang, G.Y. (1980). Requirements of carp and rainbow trout for dietary manganese and copper. *Bulletin of the Japanese Society of Scientific Fisheries*, **46**, 455–8.

Olson, K.R., Bergman, H.L. & Fromm, P.O. (1973). Uptake of methylmercuric chloride and mercuric chloride by trout: a study of uptake pathways into the whole animal and uptake by erythrocytes *in vitro*. *Journal of the Fisheries Research Board of Canada*, **30**, 1293–9.

Ondreicka, R., Kortus, R. & Ginter, E. (1971). Aluminium, its absorption, distribution and effects on phosphorus metabolism. In *Intestinal Absorption of Metal Ions, Trace Elements and Radionucleotides*. Skoryna, S.C. & Waldron-Edward, D. (eds.) pp. 297. Pergamon Press, Oxford.

Overnell, J., Fletcher, T.C. & McIntosh, R. (1988). The apparent lack of effect of supplementary dietary zinc on zinc metabolism and metallothionein concentrations in the turbot, *Scophthalmus maximus*. *Journal of Fish Biology*, **33**, 563–70.

Park, C.W. & Shimizu, C. (1989). Quantitative requirements of aluminum and iron in the formulated diets and its interrelationship with other minerals in young eel. *Nippon Suisan Gakkaishi*, **55**, 111–16.

Parry, W.H., Jackson, P.G.G., Rao, S.R.R. & Cooke, B.C. (1985). Effects of high dietary zinc on copper transport in three breeds of

Toxic metals in the diet

57

housed pregnant sheep. In *Trace Elements in Man and Animals*. Mills, C.F., Bremner, I. & Chesters, J.K. (eds.) pp. 376–78. Commonwealth Agricultural Bureau, Slough.

Pärt, P. & Lock, R.A.C. (1983). Diffusion of calcium, cadmium and mercury in a mucous solution from rainbow trout. *Comparative Biochemistry and Physiology*, **76C**, 259–63.

Patrick, F.M. & Loutit, M.W. (1978). Passage of metals to freshwater fish from their food. *Water Research*, **12**, 395–8.

Pentreath, R.J. (1973). The accumulation and retention of ^{65}Zn and ^{54}Mn by the plaice, *Pleuronectes platessa* L. *Journal of Experimental Marine Biology and Ecology*, **12**, 1–18.

Pentreath, R.J. (1976). The accumulation of mercury from food by the plaice, *Pleuronectes platessa*. *Journal of Experimental Marine Biology and Ecology*, **25**, 51–65.

Philips, G.R. & Buhler, D.R. (1978). The relative contributions of methylmercury from food or water to rainbow trout (*Salmo gairdneri*) in a controlled laboratory environment. *Transactions of the American Fisheries Society*, **107**, 853–61.

Philips, G.R. & Greogry, R.W. (1979). Assimilation efficiency of dietary methylmercury by northern pike (*Esox lucius*). *Journal of the Fisheries Research Board of Canada*, **36**, 1516–19.

Poston, H.A. (1991). Effect of dietary aluminum on growth and composition of young Atlantic salmon. *Progressive Fish-Culturist*, **53**, 7–10.

Pratap, H.B., Fu, H., Lock, R.A.C. & Wendelaar Bonga, S.E. (1989). Effect of waterborne and dietary cadmium on plasma ions of the teleost *Oreochromis mossambicus* in relation to water calcium levels. *Archives of Environmental Contamination and Toxicology*, **18**, 568–75.

Pratap, H.B. & Wendelaar Bonga, S.E. (1993). Effect of ambient and dietary cadmium on pavement cells, chloride cells, and Na$^+$/K$^+$-ATPase activity in the gills of the freshwater teleost *Oreochromis mossambicus* at normal and high calcium levels in the ambient water. *Aquatic Toxicology*, **26**, 133–50.

Recker, R.R., Blotcky, A.J., Leffler, J.A. & Rack, E.P. (1977). Evidence of aluminum absorption from the gastrointestinal tract and bone deposition by aluminum carbonate ingested with normal renal function. *Journal of Laboratory and Clinical Medicine*, **90**, 810–15.

Refsvik, T. (1983). The mechanism of biliary excretion of methyl mercury: studies with methylthiols. *Acta Pharmacologia et Toxicologica*, **53**, 153–8.

Riisgård, H.U. & Hansen, S. (1990). Biomagnification of mercury in a marine grazing food-chain: algal cells *Phaeodactylum tricornutum*, mussels *Mytilus edulis* and flounders *Platichthys flesus* studied by means of a stepwise-reduction-CVAA method. *Marine Ecology Progress Series*, **62**, 259–70.

Rodgers, D.W., Watson, T.A., Langan, J.S. & Wheaton, T.J. (1987). Effect of pH and feeding regime on methylmercury accumulation within aquatic microcosms. *Environmental Pollution*, **45**, 261–74.

Rudd, J.W., Furutani, A. & Turner, M.A. (1980). Mercury methylation by fish intestinal contents. *Applied Environmental Microbiology*, **40**, 777–82.

Rudd, J.W., Turner, M.A., Furutani, A., Swick, A.L. & Townsend, B.E. (1983). The English-Wabigoon River System: I. a synthesis of recent research with a view towards mercury amelioration. *Canadian Journal of Fisheries and Aquatic Sciences*, **40**, 2206–17.

Sandström, B., Davidson, L., Cederblad, Å. & Lönerdal, B. (1985). Effects of inorganic iron on the absorption of zinc from a test solution and a composite meal. In *Trace Elements in Man and Animals*. Mills, C.F., Bremner, I. & Chesters, J.K. (eds.) pp. 414–16. Commonwealth Agricultural Bureau, Slough.

Schmitt, C.J. & Brumbaugh, W.G. (1990). National contaminant biomonitoring program: concentrations of arsenic, cadmium, copper, lead, mercury, selenium, and zinc in US freshwater fish 1976–1984. *Archives of Environmental Contamination and Toxicology*, **19**, 731–47.

Schoenmaker, T.J.M., Klaren, P.H.M., Flik, G., Lock, R.A.C., Pang, P.K.T. & Wendelaar Bonga, S.E. (1992). Actions of cadmium on basolateral plasma membrane proteins involved in calcium uptake by fish intestine. *Journal of Membrane Biology*, **127**, 161–72.

Shears, M.A. & Fletcher, G.L. (1979). The binding of zinc to the soluble proteins of intestinal mucosa in winter flounder (*Pseudopleuronectes americanus*). *Comparative Biochemistry and Physiology*, **64B**, 297–9.

Shears, M.A. & Fletcher, G.L. (1983). Regulation of Zn^{2+} uptake from the gastrointestinal tract of a marine teleost, the Winter Flounder (*Pseudopleuronectes americanus*). *Canadian Journal of Fisheries and Aquatic Sciences*, **40** (Suppl. 2), 197–205.

Sideman, S. & Manor, D. (1982). The dialysis dementia syndrome and aluminium intoxication. *Nephron*, **31**, 1–10.

Skak, C. & Baatrup, E. (1993). Quantitative and histochemical demonstration of mercury deposits in the inner ear of trout, *Salmo trutta*, exposed to dietary methylmercury and dissolved mercuric chloride. *Aquatic Toxicology*, **25**, 55–70.

Spry, D.J. & Wood, C.M. (1985). Ion flux rates, acid–base status, and blood gases in rainbow trout, *Salmo gairdneri*, exposed to toxic zinc in natural soft water. *Canadian Journal of Fisheries and Aquatic Sciences*, **42**, 1332–41.

Spry, D.J., Hodson, P.V. & Wood, C.M. (1988). Relative contributions of dietary and waterborne zinc in the rainbow trout, *Salmo gairdneri*. *Canadian Journal of Fisheries and Aquatic Sciences*, **45**, 32–41.

Spry, D.J. & Wiener, J.G. (1991). Metal bioavailability and toxicity to fish in low-alkalinity lakes: a critical review. *Environmental Pollution*, **71**, 243–304.

Stewart, W.K. (1989). Aluminium toxicity in individuals with chronic renal disease. In *Aluminium in Food and the Environment*. Massey, R.C. & Taylor, D. (eds.) pp. 6–19. Royal Society of Chemistry Special Publication No. 73, Cambridge.

Sylva, R.N. (1976). The environmental chemistry of copper(II) in aquatic systems. *Water Research*, **10**, 789–92.

Taylor, D. (1983). The significance of the accumulation of cadmium by aquatic organisms. *Ecotoxicology and Environmental Safety*, **7**, 33–42.

Tjioe, P.S. & Van den Hamer, C.J.A. (1985). The effect of antitumor agent Cisplatin on the absorption and retention of Cu and Zn in the mouse. In *Trace Elements in Man and Animals*. Mills, C.F., Bremner, I. & Chesters, J.K. (eds.) pp. 379–81. Commonwealth Agricultural Bureau, Slough.

Turner, M.A. & Swick, A.L. (1983). The English-Wabigoon River System: IV. Interaction between mercury and selenium accumulated from waterborne and dietary sources by northern pike (*Esox lucius*). *Canadian Journal of Fisheries and Aquatic Sciences*, **40**, 2241–50.

Turnpenny, A.W.H., Dempsey, C.H., Davis, M.H. & Fleming, J.M. (1988). Factors limiting fish populations in the Loch Fleet system, an acidic drainage system in south-west Scotland. *Journal of Fish Biology*, **32**, 101–18.

Van Hattum, B., Timmermans, K.R. & Govers, H.A. (1991). Abiotic and biotic factors influencing *in situ* trace metal levels in macro invertebrates in freshwater ecosystems. *Environmental Toxicology and Chemistry*, **10**, 275–92.

Van der Voet, G.B. (1992). Intestinal absorption of aluminium. In *Aluminium in Biology and Medicine*. Ciba Foundation Symposium 169 pp. 109–122. Wiley, Chichester.

Weber, D.N., Eisch, S., Spieler, R.E. & Petering, D.H. (1992). Metal redistribution in largemouth bass (*Micropterus salmoides*) in response to restrainment stress and dietary cadmium: role of metallothionein and other metal-binding proteins. *Comparative Biochemistry and Physiology*, **101C**, 255–62.

Weisbart, M. (1973). The distribution and tissue retention of mercury-203 in the goldfish (*Carassius auratus*). *Canadian Journal of Zoology*, **51**, 143–50.

Wekell, J.C., Shearer, K.D. & Houle, C.R. (1983). High zinc supplementation of rainbow trout diets. *Progressive Fish-Culturist*, **45**, 144–7.

Wiener, J.G., Fitzgerald, W.F., Watras, C.J. & Rada, R.G. (1990). Partitioning and bioavailability of mercury in an experimentally acidified Wisconsin lake. *Environmental Toxicology and Chemistry*, **9**, 909–18.

Willis, J.N. & Sunda, W.G. (1984). Relative contributions of food and water in the accumulation of zinc by two species of marine fish. *Marine Biology*, **80**, 273–9.

Winfrey, M.R. & Rudd, J.W.M. (1990). Environmental factors affecting the formation of methylmercury in low pH lakes. *Environmental Toxicology and Chemistry*, **9**, 853–69.

Wiseman, A. & Aggett, P.J. (1985). The binding of zinc to rabbit small intestinal brush border vesicles. In *Trace Elements in Man and Animals*. Mills, C.F., Bremner, I. & Chesters, J.K. (eds.) pp. 397–9. Commonwealth Agricultural Bureau, Slough.

Wobeser, G. (1975). Prolonged oral administration of methyl mercury chloride to rainbow trout (*Salmo gairdneri*) fingerlings. *Journal of the Fisheries Research Board of Canada*, **32**, 2015–23.

Wren, C.D. & Stephenson, G.L. (1991). The effects of acidification on the accumulation and toxicity of metals to freshwater invertebrates. *Environmental Pollution*, **71**, 205–41.

Yevtushenko, N.Yu., Bren, N.V. & Sytnik, Yu. M. (1990). Heavy metal contents in invertebrates of the Danube River. *Water Science and Technology*, **22**, 119–25.

Yokel, R.A. & McNamara, P.J. (1985). Aluminum bioavailability and deposition in adult and immature rabbits. *Toxicology and Applied Pharmacology*, **77**, 344–52.

C. HOGSTRAND and C.M. WOOD

The physiology and toxicology of zinc in fish

Zinc levels in the aquatic environment

The background concentrations of Zn in aquatic environments are comparatively low. In unpolluted areas, the concentrations of total Zn in the water have been reported to be 1 μg l^{-1} or less (Spry, Wood & Hodson, 1981; Hogstrand, Lithner & Haux, 1991). Zinc has an extensive industrial use in alloys, galvanizing, pigments, and electrical equipment. On a relative basis, surface drainage and atmospheric fallout are the most important inputs of Zn to aquatic environments (Spear, 1981). Concentrations of waterborne Zn in industrialized areas rarely exceed 50 μg l^{-1} (Coombs, 1980; Spear, 1981; Hogstrand and Haux, 1991), although concentrations of dissolved Zn far above 100 μg l^{-1} have been reported (Abdullah et al., 1976; Roch & McCarter, 1984). There is also a coupling between acidification and increased waterborne concentrations of Zn, probably caused by leaching of Zn from rocks and sediments (Baker, 1982).

Zinc is a micronutrient

A total absence of Zn is not compatible with life. Although the involvement of Zn in biological systems has been suspected for a very long time, it was not until the middle of this century that evidence for the biochemical functions of Zn started to emerge. Keilin and Mann (1940) were the first to recognize the involvement of Zn in enzymes by their isolation of carbonic anhydrase. To date, over 300 proteins have been identified that need Zn for their functions and the number is rapidly increasing (Vallee & Falchuk, 1993). The biological activities of these proteins include steps in the metabolism of nucleic acids, proteins, carbohydrates, and fatty acids. Perhaps even more important are the several proteins involved in replication and transcription of DNA. The role of Zn in gene regulation is presently a rapidly expanding area of research (Wu & Wu, 1987; Vallee & Falchuk, 1993). The chemical properties of Zn make the element uniquely suited for interaction with biological macromolecules (Vallee & Falchuk, 1993). For example, the large number of coordination numbers and geometries that Zn can assume in proteins allows the protein to have structural effects on Zn, and conversely the bound Zn can modulate the shape of the interacting protein. Another important feature is that Zn is not reduced or oxidized under physiological conditions. The majority of Zn proteins are enzymes that use one single Zn atom as prosthetic group. However in some, two or more Zn atoms act as catalytic unit. The second Zn site (coactive site) modulates the activity, but is dispensable for catalysis (Vallee & Auld, 1992). Finally, a few proteins use Zn to stabilize or modulate the structure of the protein (Williams, 1984). Among these are some DNA binding proteins, including the glucocorticoid receptor (Hollenberg et al., 1987; Freedman et al., 1988), the oestrogen receptor (Schwabe, Neuhaus & Rhodes, 1990), and the trans-acting factors responsible for metallothionein (MT) induction (Fig. 1) (Imbert et al., 1989; Searle, 1990; Thiele, 1992; Koizumi et al., 1992).

The average fish contains about 10 to 40 µg Zn g^{-1} wet weight, depending on species (Johnson, 1987; Spry, Hodson & Wood, 1988; Shearer, 1984; Shearer et al., 1992). Most of the Zn is found in skin, muscle and bone, which together constitute some 50–60% of the whole body Zn (Pentreath, 1973, 1976; Wicklund Glynn, 1991). In most tissues, the concentrations of Zn are between 5 and 100 µg g^{-1} wet weight (Table 1). However, some tissues, particularly in a few species, possess amazingly high levels of Zn. Most notable is the liver of squirrelfish (Holocentrus rufus; Hogstrand & Haux, 1990). Although there is a pronounced variation amongst individuals, the hepatic Zn levels in seemingly non-exposed squirrelfish (average 2600 µg g^{-1} wet weight) are among the highest found in any living tissue (Hogstrand & Haux, 1990). The highest Zn levels known in mammals are those in the tapetum luicidum of the choroid of the eye in dog and fox where average Zn levels between 18 000 and 33 000 µg g^{-1} wet weight have been reported (Weizel et al., 1954). Also in most fish, various parts of the eye – particularly the retina and the choroid – have unusually high levels of Zn (Table 1) albeit not as spectacular as those of the tapetum luicidum (Eckhert, 1983). The physiological significance of these extraordinary Zn accumulations is yet to be elucidated.

Routes of zinc uptake by fish

In both seawater and freshwater fish, uptake of Zn from the environment can occur both through the gills and the gastrointestinal system

Table 1. *Levels of Zn in various tissues of fish*

Tissue	Teleosts	Elasmobranchs
Blood plasma	8–24 (2)[a,b,c]	–
Eye (choroid)	340–2700 (3)[d]	–
Eye (retina)	390–4900 (3)[d]	–
Gill	25–50 (2)[e,f,g]	–
Intestine	14–26 (4)[h]	23–51 (5)[m]
Kidney	22–110 (6)[e,g,h]	10–19 (5)[m]
Liver	15–2600 (11)[e,f,g,h,i,j,k]	9–16 (5)[m]
Muscle	4–19 (6)[h,i,k]	3–5 (5)[m]
Ovary	20–100 (2)[i,k]	–
Scale	47–100 (2)[i,l]	–
Skin	23 (1)[i]	9–13 (2)[m]

The values shown are the means, expressed as µg g^{-1} wet weight, reported by the authors. The number of species that the figures are based upon is given within parentheses. In some cases where the Zn content has been presented on a dry weight basis, a conversation factor of ×0.25 has been used.

[a]Bettger *et al.* (1987), [b]Knox *et al.* (1984), [c]Shearer *et al.* (1992), [d]Eckhert (1983), [e]Wicklund Glynn *et al.* (1992), [f]Hogstrand *et al.* (1994), [g]Olsson *et al.* (1989a), [h]Takeda & Shimizu (1982), [i]Shearer (1984), [j]Hogstrand & Haux (1990), [k]Hellou *et al.* (1992), [l]Abdullah *et al.* (1976), [m]Hornung *et al.* (1993)

(Renfro *et al.*, 1975; Spry *et al.*, 1988). The uptake of waterborne Zn by the gills is dependent on the chemical composition of the water. A high Ca concentration of the water greatly reduces the rate of Zn uptake and the accumulation of Zn by gill tissue (Bradley & Sprague, 1985a; Spry & Wood, 1988, 1989a; Wicklund, 1990; Bentley, 1992). Also, the branchial influx of Zn seems to be reduced if elevated levels of Cd are present in the water (Bentley, 1992). There is evidence that a reduced water pH decreases the accumulation rate of Zn in gills, while the influx of Zn to the whole animal may actually be increased (Bradley & Sprague, 1985a; Bentley, 1992). A possible explanation for this apparent paradox could be that the basolateral transfer of Zn in the gill epithelium is stimulated by low water pH. The uptake rate of Zn through the gills is comparatively slow compared to that of the more abundant hydrominerals. In freshwater at a Ca concentration of 1 mM, a Na concentration of 0.6 mM, and a Zn concentration of 150 µg l^{-1} (2.3 µM), the branchial influx of Zn in rainbow trout is approximately 1000 times slower than the influx of Ca and 6000 times slower than that of Na (Laurén & McDonald, 1985; Spry & Wood, 1989a,b; Hogstrand et al., 1994, 1995). However, Zn passes more quickly through the gills than Cd, which might be expected since Zn is a nutrient and Cd is probably not (Wicklund Glynn, 1991).

There is limited information on the nature of the intestinal Zn uptake in fish. The absorption of Zn in the marine winter flounder (*Pseudopleuronectes americanus*) and plaice (*Pleuronectes platessa*) seems to be highest in the anterior part of the intestine (Pentreath, 1976; Shears & Fletcher, 1983). The uptake of Zn to the enterocytes became saturated at high Zn loads, suggesting that binding of Zn to more or less specific sites on the surface of, or inside, the enterocytes is involved (Shears & Fletcher, 1983). The entry step through the brushborder membrane was inhibited by the presence of Cu, Cd, Co, Cr, Ni, Mg, and Hg. The uptake of Zn was unaffected by the Ca content of the lumen, which would indicate that the mechanism of Zn uptake in winter flounder may differ from that in mammals where high Ca reduces the absorption (see Underwood, 1977; Vallee & Falchuk, 1993). The basolateral transfer of Zn in winter flounder enterocytes showed no saturation kinetics, suggesting a passive diffusion process (Shears & Fletcher, 1983). In mammals, Zn is absorbed mainly from the duodenum, ileum and jejunum (Lee et al., 1989). Despite a number of studies on the topic, the underlying mechanisms of Zn uptake in mammals are poorly understood (see Underwood, 1977; Vallee & Falchuk, 1993). However, the discovery of a cysteine-rich intestinal protein, CRIP, has led to a new conceptual model (Hempe & Cousins, 1992). According to this model, Zn passes through the brushborder membrane by an as yet unidentified mechanism, then combines with CRIP, which functions as a diffusible intracellular transport protein that brings Zn over to the basolateral membrane, where Zn is transferred to a putative transmembrane transporter. Metallothionein (MT), a low molecular weight metal binding protein (see below), may exert a negative feedback on the Zn absorption by limiting the amount of Zn that binds to CRIP. This model has yet to be evaluated in fish.

The relative importance of gills versus intestine for uptake of Zn varies considerably according to the external conditions. In seawater, at environmentally realistic levels of waterborne Zn, there is little doubt that the intestine is more important than the gill as uptake route for nutritional Zn (Pentreath, 1973, 1976; Renfro et al., 1975; Milner, 1982; Willis and Sunda, 1984). Zn is absorbed from the food, and possibly from the water in the intestine because marine fish drink considerable volumes. At higher concentrations of waterborne Zn ($600 \ \mu g \ l^{-1}$), the uptake of dissolved Zn becomes more important and up to 50% of the accumulated Zn may be contributed by the water

(Milner, 1982; Willis & Sunda, 1984). It is not clear, however, whether this increased uptake from water is a result of an increased intestinal absorption of Zn, if more Zn is taken up by the gills, or both. In most cases, the intestine is responsible for most of the Zn uptake also in freshwater fish, even though freshwater fish do not drink to any great extent, but there are conditions where the gills play an important role. Spry *et al.* (1988) found that, in hard water, an increased concentration of waterborne Zn (90 μg l^{-1}) was sufficient to reverse symptoms of Zn deficiency in rainbow trout fed on a Zn-deprived diet. Even when dietary Zn was adequate, the same concentration of waterborne Zn still contributed 57% of the Zn accumulation. Thus, the intestine is the dominating pathway for normal Zn uptake in fish, while the input from the gills will become increasingly important at higher concentrations of waterborne Zn and/or when dietary Zn is deficient.

Internal Zn regulation

There is an important difference between Zn and many other so-called heavy metals in that Zn is an essential element. This fact is very well reflected in the internal handling of Zn, which is exquisitely poised. Processes of uptake, distribution, intracellular metabolism, and excretion are so well regulated that Zn deficiency is a next-to-unknown condition in fish, and Zn toxicity is uncommon. However, in neither fish nor mammals are the actual mechanisms for Zn homeostasis particularly well understood, and the research in this area has mostly been descriptive.

From the primary uptake organs, the intestine and the gills, Zn is transported in the blood to various parts of the body. The plasma level of Zn in fish (Table 1) is 10 to 20 times higher than that in most mammals (see Underwood, 1977). This is something that seems to be specific to plasma, because the erythrocyte Zn concentration is similar in rainbow trout to that in mammals, and the whole body content of Zn does not differ between fish and man (Bettger *et al.*, 1987; Shearer *et al.*, 1992; see Vallee & Falchuk, 1993). In rainbow trout plasma 99.8% of the total Zn content is bound to plasma proteins (Bettger *et al.*, 1987). Most of this Zn seems to be relatively loosely associated. In mammals, 60–70% of the plasma Zn is loosely bound to a protein that co-elutes with albumin on size-exclusion chromatography and 30–40% is tightly bound to α_2-globulin (Parisi & Vallee, 1970; Falchuk, 1977).

In rainbow trout, Zn that is taken up from the environment equilibrates with the rapidly exchanging body pool of Zn in about two days (Spry & Wood, 1989*a*). Based on the whole body content of Zn (Spry *et al.*, 1988), the plasma concentration of Zn (Bettger *et al.*, 1987), and the radiospace of exogenous ^{65}Zn (Spry & Wood, 1989a), the size of the pool rapidly exchanging Zn in rainbow trout can be calculated to be 40% of the whole body pool at the most. Renfro *et al.* (1975) treated a gobiid with tracer levels of ^{65}Zn for 96 days and within this period, only 18% of the Zn pool exchanged with the exogenous Zn. Thus, most of the Zn in fish is slowly exchanging.

Fish that are exposed to excessive amounts of Zn accumulate most of the metal in skin, scales, muscle, and bone, which are the organs that normally contain most Zn on a relative basis in fish from pristine environments (see Phillips & Russo, 1978). That is, the relative distribution of Zn does not seem to be markedly altered by Zn exposure. The overall conclusion is that there may be no such a thing as a special storage organ for Zn.

The liver is the 'classic' organ for metal metabolism. Quite frequently, it is found that fish exposed to elevated levels of Zn do not accumulate the metal in liver (Roch *et al.*, 1982; Wicklund Glynn *et al.*, 1992; Hogstrand *et al.*, 1994). However, the presence of increased hepatic Zn levels in feral fish from Zn contaminated areas, in both freshwater and marine environments, demonstrates that Zn like Cd and Cu may accumulate in liver during some exposures (Hogstrand & Haux, 1990; Hogstrand *et al.*, 1991).

Compared to Cd, Zn is rapidly eliminated from gills, liver and kidney (Wicklund Glynn, 1991; Wicklund Glynn *et al.*, 1992). In minnows, the half-times of the rapidly eliminated pool of Zn in the gills and liver ($\sim$ 70 to 80% of the total Zn in these organs) are two days and one week, respectively (Wicklund Glynn, 1991; Wicklund Glynn *et al.*, 1992). However, on a whole body basis, most of the Zn is eliminated very slowly. During a two month period, no elimination of Zn from the fish could be detected (Wicklund Glynn, 1991; Wicklund Glynn *et al.*, 1992). Other studies report the biological half-life of the pool of slowly exchanging Zn pool to be 100 to 300 days (Pentreath, 1973, 1976; see also Phillips & Russo, 1978).

The intracellular localization of Zn is dynamic and seems to be under considerable physiological control, although the processes involved are not clear (Olsson, Haux & Förlin, 1987; Olsson *et al.*, 1990). However, some generalizations are possible. Zinc is abundant in the nucleus (Julshamn *et al.*, 1988; Baatrup, 1989), which is not surprising considering the large number of Zn proteins involved in replication and transcription of DNA (Wu & Wu, 1987). Another organelle with high Zn content is the lysosome. Sauer and Watabe (1989) found that 'lysosome-like structures' in the scale osteoblast had high contents of Zn, which increased considerably during exposure. It is quite possible that these 'lysosomes' are responsible for the deposition of Zn in scales. Typically, about 50% of the cellular Zn is found in the cytosol although this number may vary from 15 to 90% depending on organ, physiological state, and metal burden (Kito *et al.*, 1982*a,b*; Olsson *et al.*, 1987, 1990; Hogstrand & Haux, 1990; Hogstrand *et al.*, 1991; Wicklund Glynn, 1991).

Since Zn is a potentially toxic metal, the intracellular handling of the metal must ensure that there is virtually no free Zn present in the cell. Movements of Zn between different intracellular compartments are therefore likely to take place by direct interaction beween Zn ligands. Probably the best known intracellular Zn ligand is metallothionein (MT). MT is a small ubiquitous protein of about 60 amino acid residues (Kägi & Schäffer, 1988). Typically, MT contains 30% cysteinyl residues, but no aromatic amino acids and no histidine (Kojima *et al.*, 1976). There are no disulphide bridges and all cysteinyl groups are involved in the chelation of metals. In addition to Zn, a variety of transition elements, such as Cu, Ag, Cd, and Hg can bind to MT. However, under 'non-exposure conditions', the metals that are found in the protein *in vivo* are Zn and Cu. In most species and the majority of tissues, Zn is the predominant metal in MT (Krezoski *et al.*, 1988). A total of 7 atoms of bivalent metals, such as Zn or Cd, or 12 atoms of univalent metals, such as Cu and Ag, or a combination of the two, bind to MT via metal–thiolate linkages (Kägi & Schäffer, 1988). The affinity of metals to MT differs between elements. The stability constants have not been calculated for the binding of Cu and Zn to MT from fish, but the conserved positions of the cysteinyl residues make it probable that the stability constants are close to those determined for MT from mammals (10^{11}–10^{14} M for Zn, and 10^{17}–10^{19} M for Cu; see Hamer, 1986; Kägi & Kojima, 1987). One or two isoforms of MT have been isolated from fish, whereas the number of mammalian isoforms varies between two and ten, depending on species (Noël-Lambot, Gerday & Disteche, 1978; Krezoski *et al.*, 1988; Olsson & Hogstrand, 1987; Hunziker & Kägi, 1988; George *et al.*, 1992). A central and remarkable feature of MT is that the synthesis of the protein is directly induced by metals (Fig. 1). In fish, metal stimulated transcription seems to be the main pathway of MT induction since the MT gene(s) are not particularly responsive to hormones that are known to elicit strong induction in mammals (Chan *et al.*, 1989; Hyllner *et al.*, 1989; Olsson *et al.*, 1990; George *et al.*, 1992; Kille *et al.*, 1992). The metals that have been shown to induce MT production in fish are Cu, Zn, Cd, Hg, and Ag, essentially the same as those which bind to MT (Olafson & Thompson, 1974; McCarter & Roch, 1983; Chan *et al.*, 1989; Olsson *et al.*, 1989*a*,*b*; Zafarullah, Olsson & Gedamu, 1989; George *et al.*, 1992; Hogstrand & Wood, 1993).

The mechanisms, control, and sites of MT degradation are less known and there is basically no information on these processes in *piscine* systems. The presence of mammalian MT – which is structurally very similar to MT from fish – in lysosomes and the potency of lysosomal proteinases to degrade Zn–MT suggest that the lysosomes are involved in the breakdown of Zn-MT (Fig. 1) (Mehra & Bremner, 1985; Chen & Failla, 1989; Kershaw & Klaassen, 1992; Choudhuri, McKim & Klaassen, 1992). The

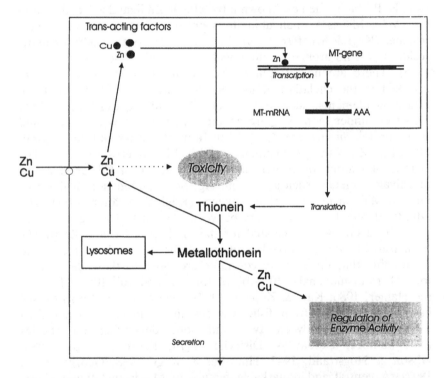

Fig. 1. Hypothetical model of synthesis, elimination, and possible functions of MT in fish. The induction of MT synthesis is mediated by trans-acting factors that bind to the MT-gene(s) and stimulate transcription. Newly synthesized thionein (apo-MT) binds metals and becomes MT. The elimination of MT from the cell involves lysosomal degradation and secretion. Functions of MT may include protection against potential metal toxicity, and activation and/or inactivation of Zn-proteins.

degree of metal saturation of MT markedly affects the degradation such that each Zn atom added to the protein adds to the resistance against degradation (Choudhuri *et al.*, 1992). It is also believed that MT is secreted from the cell, based on the findings that MT is present in plasma, urine, and bile (see Bremner & Beattie, 1990).

After fractionation of the cell, most MT is found in the cytosolic fraction (see Cousins, 1985). However, immunohistochemistry has revealed that MT is also abundant in the nucleus (Kito, Ose & Sato, 1986). In liver of unexposed fish, 20 to 40% of the cytosolic Zn is bound to MT (Noël-Lambot *et al.*, 1978; Hogstrand & Haux, 1990; Hogstrand *et al.*, 1991). During situations of metal exposure, the portion bound to MT often increases and fractions of 30 to 80% of the cytosolic Zn have been found in MT of Zn exposed fish (Kito *et al.*, 1982*a*,*b*; Hogstrand & Haux, 1990; Hogstrand *et al.*, 1991). Thus during Zn exposure, MT may therefore bind up to 40% of the Zn in the liver. Less of the Zn present in the gills seems to bind to MT and for rainbow trout, exposed for 60 days to 150 µg Zn l^{-1} (2.3 µM), it was calculated that no more than 2% of the total Zn in the gills was associated with MT (Hogstrand *et al.*, 1994).

In spite of intensive research on MT for 30 years, the exact function(s) of MT are yet to be demonstrated. The ability of MT to be induced by and to bind metals suggests a function of MT in the regulation of metals during both normal and exposure conditions (Fig. 1). In fish, efforts to establish a role for MT have mainly focused on the capacity of MT to detoxify metals. MT is induced when the cell experiences excessive concentrations of the essential elements Cu and Zn, or if non-essential Cd or Hg enters the cell. The induction of MT results in an increased binding of these elements, thus rendering them less toxic. There is no direct experimental evidence for a protective role of MT against Zn toxicity, although any strong chelator of Zn in theory would act as such. However, there are convincing arguments for a decreased toxicity of copper and cadmium when these elements are bound to MT (see Hogstrand & Haux, 1991).

Pretreatment of animals with sublethal concentrations of metal often increases the tolerance to subsequent challenges to the same metal (see McDonald & Wood, 1993). Tolerance is usually referred to as the ability to survive in acute toxicity tests. There is a general agreement that MT is a factor responsible for acquired tolerance to metals in mammals (see Cherian & Nordberg, 1983). However, there is a crucial difference between mammals and fish in that acute toxicity caused by many metals, including Zn, is the direct consequence of insult to the gills (McDonald & Wood, 1993). Consequently, if MT is able to influence the tolerance to

Zn in acute toxicity tests, it would be the MT in the gills that is responsible for the effect. There is some experimental evidence that MT in the gills is involved the development of tolerance to Zn. Bradley, DuQuesnay & Sprague (1985) showed that the levels of MT in the gills and the LC_{50} for Zn covaried in rainbow trout during exposure to waterborne Zn. However, in a recent study it was found that rainbow trout can develop a considerable degree of tolerance to a lethal concentration of Zn without any induction of MT in gills or liver (Hogstrand et al., 1994). Thus, development of increased tolerance to waterborne Zn may have little, if anything, to do with MT induction.

MT seems to be involved in the regulation of Zn during the reproduction in fish. In rainbow trout, the hepatic MT content increased during sexual maturation of females (Olsson et al., 1987). It was possible to mimic the induction of hepatic MT during female sexual maturation by the injection of oestradiol-17β into juvenile rainbow trout (Olsson et al., 1989b). The increase in hepatic MT during sexual maturation is not unique for rainbow trout. High levels of hepatic MT and zinc have been observed in plaice during the reproductive period (Overnell, McIntosh & Fletcher, 1987). During the sexual maturation of female fish, there is an extensive synthesis of the protein vitellogenin (Haux & Norberg, 1985). This is one example of many in the animal kingdom where an increased formation of Zn–MT is observed during a period of high metabolic activity (Kägi & Schäffer, 1988). Given that Zn is required for DNA- , RNA-polymerases and some transcription factors (Wu & Wu, 1987), it is possible that MT can regulate the activities of these proteins by donating or accepting Zn. MT may also serve as a storage site for excess hepatic Zn during and after the synthesis of vitellogenin (Fig. 1) (Olsson et al., 1989b). Clearly, the exact function of MT during the reproduction in fish remains to be established.

Apart from MT, there are indications of other inducible less known Zn-binding proteins in fish (Davies, 1985; Pierson, 1985a,b). For example, Spry & Wood (1989b) were not able to show induction of MT from exposure of rainbow trout to waterborne or dietary Zn. There was, however, evidence for increased levels of a heat-stable protein with a molecular weight of <3000, in intestine and gills, that was able to bind Cd.

Zinc as a toxicant

There is a wealth of data on the toxicity of waterborne Zn to aquatic organisms (for review see Spear, 1981), whereas very few studies have investigated the toxic effects of dietary Zn on fish. Knox, Cowey and

Adron (1984) found no effects when rainbow trout were fed a diet with 500 mg Zn kg^{-1} for 20 weeks. Wekell, Shearer and Houle (1983) reported that there was no growth inhibition in fingerling rainbow trout given dietary Zn at a level of 1700 mg Zn kg^{-1}. Similarly, feeding juvenile rainbow trout a diet of dry pellets, containing 529 mg Zn kg^{-1}, for 16 weeks did not have any adverse effects on growth or whole body element composition though it greatly elevated the whole body Zn burden (Spry *et al.*, 1988). If anything, the high-Zn diet stimulated growth. Thus, only Zn dissolved in water is of toxic concern.

Reported values of 96 h LC$_{50}$ for waterborne Zn ranges from 90 µg l^{-1} to 40 mg l^{-1} (Spear, 1981; Bradley & Sprague, 1985*b*). Typical 96 h LC$_{50}$ concentrations are 1 to 10 mg l^{-1} in soft water and 3 to 20 mg l^{-1} in hardwater (Spear, 1981). The major modifying physicochemical factors of Zn toxicity are the hardness, and pH of the water (Spear, 1981; Bradley & Sprague, 1985*a,b*). Acute lethality of dissolved Zn increases with decreasing water hardness and [H^{+}]. Carbonate alkalinity, on the other hand, has limited influence on the toxicity of Zn (Bradley & Sprague, 1985*b*). A possible explanation for the protective effect of low pH is that less Zn accumulates in the gills when the water pH is low (Bradley & Sprague, 1985*a*). This, in turn, may be explained by a stimulated transfer of Zn across the basolateral membrane of the epithelium (see pp. 72–74). There are marked differences in Zn sensitivity between species, the order *Perciformes* being the most resistant group and the order *Clupeiformes* the most sensitive (Spear, 1981).

There are two principal mechanisms of Zn toxicity and both of these can be fatal. At very high concentrations of Zn the fish dies from hypoxia, caused by gross morphological damage to the gills – oedema, inflamation, cell sloughing, and fusion – effects common for heavy metals at 'industrial' concentrations (Skidmore, 1970; Spry & Wood, 1984). The second mechanism, which operates at lower Zn concentrations, is an impairment of the branchial Ca uptake, which leads to a hypocalcemia that may become terminal (Spry & Wood, 1985). The effect of Zn on the branchial Ca uptake is present at concentrations of Zn that may occur in polluted environments and that are far below lethal (Hogstrand *et al.*, 1994, 1995). Furthermore, Zn (150 µg l^{-1} = 2.3 µM; [Ca] = 1 mM) chronically inhibited the Ca influx without any long-term effects on plasma [Ca], growth, or whole body element composition in rainbow trout exposed for one month. The action that Zn has on the Ca homeostasis may be the most sensitive effect of elevated waterborne Zn. The most sensitive gross effect of Zn exposure is an impairment of reproduction. In soft water, concentrations down to 50 µg Zn l^{-1} has been found to reduce the number of spawnings and eggs produced per female (Bengsson, 1974; Spehar, 1976). Since oocyte production is dependent on vitellogenin synthesis, which requires large amounts of Ca, an impaired spawning may be directly linked to the interference by Zn with the branchial Ca uptake.

Evidence for a shared uptake pathway of zinc and calcium across the gills

Waterborne Ca is a competitive inhibitor of branchial Zn influx (Fig. 2(a); Spry & Wood, 1989a). Conversely, three recent studies have shown that Zn inhibits the influx of Ca across the gills (Fig. 2(b); Sayer, Reader & Morris, 1991; Hogstrand et al., 1994, 1995). The inhibition was found to be mainly of a competitive nature although a minor non-competitive component was involved (Fig. 2(b); Hogstrand et al., 1994, 1995). These interactions not only provide an explanation for the protective effect of Ca against Zn poisoning, but also suggest that these two elements partially or fully share a common uptake route across the gills. The uptake of Ca at the gills has been intensively examined by a number of investigators (Perry & Wood, 1985; Perry & Flik, 1988; Flik et al., 1993; Perry, Goss & Fenwick, 1992; Marshall, Bryson & Wood, 1992; McCormick, Hasegawa & Hirano, 1992; Verbost et al., 1993, 1994). The current consensus is that the mitochondria-rich 'chloride-cells' are the sites of uptake. Ca is believed to pass passively through the apical membrane of the chloride cells via a voltage-insensitive Ca channel. Inside the chloride cells, Ca binds to Ca sequestering proteins and is also transported into the endoplasmic reticulum, thus, ensuring that the intracellular activity of Ca is kept very low (10^{-7} M). The basolateral transfer of Ca is mediated by a high-affinity Ca–ATPase and a low-affinity Na/Ca-exchanger. Zinc may share this entire pathway with Ca, but more likely, a part of the uptake route is common between the elements (Fig. 3).

In a recent study there was a scattered, but highly significant linear relationship between the total influx of ^{65}Zn into the fish and the accumulation of ^{65}Zn in the gill epithelium (Hogstrand et al., 1995). This relationship would indicate that the rate limiting step for Zn influx is the transport through the apical membrane and that the Zn influx may be regulated by changes in the permeability of the apical membrane to Zn. If the apical transfer of Zn occurs via the Ca channel, the entry step of Zn could be regulated by the hypocalcemic hormone, stanniocalcin, which is thought to act on this channel (Lafeber et al., 1988; Verbost et al., 1989, 1993; Flik, 1990). It has been shown that Cd enters the chloride cells through the same route as Ca and the entry of Cd is reduced in stanniocalcin-injected fish

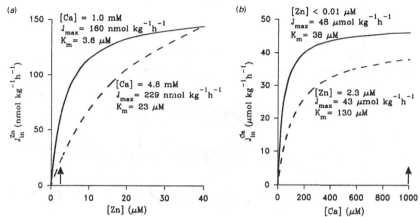

Fig. 2. (*a*) The effect of waterborne Ca on branchial Zn influx (left; data from Spry and Wood, 1989) and (*b*) the effect of waterborne Zn on branchial Ca influx (right; data from Hogstrand *et al.*, 1994) in the rainbow trout (*Oncorhynchus mykiss*). Ca competitively inhibits the influx of Zn (increased K_m), while the maximum influx capacity for Zn (J_{max}) is slightly increased by Ca. Zn (2.3 μM) exerts a pronounced competitive inhibition on the Ca influx (increased K_m), but there is also a mild non-competitive component involved (reduced J_{max}). The arrows indicate the concentrations of Ca and Zn in the water during the experiments described by Hogstrand *et al.* (1994, 1995). In most waters, the concentration of waterborne Ca yields an influx of Ca that is on the 'saturated' portion of the kinetic curve, whereas the concentration of waterborne Zn rarely reaches the K_m of the influx. Thus, a change in the affinity (K_m) for a mutual Ca/Zn carrier would have a marked effect on the Zn influx, but only slightly influence the influx of Ca.

(Verbost *et al.*, 1989). Given the chemical similarities between Zn and Cd, and the competitive interactions between Zn and Ca influx (Spry & Wood, 1989*a*; Hogstrand *et al.*, 1994, 1995), and between Zn and Cd influx (Bentley, 1992), it seems quite possible that Zn influx is regulated by stanniocalcin at the level of the apical membrane. Studies of the temporal changes in Zn influx, in relation to fluctuations in the kinetics of Ca influx, provide further evidence for a coupling between the branchial uptake of Zn and Ca. This will be dealt with in the following section.

Mechanism of adaptation to waterborne zinc

On pp. 69–70, we concluded that induction of MT is not a prerequisite for the development of an increased tolerance to lethal concentrations

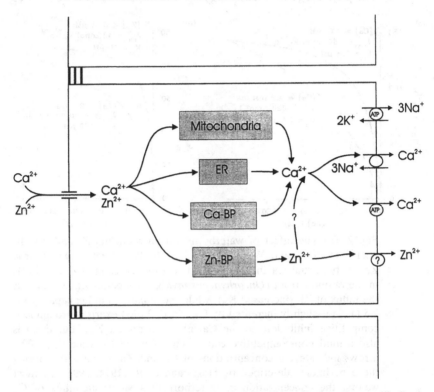

Fig. 3. Model for uptake of Ca and Zn through the teleost gill. Both Ca and Zn may pass across the apical membrane of the chloride cell via a voltage-independent Ca/Zn channel, driven by the electrochemical gradient. Inside the chloride cell, both elements are sequestered by various compartments (ER = endoplasmic reticulum, Ca–BP = calcium-binding protein, Zn–BP = zinc-binding protein) to keep the cytosolic activities of the free ions at a minimum. The basolateral transfer of Ca is mediated by a high-affinity Ca–ATPase and a Na/Ca antiporter. The route for Zn transfer across the basolateral membrane is not known, but hypothetically, one of the Ca transporters could be used or there might be a totally separate transport mechanism for Zn. The competitive interactions between Zn and Ca influx could occur at any of these levels.

of Zn (Hogstrand *et al.*, 1994). In a sequel experiment, during similar conditions, the same Zn treatment did not lead to an elevated tolerance to lethal Zn (Hogstrand *et al.*, 1995). However, the fish showed physiological evidence of adaptation to the elevated sublethal Zn. In the beginning of this experiment, Zn caused hypocalcemia, an increased rate of protein degradation in the gills, and impaired growth. None of these effects was present by the end of the exposure period (one month) and, furthermore, there were no differences in whole body contents of Zn and Ca between the Zn exposed fish and the controls. Surprisingly, after up to two months of exposure to waterborne Zn, the affinity of the Ca carrier mechanism (inverse of K_m for unidirectional Ca influx) was still markedly decreased, which is indicative of continuous competitive inhibition by Zn (Hogstrand *et al.*, 1994; 1995). Moreover, there was evidence that the intrinsic properties of the carrier were changed so that the affinity for Ca was reduced (that is K_M increased) even if Zn was removed from the water. This pattern is quite opposite to that found during acclimation to other metals, such as Al and Cu, where acclimation is conferred by the restoration of the branchial fluxes of Na (McDonald & Wood, 1993). The significance of the long-term response to Zn, compared with that to Al and Cu, is probably that Zn utilizes the Ca uptake pathway. The results suggest that the fish were able to reduce the influx of Zn by 'deliberately' keeping the affinity of the common Ca/Zn carrier low (Hogstrand *et al.*, 1994; 1995). Such an approach would limit the accumulation of Zn in gills and the influx of Zn into the whole animal. In fact, the time-course of the changes in Zn influx, during the experiment, could be explained by the simultaneous alterations in the kinetic properties of the Ca transporter. These experiments were carried out in hard water ([Ca] = 1.0 mM) and at this Ca concentration, the influx of Ca is already on the 'saturated' portion of the kinetic curve (Fig. 2(*b*); Hogstrand *et al.*, 1994a). The concentration of Zn (150 μg l^{-1} = 2.3 μM), on the other hand, was well below the K_m for Zn influx (Fig. 2(*a*); Spry & Wood, 1989a). Thus, the persistently low affinity of the proposed common Ca/Zn carrier only marginally affected the Ca influx, but reduced the influx of Zn by half (Hogstrand *et al.*, 1994; 1995). We propose that this may be the most significant mechanism of adaptation to elevated but sub-lethal levels of waterborne Zn.

Acknowledgements

This study was supported by a DFO/NSERC Science Subvention Program Grant to CMW, and a PDF to CH from the Swedish Council for Forestry and Agricultural Research.

Note added in proof

Since this chapter was written, new information has been made available concerning the interactions between Ca and Zn in gills of freshwater

76   C. HOGSTRAND AND C.M. WOOD

rainbow trout (see pp. 72–75; Hogstrand, Verbost et al., 1996). First, pharmacological blockage of the apical Ca channel with La inhibits influx of both Ca and Zn across the gills. Moreover, endogenous release of the calciostatic hormone, stanniocalcin, also reduces the branchial influx of Ca and Zn. These results strongly support our interpretation that the apical entry of Zn occurs through the so-called Ca channel, and that this channel may be a relatively non-specific channel for divalent ions. Secondly, Zn is not transported by the basolateral high-affinity Ca–ATPase of the chloride cells. Indeed, the activity of this basolateral Ca-pump is severely inhibited by picomolar concentrations of free Zn^{2+}.

References

Abdullah, M.I., Banks, J.W., Miles, D.L. & O'Grady, K.T. (1976). Environmental dependence of manganese and zinc in the scales of Atlantic salmon, Salmo salar (L), and brown trout, Salmo trutta (L). Freshwater Biology, 6, 161–6.

Baatrup, E. (1989). Selenium-induced autometallographic demonstration of endogenous zinc in organs of the rainbow trout, Salmo gairdneri. Histochemistry, 90, 417–25.

Baker, J.P. (1982). Effects on fish of metals associated with acidification. In Acid rain/Fisheries. Johnson, R.E. (ed.), pp. 165–75. Americal Fisheries Society, Bethesda, MD.

Bengsson, B.E. (1974). Effects of zinc on growth of the minnow, Phoxinus phoxinus. Oikos, 25, 370–3.

Bentley, P.J. (1992). Influx of zinc by channel catfish (Ictalurus punctatus): uptake from external environment solutions. Comparative Biochemistry and Physiology, 101C, 215–17.

Bettger, W.J., Spry, D.J., Cockell, K.A., Cho, C.H. & Hilton, J.W. (1987). The distribution of zinc and copper in plasma, erythrocytes and erythrocyte membranes of rainbow trout (Salmo gairdneri). Comparative Biochemistry and Physiology, 87C, 445–51.

Bradley, R.W., DuQuesnay, C. & Sprague, J.B. (1985). Acclimation of rainbow trout, Salmo gairdneri Richardson, to zinc: kinetics and mechanism of enhanced tolerance induction. Journal of Fish Biology, 27, 367–79.

Bradley, R.W. & Sprague, J.B. (1985a). Accumulation of zinc by rainbow trout as influenced by pH, water hardness and fish size. Environmental Toxicology and Chemistry, 4, 685–94.

Bradley, R.W. & Sprague, J.B. (1985b). The influence of pH, water hardness, and alkalinity on the acute lethality of zinc to rainbow trout (Salmo gairdneri). Canadian Journal of Fisheries and Aquatic Sciences, 42, 731–6.

Bremner, I. & Beattie, J.H. (1990). Metallothionein and the trace minerals. *Annual Review of Nutrition*, **10**, 63–83.

Chan, K.M., Davidson, W.S., Hew, C.L. & Fletcher, G.L. (1989). Molecular cloning of metallothionein cDNA and analysis of metallothionein gene expression in winter flounder tissues. *Canadian Journal of Zoology*, **67** 2520–7.

Chen, M.L. & Failla, M.L. (1989). Degradation of zinc–metallothionein in monolayer cultures of rat hepatocytes (42898). *Proceedings of the Society for Experimental Biology and Medicine*, **191**, 130–8.

Cherian, M.G. & Nordberg, M. (1983). Cellular adaptation in metal toxicology and metallothionein. *Toxicology*, **28**, 1–15.

Choudhuri, S., McKim, J.M. & Klaassen, C.D. (1992). Role of hepatic lysosomes in the degradation of metallothionein. *Toxicology and Applied Pharmacology*, **115**, 64–71.

Coombs, T.L. (1980). Heavy metal pollutants in the aquatic environment. In *Animals and Environmental Fitness: Physiological and Biochemical Aspects of Adaptation and Ecology*. Gilles, R. (ed.), pp. 283–302. Pergamon Press, Oxford, New York.

Cousins, R.J. (1985). Absorption, transport, and hepatic metabolism of copper and zinc: special reference to metallothionein and ceruloplasmin. *Physiological Reviews*, **65**, 238–310.

Davies, P.E. (1985). The toxicology and metabolism of chlorothalonil in fish. iv. Zinc exposure and the significance of metallothionein in detoxification in *Salmo gairdneri*. *Aquatic Toxicology*, **7**, 301–6.

Eckhert, C.D. (1983). Elemental concentrations in ocular tissues of various species. *Experimental Eye Research*, **37**, 639–47.

Falchuk, K.H. (1977). Effects of acute disease and ACTH on serum zinc proteins. *New England Journal of Medicine*, **269**, 1129–34.

Flik, G. (1990). Hypocalcin physiology. In *Progress in Comparative Endocrinology*. Epple, A., Scanes, C.G. & Stetson, M.H. (eds.), pp. 578–85, Wiley-Liss, Inc., New York.

Flik, G., Van Der Welden, J.A., Dechering, K.J., Verbost, P.M., Schoenmakers, T.J.M., Kolar, Z.I. & Wendelaar Bonga, S.E. (1993). Ca²⁺ and Mg²⁺ transport in gills and gut of tilapia, *Oreochromis mossambicus*: a review. *Journal of Experimental Zoology*, **265**, 356–65.

Freedman, L.P., Luisi, B.F., Korszun, Z.R., Basavappa, R., Sigler, P.B. & Yamamoto, K.R. (1988). The function and structure of the metal coordination sites within the glucocorticoid receptor DNA binding domain. *Nature*, **334**, 543–6.

George, S., Burgess, D., Leaver, M. & Frerichs, N. (1992). Metallothionein induction in cultured fibroblasts and liver of a marine flatfish, the turbot, *Scophtalamus maximus*. *Fish Physiology and Biochemistry*, **10**, 43–54.

Hamer, D.H. (1986). Metallothionein. *Annual Review of Biochemistry*, **55**, 913–51.

Haux, C. & Norberg, B. (1985). The influence of estradiol-17β on the liver content of protein, lipids, glycogen and nucleic acids in juvenile rainbow trout, *Salmo gairdneri*. *Comparative Biochemistry and Physiology*, **81B**, 275–9.

Hellou, J., Warren, W.G., Payne, J.F., Belkhode, S. & Lobel, P. (1992). Heavy metals and other elements in three tissues of cod, *Gadus morhua* from the northwest Atlantic. *Marine Pollution Bulletin*, **24**, 452–8.

Hempe, J.M. & Cousins, R.J. (1992). Cysteine-rich intestinal protein and intestinal metallothionein: an inverse relationship as a conceptual model for zinc absorption in rats. *Journal of Nutrition*, **122**, 89–95.

Hogstrand, C. & Haux, C. (1990). Metallothionein as an indicator of heavy-metal exposure in two subtropical fish species. *Journal of Experimental Marine Biology and Ecology*, **138**, 69–84.

Hogstrand, C. & Haux, C. (1991). Binding and detoxification of heavy metals in lower vertebrates with reference to metallothionein. *Comparative Biochemistry and Physiology*, **100C**, 137–41.

Hogstrand, C., Lithner, G. & Haux, C. (1991). The importance of metallothionein for the accumulation of copper, zinc and cadmium in environmentally exposed perch, *Perca fluviatilis. Pharmacology and Toxicology*, **68**, 492–501.

Hogstrand, C., Reid, S.D. & Wood, C.M. (1995). Calcium *versus* zinc transport in the gills of freshwater rainbow trout, and the cost of adaptation to waterborne zinc. *Journal of Experimental Biology* **198**, 337–48.

Hogstrand, C., Wilson, R.W., Polgar, D. & Wood, C.M. (1994). Effects of zinc on the branchial calcium uptake in freshwater rainbow trout during adaptation to waterborne zinc. *Journal of Experimental Biology* **186**, 55–73.

Hogstrand, C. & Wood, C.M. (1993). Hazard assessment of metal pollution. In *Proceedings of the First International Conference on Transport, Fate, and Effects of Silver in the Environment*. Andren, A.W. (ed.), pp. 75–84, University of Wisconsin, Wisconsin.

Hogstrand, C., Verbost, P.M., Wendelaar Bonga, S.J. & Wood, C.M. (1996). Mechanisms of zinc uptake in gills of freshwater rainbow trout: interplay with calcium transport. *American Journal of Physiology* (in press).

Hollenberg, S.M., Gigure, V., Segui, P. & Evans, R.M. (1987). Co-localization of DNA-binding and transepithelial activation functions in the human glucocorticoid receptor. *Cell*, **49**, 39–46.

Hornung, H., Krom, M.D., Cohen, Y. & Bernhard, M. (1993). Trace metal content in deep-water sharks from the eastern Mediterranean Sea. *Marine Biology*, **115**, 331–8.

Hunziker, P.E. & Kägi, J.H.R. (1988). Metallothionein: a multigene protein. In *Essential and Toxic Trace Elements in Human Health* and Disease. Prasad, A.S. (ed.), pp. 349–63. Alan R. Liss, Inc., New York.

Hyllner, S.J., Andersson, T., Haux, C. & Olsson, P.-E. (1989). Cortisol induction of metallothionein in primary culture of rainbow trout hepatocytes. *Journal of Cell Physiology*, **139**, 24–8.

Imbert, J., Zafarullah, M., Culotta, V.C., Gedamu, L. & Hamer, D. (1989). Transcription factor MBF-I interacts with metal regulatory elements of higher eucaryotic metallothionein genes. *Molecular Cell Biology*, **9**, 5315–23.

Johnson, M.G. (1987). Trace element loadings to sediments of fourteen Ontario lakes and correlations with concentrations in fish. *Canadian Journal of Fisheries and Aquatic Sciences*, **44**, 3–13.

Julshamn, K., Andersen, K.-J., Ringdal, O. & Brenna, J. (1988). Effect of dietary copper on the hepatic concentration and subcellular distribution of copper and zinc in the rainbow trout (*Salmo gairdneri*). *Aquaculture*, **73**, 143–55.

Kägi, J.H.R. & Kojima, Y. (1987). Chemistry and biochemistry of metallothionein. In *Experientia, Suppl., Metallothionein II*, vol. 52. Kägi, J.H.R. and Kojima, Y. (eds.), pp. 25–61, Birkhäuser Verlag, Basel.

Kägi, J.H.R. & Schäffer, A. (1988). Biochemistry of metallothionein. *Biochemistry*, **27**, 8509–15.

Keilin, D. & Mann, T. (1940). Carbonic anhydrase. Purification and nature of the enzyme. *Biochemical Journal*, **34**, 1163–76.

Kershaw, W.C. & Klaassen, C.D. (1992). Degradation and metal composition of hepatic isometallothioneins in rats. *Toxicology and Applied Pharmacology*, **112**, 24–31.

Kille, P., Kay, J., Leaver, M. & George, S. (1992). Induction of piscine metallothioneins as a primary response to heavy metal pollutants: applicability of new sensitive molecular probes. *Aquatic Toxicology*, **22**, 279–86.

Kito, H., Tazawa, T., Ose, Y., Sato, T. & Ishikawa, T. (1982*a*). Formation of metallothionein in fish. *Comparative Biochemistry and Physiology*, **73C**, 129–34.

Kito, H., Tazawa, T., Ose, Y., Sato, T. & Ishikawa, T. (1982*b*). Protection by metallothionein against cadmium toxicity. *Comparative Biochemistry and Physiology*, **73C**, 135–9.

Kito, H., Ose, Y. & Sato, T. (1986). Cadmium-binding protein (metallothionein) in carp. *Environmental Health Perspectives*, **65**, 117–24.

Knox, D., Cowey, C.B. & Adron, J.W. (1984). Effects of dietary zinc intake upon copper metabolism in rainbow trout (*Salmo gairdneri*). *Aquaculture*, **40**, 199–207.

Koizumi, S., Yamada, H., Suzuki, K. & Otsuka, F. (1992). Zinc-specific activation of a HeLa cell nuclear protein which interacts with a metal responsive element of the human metallothionein-II$_A$ gene. *European Journal of Biochemistry*, **210**, 555–60.

Kojima, Y., Berger, C., Vallee, B.L. & Kägi, J.H.R. (1976). Amino-acid sequence of equine renal metallothionein-1B. *Proceedings of the National Academy of Sciences, USA*, **73**, 3413–17.

Krezoski, S., Laib, J., Onana, P., Hartmann, T., Chen, P., Shaw III, C.F. & Petering, D.H. (1988). Presence of Zn, Cu-binding protein in liver of freshwater fishes in the absence of elevated exogenous metal: Relevance to toxic metal exposure. *Marine Environmental Research*, **24**, 147–50.

Lafeber, F.P.J.G., Flik, G., Wendelaar Bonga, S.E. & Perry, S.F. (1988). Hypocalcin from Stannius corpuscles inhibits gill calcium uptake in trout. *American Journal of Physiology*, **254**, R891–6.

Laurén, D.J. & McDonald, D.G. (1985). Effects of copper on branchial ionoregulation in the rainbow trout, *Salmo gairdneri* Richardson. *Journal of Comparative Physiology B*, **155**, 635–44.

Lee, H.H., Prasad, A.S., Brewer, G.J. & Owyang, C. (1989). Zinc absorption in human small intestine. *American Journal of Physiology*, **256**, G687–91.

McCarter, J.A. & Roch, M. (1983). Hepatic metallothionein and resistance to copper in juvenile coho salmon. *Comparative Biochemistry and Physiology*, **74C**, 133–7.

McCormick, S.D., Hasegawa, S. & Hirano, T. (1992). Calcium uptake in the skin of a freshwater teleost. *Proceedings of the National Academy of Sciences, USA*, **89**, 3635–8.

McDonald, D.G. & Wood, C.M. (1993). Branchial mechanisms of acclimation to metals in freshwater fish. In *Fish Ecophysiology*. Rankin, J.C. and Jensen, F.B. (eds.), pp. 297–321, Chapman and Hall, London.

Marshall, W.S., Bryson, S.E. & Wood, C.M. (1992). Calcium transport by isolated skin of rainbow trout. *Journal of Experimental Biology*, **166**, 297–316.

Mehra, R.K. & Bremner, I. (1985). Studies on the metabolism of rat liver copper–metallothionein. *Biochemical Journal*, **227**, 903–38.

Milner, N.J. (1982). The accumulation of zinc by O-group plaice, *Pleuronectes platessa* (L.), from high concentrations in sea water and food. *Journal of Fish Biology*, **21**, 325–36.

Noël-Lambot, F., Gerday, C. & Disteche, A. (1978). Distribution of Cd, Zn, and Cu in liver and gills of the eel *Anguilla anguilla* with special reference to metallothioneins. *Comparative Biochemistry and Physiology*, **61C**, 177–87.

Olafson, R.W. & Thompson, J.A.J. (1974). Isolation of heavy metal binding proteins from marine vertebrates. *Marine Biology*, **28**, 83–6.

Olsson, P.-E. & Hogstrand, C. (1987). Improved separation of perch liver metallothionein by fast protein liquid chromatography. *Journal of Chromatography*, **402**, 293–9.

Olsson, P.-E., Haux, C. & Förlin, L. (1987). Variations in hepatic metallothionein, zinc and copper levels during an annual reproductive cycle in rainbow trout, *Salmo gairdneri*. *Fish Physiology and Biochemistry*, **3**, 39–47.

Olsson, P.-E., Larsson, A., Maage, A., Haux, C. & Bonham, K. (1989*a*). Induction of metallothionein synthesis in rainbow trout, *Salmo gairdneri*, during long-term exposure to waterborne cadmium. *Fish Physiology and Biochemistry*, **6**, 221–9.

Olsson, P.-E., Zafarullah, M., Foster, R., Hamor, T. & Gedamu, L. (1990). Developmental regulation of metallothionein mRNA, zinc and copper levels in rainbow trout, *Salmo gairdneri*. *European Journal of Biochemistry*, **193**, 229–36.

Olsson, P.-E., Zafarullah, M. & Gedamu, L. (1989*b*). A role of metallothionein in zinc regulation after oestradiol induction of vitellogenin synthesis in rainbow trout, *Salmo gairdneri*. *Biochemical Journal*, **257**, 555–60.

Overnell, J., McIntosh, R. & Fletcher, T.C. (1987). The levels of liver metallothionein and zinc in plaice, *Pleuronectes platessa* L., during the breeding season, and the effect of oestradiol injection. *Journal of Fish Biology*, **30**, 539–46.

Parisi, A.F. & Vallee, B.L. (1970). Isolation of a zinc alpha-2 macroglobulin from human serum. *Biochemistry*, **9**, 2421–6.

Pentreath, R.J. (1973). The accumulation and retention of ^{65}Zn and ^{54}Mn by the plaice *Pleuronectes platessa* L. *Journal of Experimental Marine Biology and Ecology*, **12**, 1–18.

Pentreath, R.J. (1976). Some further studies on the accumulation of ^{65}Zn and ^{54}Mn by the plaice, *Pleuronectes platessa*. *Journal of Experimental Marine Biology and Ecology*, **21**, 179–89.

Perry, S.F. & Flik, G. (1988). Characterization of branchial transepithelial calcium fluxes in freshwater rainbow trout, *Salmo gairdneri*. *American Journal of Physiology*, **23**, R491–8.

Perry, S.F., Goss, G.G. & Fenwick, J.C. (1992). Interrelationships between gill chloride cell morphology and calcium uptake in freshwater teleosts. *Fish Physiology and Biochemistry*, **10**, 327–37.

Perry, S.F. & Wood, C.M. (1985). Kinetics of branchial calcium uptake in the rainbow trout: Effects of acclimation to various external calcium levels. *Journal of Experimental Biology*, **116**, 411–33.

Phillips, G.R. & Russo, R.C. (1978). Metal Bioaccumulation in Fishes and Aquatic Invertebrates: A Literature Review. EPA–600/3–78–

82   C. HOGSTRAND AND C.M. WOOD

103, National Technical Information Service, US Department of Commerce, Springfield, VA.

Pierson, K.B. (1985a). Isolation and partial characterization of a non-thionein, zinc-binding protein from the liver of rainbow trout (*Salmo gairdneri*). *Comparative Biochemistry and Physiology*, **80C**, 299–304.

Pierson, K.B. (1985b). Occurrence and synthesis of a non-thionein, zinc-binding protein in the rainbow trout (*Salmo gairdneri*). *Comparative Biochemistry and Physiology*, **81C**, 71–5.

Renfro, W.C., Fowler, S.W., Heyraud, M. & La Rosa, J. (1975). Relative importance of food and water in long-term zinc^{-65} accumulation by marine biota. *Journal of Fisheries Research Board of Canada*, **32**, 1339–45.

Roch, M. & McCarter, J.A. (1984). Hepatic metallothionein production and resistance to heavy metals by rainbow trout (*Salmo gairdneri*). II. Held in a series of contaminated lakes. *Comparative Biochemistry and Physiology*, **77C**, 77–82.

Roch, M., McCarter, J.A., Matheson, A.T., Clark, M.J.R. & Olafson, R.W. (1982). Hepatic metallothionein in rainbow trout (*Salmo gairdneri*) as an indicator of metal pollution in the Campbell River system. *Canadian Journal of Fisheries and Aquatic Sciences*, **39**, 1596–601.

Sauer, G.R. & Watabe, N. (1989). Ultrastructural and histochemical aspects of zinc accumulation by fish scales. *Tissue and Cell*, **21**, 935–43.

Sayer, M.D.J., Reader, J.P. & Morris, R. (1991). Effects of six trace metals on calcium fluxes in brown trout (*Salmo truta* L.) in soft water. *Journal of Comparative Physiology B*, **161**, 537–42.

Schwabe, J.W.R., Neuhaus, D. & Rhodes, D. (1990). Solution structure of the DNA-binding domain of the oestrogen receptor. *Nature*, **348**, 458–61.

Searle, P. (1990). Zinc dependent binding of a liver nuclear factor to metal response element MRE-a of the mouse metallothionein-l gene and variant sequences. *Nucleic Acids Research*, **18**, 4683–90.

Shearer, K.D. (1984). Changes in elemental composition of hatchery-reared rainbow trout, *Salmo gairdneri*, associated with growth and reproduction. *Canadian Journal of Fisheries and Aquatic Sciences*, **41**, 1592–600.

Shearer, K.D., Maage, A., Opstvedt, J. & Mundheim, H. (1992). Effects of high-ash diets on growth, feed efficiency, and zinc status of juvenile Atlantic salmon (*Salmo salar*). *Aquaculture*, **106**, 345–55.

Shears, M.A. & Fletcher, G.L. (1983). Regulation of Zn^{2+} uptake from the gasterointestinal tract of a marine teleost, the winter flounder (*Pseudopleuronectes americanus*). *Canadian Journal of Fisheries and Aquatic Sciences*, **40** (suppl. 2), 197–205.

Skidmore, J.F. (1970). Respiration and osmoregulation in rainbow trout with gills damaged by zinc sulphate. *Journal of Experimental Biology*, **52**, 481–94.

Spear, P.A. (1981). *Zinc in the Aquatic Environment: Chemistry, Distribution, and Toxicology*. National Research Council of Canada, Environmental Secretariat publication 17589, Publications NRCC/ CNRC, Ottawa.

Spehar, R.L. (1976). Cadmium and zinc toxicity to flagfish, *Jordanella floridae*. *Journal of Fisheries Research Board of Canada*, **33**, 1939–45.

Spry, D.J., Hodson, P.V. & Wood, C.M. (1988). Relative contributions of dietary and waterborne zinc in the rainbow trout, *Salmo gairdneri*. *Canadian Journal of Fisheries and Aquatic Sciences*, **45**, 32–41.

Spry, D.J. & Wood, C.M. (1984). Acid–base, plasma ion, and blood gas changes in rainbow trout during short term toxic zinc exposure. *Journal of Comparative Physiology B*, **154**, 149–58.

Spry, D.J. & Wood, C.M. (1985). Ion flux rates, acid–base status, and blood gases in rainbow trout, *Salmo gairdneri*, exposed to toxic zinc in natural soft water. *Canadian Journal of Fisheries and Aquatic Sciences*, **42**, 1332–41.

Spry, D.J. & Wood, C.M. (1988). Zinc influx across the isolated, perfused head preparation of the rainbow trout (*Salmo gairdneri*) in hard and soft water. *Canadian Journal of Fisheries and Aquatic Sciences*, **45** 2206–15.

Spry, D.J. & Wood, C.M. (1989*a*). A kinetic method for the measurement of zinc influx *in vivo* in the rainbow trout, and the effects of waterborne calcium on flux rates. *Journal of Experimental Biology*, **142**, 425–46.

Spry, D.J. & Wood, C.M. (1989*b*). The influence of dietary and waterborne zinc on heat-stable metal ligands in rainbow trout, *Salmo gairdneri* Richardson: quantification by ^{109}Cd radioassay and evaluation of the assay. *Journal of Fish Biology*, **35**, 557–76.

Spry, D.J., Wood, C.M. & Hodson, P.V. (1981). *The Effects of Environmental Acid on Freshwater Fish with Particular Reference to the Softwater Lakes in Ontario and the Modifying Effects of Heavy Metals. A Literature Review*. Canadian Technical Reports of Fisheries and Aquatic Sciences, 999.

Takeda, H. & Shimizu, C. (1982). Existence of the metallothionein-like protein in various fish tissues. *Bulletin of the Japanese Society of Scientific Fisheries*, **48**, 711–15.

Thiele, D.J. (1992). Metal-regulated transcription in eucaryotes. *Nucleic Acids Research*, **20**, 1183–91.

Underwood, E.J. (1977). *Trace Elements in Human and Animal Nutrition*. Academic Press, New York.

Vallee, B.L. & Auld, D.S. (1992). Active zinc binding sites of zinc metalloenzymes. In *Matrix Metalloproteinases and Inhibitors. Matrix*

84 C. HOGSTRAND AND C.M. WOOD

Supplement 1. Birkedal-Hansen, H., Werb, Z., Welgus, H. and Van Wart, H. (eds.), p. 5, Fisher, Stuttgart, Germany.

Vallee, B.L. & Falchuk, K.H. (1993). The biochemical basis of zinc physiology. *Physiological Reviews,* **73,** 79–118.

Verbost, P.M., Flik, G., Pang, P.K.T., Lock, R.A.C. & Wendelaar Bonga, S.E. (1989). Cadmium inhibition of the erythrocyte Ca^{2+} pump. *Journal of Biological Chemistry,* **264,** 5613–15.

Verbost, P.M., Flik, G., Fenwick, J.C., Greco, A.-M., Pang, P.K.T. & Wendelaar Bonga, S.E. (1993). Branchial calcium uptake: possible mechanisms of control by stanniocalcin. *Fish Physiology and Biochemistry,* **11,** 205–15.

Verbost, P.M., Schoenmakers, Th.J.M., Flik, G. & Wendelaar Bonga, S.E. (1994). Kinetics of ATP- and Na^+-gradient driven Ca^{2+} transport in basolateral membranes from gills of freshwater- and seawater-adapted tilapia. *Journal of Experimental Biology,* (in press).

Weizel, G., Strecker, F.-J., Roester, U., Buddeke, E. & Fretzdorff, A.-M. (1954). Zink im tapetum lucidum. *Hoppe-Seyler's Zeitschrift der Physiologische Chemie,* **296,** 19–30.

Wekell, J.C., Shearer, K.D. & Houle, C.R. (1983). High zinc supplementation of rainbow trout diets. *Progressive Fish-Culturist,* **45,** 144–7.

Wicklund, A. (1990). Metabolism of cadmium and zinc in fish. *PhD Thesis, ISBN 91-554-2530-5,* Uppsala University, Uppsala, Sweden.

Wicklund Glynn, A. (1991). Cadmium and zinc kinetics in fish: studies on water-borne ^{109}Cd and ^{65}Zn turnover and intracellular distribution in minnows, *Phoxinus phoxinus. Pharmacology and Toxicology,* **69,** 485–91.

Wicklund Glynn, A., Haux, C. & Hogstrand, C. (1992). Chronic toxicity and metabolism of Cd and Zn in juvenile minnows (*Phoxinus phoxinus*) exposed to a Cd and Zn mixture. *Canadian Journal of Fisheries and Aquatic Sciences,* **49,** 2070–9.

Williams, R.J.P. (1984). Zinc: what is its role in biology? *Endeavour, New Series,* **8,** 65–70.

Willis, J.N. & Sunda, W.G. (1984). Relative contributions of food and water in the accumulation of zinc by two species of marine fish. *Marine Biology,* **80,** 273–9.

Wu, F.Y. & Wu, C.-W. (1987). Zinc in DNA replication and transcription. *Annual Review of Nutrition,* **7,** 251–72.

Zafarullah, M., Olsson, P.-E. & Gedamu, L. (1989). Endogenous and heavy-metal-ion-induced metallothionein gene expression in salmonid tissues and cell lines. *Gene,* **83,** 85–93.

E.W. TAYLOR, M.W. BEAUMONT,
P.J. BUTLER, J. MAIR and M.S.I. MUJALLID

Lethal and sub-lethal effects of copper upon fish: a role for ammonia toxicity?

Copper in the aquatic environment

Although traces of copper are essential constituents of some enzymes, copper is toxic to both animals and plants at levels only just in excess of those found in many unpolluted aquatic environments. Levels of dissolved copper are often increased from anthropogenic origins such as mine washings and direct application as an algicide, molluscicide or anti-fouling agent. Natural rock weathering adds about 2×10^5 tonnes yr^{-1} of copper into the river systems of the world. In comparison Bowen (1979) estimated an annual addition of 6.19×10^6 tonnes of copper from mining sources. Global copper emissions tripled between 1950 and 1980 (Moore and Ramamoothy, 1984) and this order of magnitude difference between anthropogenic and natural inputs has led copper to be classified as one of the more potentially hazardous heavy metals (Sposito, 1986). While copper concentration in surface waters in general rarely rises above 5 $\mu g\ l^{-1}$ or 0.1 $\mu mol\ l^{-1}$ (Spear & Pierce, 1979), in common with other heavy metals, copper is mobilized during acid episodes. Thus, particularly in those soft, poorly buffered waters that are prone to acidification, copper concentrations can be significantly elevated by acid rain. Turnpenny et al. (1987) found that in mildly acidic streams in Wales, copper and zinc were more important determinants of fisheries status than the acidity itself.

Fortunately, copper can exist in natural waters in a wide variety of physico-chemical forms many of which are of low bioavailability and/ or toxicity. Of the divalent ions in the first transition series, irrespective of the ligand type or concentration, copper forms the most stable organic complexes (Irving & Williams, 1953). A relatively low concentration of organic complexing agents may effectively bind all the available copper (Lerman & Childs, 1973; Morel, McDuff & Morgan, 1973). In natural waters, organic ligands include humic, fulvic and amino acids and polypeptides, all of which can be derived from decomposition processes and anthropogenic inputs such as sewage. Due to the size and shape of these molecules and to the coordination of copper with more than one ligand of the complexing agent, such complexes are predicted to be generally innocuous to aquatic organisms (Spear & Pierce, 1979). Copper may also be found in particulate and colloidal states arising from adsorption of copper by, for example, hydrous metal oxides and clays. Little is known of the effects of these forms of copper once they become incorporated into the food chain but, at least in the short term, they will also be non-toxic (Shaw & Brown, 1974). However, the capacity of organics to form complexes with copper is reduced as pH declines (Shapiro, 1964; Culp, 1975). The acidification of surface waters will therefore increase the availability of copper to aquatic biota.

Copper ions coordinate with water molecules to form aquo ions $[Cu(H_2O)_6]^{2+}$. Complexes are formed by the successive displacement of water molecules. In the absence of organics and in freshwater, bicarbonate and carbonate are the only inorganic ions that need to be considered and the chemistry of copper is basically that of copper in calcium bicarbonate solutions (Stiff, 1971). A number of authors have produced models to predict the copper species present in freshwater systems (e.g. Shaw & Brown, 1974; Ernst, Allen & Mancy, 1975; Howarth & Sprague, 1978; Waiwood & Beamish, 1978). Two factors are important, pH and bicarbonate concentration. At neutral pH, $CuCO_3$ is the predominant form over a total carbonate range of three orders of magnitude but decreasing alkalinity increases the contributions of the $CuOH^+$ and Cu^{2+} species at the expense of $Cu(CO_3)_2^{2-}$ (Ernst *et al.*, 1975). Below pH 5, the contribution of the cupric ion to the total soluble copper is greatest regardless of the alkalinity and may in fact be the only form present (Howarth & Sprague, 1978). As acidity declines, the concentration of the cupric ion also falls in favour of carbonate and hydroxides so that at pH 8 to 9, Cu^{2+} is almost entirely absent (Howarth & Sprague, 1978).

Such analysis is important since only certain copper species have a toxic effect. Shaw and Brown (1974) found both the concentration of Cu^{2+} and $CuCO_3$ to be determinants of toxicity. However, these authors used a rather limited model and did not consider the formation of copper hydroxides to be important. Other authors, using models that included a wider range of species, found evidence to indicate that a given level of copper toxicity is best explained by the concentration of Cu^{2+} and perhaps also of $CuOH^+$ but not $CuCO_3$ (Pagenkopf, Russo & Thurston, 1974; Howarth & Sprague, 1978; Waiwood & Beamish, 1978; Chakoumakos, Russo & Thurston, 1979).

The toxic action of copper

In common with many other waterborne pollutants and in particular with other heavy metals and low pH, the primary target of the toxic actions of copper on fish is their gills (Evans, 1987). Fish gills are structurally highly specialized for their role in respiratory gas exchange, presenting a large surface area to the ventilatory water stream and having thin epithelial layers covering the fragile lamellae. Comprehensive structural damage is caused by severe copper exposure, characterized by the collapse and even fusing of lamellae, the lifting of lamellar epithelium away from pillar cells and swelling of epithelial cells forming ridges in the epithelium (Kirk & Lewis, 1993). The effects of acute exposure to lethal levels of copper on gill histology were described by Wilson and Taylor (1993*a*) and are illustrated in Fig. 1. This injury is similar to that caused by both zinc (Skidmore & Tovell, 1972) and acid (Westfall, 1945; Daye and Garside, 1976; Ultsch & Gros, 1979). In addition to ultrastructural damage, Kirk and Lewis (1993) observed an increase in the secretion of mucus and a concomitant swelling of the mucus layer around the gill which also contained a significant amount of cellular debris. The secretion of mucus is a common stress response to a variety of environmental factors including heavy metals, ammonia and salinity changes (Eddy & Fraser, 1982). Miller and Mackay (1982) suggest that secretion of mucus is actually a specific response to only certain pollutants, including exposure to low pH, and that in fact the elevation of secretion following copper exposure is a short-term phenomenon due to the evacuation of existing mucous cells which then remain empty.

Such extreme damage undoubtedly has a strong effect upon the ability of the gill – to function effectively as exchange organs. The structural damage is likely to disrupt the perfusion of blood through the gills and their ventilation with water and to increase diffusion distances (Hughes, 1976; Tuurala & Soivio, 1982). An expanded mucus layer may exacerbate the problem (Ultsch & Gros, 1979).

In freshwater, where fish are hyperosmotic to the surrounding water, copper causes a net loss of ions such as Na^+ and Cl^- (Lewis & Lewis, 1971; Christiansen *et al.*, 1972; Schreck & Lorz, 1978; Stagg & Shuttleworth, 1982). Laurén and McDonald (1985) found that in rainbow trout this net loss comprises two components. At copper concentrations as low as 0.2 μmol l^{-1}, the active uptake of ions was inhibited. At copper levels of 0.8 μmol l^{-1}, there was a second contributing factor to the net ion loss, an increase in the permeability of the gill membrane to these ions and hence a greater passive ion efflux. These effects are

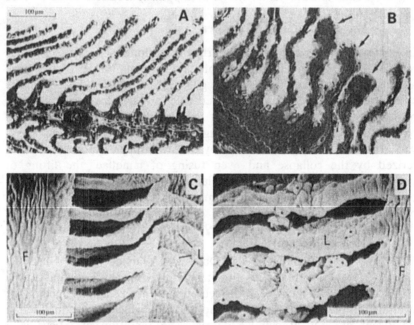

Fig. 1. Changes in gill structure following damage by exposure to copper in freshwater trout. Light micrographs of gills from (a) control fish, and (b) trout exposed to 4.9 μmol l⁻¹ copper. In the control fish, lamellae are thin and uniform with a minimal blood–water diffusion distance. After exposure to copper (b). Cells on the surface of lamellae appear swollen, causing a thickening of the respiratory epithelium and hence an increase in the blood–water diffusion distance. Arrows indicate lamellae where curling has occurred, leading to a further reduction in the effective surface area for gas exchange. Scanning electron micrographs of gills from (c) control fish, and (d) trout exposed to 4.9 μ mol l⁻¹ copper. Asterisks in (d) mark respiratory pavement cells (as indicated by their microridged surface under higher

similar to those observed during exposure to low pH and it has been suggested that acid and copper share a similar mechanism of toxicity (Laurén & McDonald, 1985). Wood (1989) reported that death in rainbow trout acutely exposed to acid occurs once sodium or chloride concentrations ([Na⁺], [Cl⁻]) fall below 70% of normal. However, he did not suggest that the absolute level of these ions is the critical factor, since salmonids tolerate this sort of dilution in a number of situations, including migration (Miles, 1971). Mortality is related instead to the rapidity of ion loss (McDonald, 1983) and there is evidence for a series of physiological events leading to death that is associated with

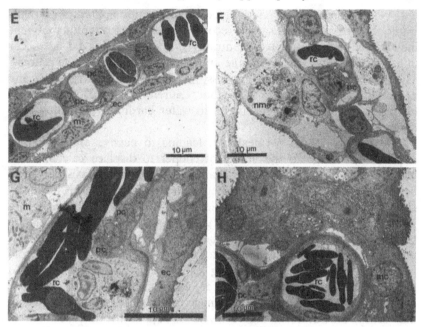

magnification) which have become almost completely detached from
the surrounding lamellar epithelium on exposure to copper. L, lamellae; F, filament. (From Wilson & Taylor, 1993*a*.) Transmission electron micrographs of gill lamellae from brown trout (e) in unpolluted
water, and (f–h) after exposure to low levels (0.7 μmol l^{-1}) of dissolved
copper. Exposure to copper has caused proliferation of mucocytes,
degeneration of chloride cells and epithelial and pillar cells to become
detached, resulting in increased diffusion distances, together with
aggregation of red blood cells in the disrupted blood spaces; ec,
epithelial cell; mc, mucocyte; m, mitochondrial rich (chloride) cell;
nm, necrotic chloride cell; pc, pillar cell; rc, red blood cell.
(Unpublished data from M.S.I. Mujallid, E.W. Taylor & P.J. Butler.)

rapid ion loss, which causes a decline in plasma osmotic potential not
compensated for by ion loss from the intracellular fluid (Milligan &
Wood, 1982). Instead, an osmotic flux of water occurs from the plasma
to the intracellular fluid so that plasma volume falls by up to 27%
(Wood, 1989).

The mechanisms of copper toxicity have, in general, been elucidated
from the observed influences on relative toxicity of changes in water
quality, in particular its pH, hardness and alkalinity (Miller & MacKay,
1980). The toxicity of copper is dependent on the ambient calcium
levels during exposure (e.g. Alabaster & Lloyd, 1980; Reid & McDon-

ald, 1988). Rainbow trout acclimated to, and experimented on, in hard water showed a compensatory recovery in the passive efflux of ions induced by higher copper concentrations. Such recovery did not occur in fish in soft water, indicating the importance of calcium in the process. The two components of net ion loss, an increase in passive efflux and an inhibition of active uptake, that Laurén and McDonald (1985) observed, had different sensitivities to water hardness, indicating their different origins.

The increase in gill permeability, leading to passive efflux of ions, is probably related to the ability of copper to displace calcium from biological ligands (Nieboer & Richardson, 1980). Calcium is known to be important in controlling the integrity and permeability of the branchial epithelium in fishes (Potts & Fleming, 1971; McDonald & Rogano, 1986) and Laurén & McDonald (1985) proposed that the stimulation of ionic effluxes by copper may be a result of increased ionic permeability of the gill, secondary to the displacement of Ca^{2+} from paracellular tight junctions, causing a breakdown in cell adhesion. At low enough concentrations of copper, ionoregulatory disturbances can occur without apparent physical damage to the gills (Laurén & McDonald, 1985). However, at higher levels of copper the displacement of intercellular and membrane-bound calcium may lead to a breakdown in branchial cell volume control and epithelial organization. Indeed, exposure of freshwater trout to 4.9 mmol l^{-1} copper caused severe gill histopathologies which included cell adhesion failure, swelling, and pillar cell detachment leading to numerous haematomas (red cells pooling in regions of lamellae) (Wilson & Taylor, 1993a). These effects are illustrated in Fig. 1.

The second component of net ion loss described by Laurén and McDonald (1985) was an inhibition of active ion uptake which occurred at copper concentrations as low as 0.2 μmol l^{-1}. There was no recovery of ion uptake over time, regardless of calcium concentration. This effect is most probably derived from an inhibition of transport enzymes in the branchial epithelium, in particular of Na^+/K^+ ATPase (Lorz & McPherson, 1976). It has been demonstrated that copper inhibits branchial ATPases both *in vitro* (Shephard & Simkiss, 1978; Beckman & Zaugg, 1988), in isolated gill or opercular membrane preparations (Stagg & Shuttleworth, 1982; Crespo & Karnaky, 1983), and *in vivo* in freshwater rainbow trout (Laurén & McDonald, 1987). The inhibition of active NaCl uptake is most likely due to the high affinity of the cupric ion for sulphydryl groups (-SH) on transport enzymes such as Na^+ K^+-ATPase located in the gill epithelium (Stagg & Shuttleworth, 1982). Exposure of brown trout to 0.47 μmol l^{-1} copper in acid, soft water (pH 5) caused gill Na^+/K^+ ATPase activity to fall from a control (pH 7.0 no copper) value of 1.31 ± 0.38 to 0.43 ± 0.06 µmol Pi mg^{-1} protein h^{-1}. (J. Mair, unpubl. obs.).

Crespo and Karnaky (1983) found copper to have an inhibitory effect upon chloride ion secretion across the opercular epithelium of killifish (*Fundulus heteroclitus*) only when applied to its serosal side. This epithelium contains chloride cells identical in structure to those of the gill epithelium. Enzymes for ion secretion are situated exclusively on the basolateral membrane and this experiment therefore adds further evidence for a specific effect of copper on transport enzymes.

Responses of freshwater trout to copper

Exposure to copper is acutely lethal to rainbow trout in freshwater (Wilson & Taylor, 1993a). Fish die within 24 hours of exposure to 4.9 µmol l^{-1} copper (levels measured in the Tees estuary) and the changes recorded during this period provide clues to the physiological effects of the toxic metal ion. These are illustrated in Fig. 2(a)–(i). The fish's ability to ionoregulate broke down so that sodium and chloride levels in plasma progressively fell (Fig. 2(a)). Plasma calcium levels were regulated at first but rose precipitously at the point of death. In contrast potassium levels increased progressively (Fig. 2(b)), and these changes were interpreted as indicating potassium efflux from damaged tissues, secondary to plasma ionic dilution.

The structural disruption of the gill epithelium in freshwater trout exposed to copper has drastic effects on respiratory gas exchange. The fish become progressively hypoxic (Fig. 2(c)) and as a consequence haemoglobin oxygenation is reduced so that oxygen supply to tissues is reduced (Fig. 2(d)). Carbon dioxide accumulates (Fig. 2(e)) and the fish switch to anaerobic metabolism, so that lactic acid levels in the blood rise (Fig. 2(f)). The blood becomes progressively acidotic (Fig. 2(g)) due to a combined respiratory (CO_2) and metabolic (lactic acid) acidosis. This, in turn, may cause a rightward Bohr or Root shift on the fish's haemoglobin, reducing its affinity for oxygen, and further curtailing oxygen transport. Physical damage to the gills (and reduced diffusion of NH_3) may also be responsible for the progressive accumulation of plasma ammonia (Fig. 2(h)). However, inhibition of active ammonium ion transport NH_4^+/Na^+ or H^+ exchange may also contribute to this effect.

Possibly, as part of its attempt to compensate for these changes by increasing oxygen transport, the fish exhibit an increase in haematocrit to values as high as 50% (Fig. 2(i)). Part of this increase is due to

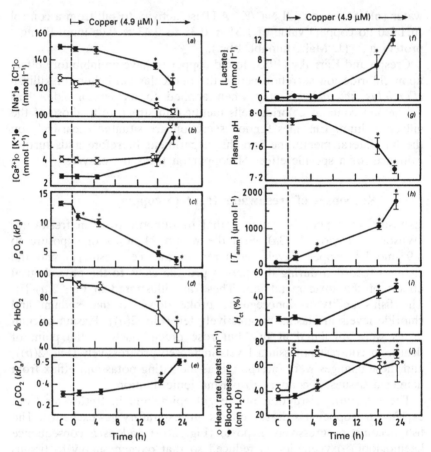

Fig. 2. The toxic effects of 24 h exposure to 4.9 μ mol^{-1} copper on trout measured as changes in: (*a*) concentration of sodium and chloride ions in blood plasma, (*b*) concentration of calcium and potassium ions in blood plasma, (*c*) P_aO_2, (*d*) percentage saturation of haemoglobin, (*e*) P_aCO_2, (*f*) concentration of lactate in blood plasma, (*g*) blood plasma pH, (*h*) ammonia accumulation in blood plasma, (*i*) haematocrit, and (*j*) heart rate and blood pressure. Values are means ±SEM with asterisk denoting significant deviation ($P<0.05$) from the mean values prior to copper exposure (C). (From Wilson & Taylor, 1993*a*.)

the osmotic movement of water into erythrocytes from the diluted plasma, causing them to swell (Soivio & Nikinmaa, 1981; Milligan & Wood, 1982). However, there was also an increase in red cell numbers. This compares with the earlier observations of McKim *et al.* (1970)

who recorded a rise in haematocrit and red blood cell count during exposure of brook trout (*Salvelinus fontinalis*) to copper and of Beckman and Zaugg (1988) who observed an elevated haematocrit in copper-exposed Chinook salmon (*Oncorhynchus tshawytscha*). The increased haematocrit is probably due to a catecholamine-mediated release of red cells from the spleen (Yamamoto, Itazawa & Kobayashi, 1980). Ye and Randall (1991) observed a release of catecholamines into the plasma of acid-exposed rainbow trout just prior to death.

The elevated haematocrit ensuing from reduced plasma volume, red cell swelling and mobilization, together with the increase in plasma proteins will contribute to an increased blood viscosity (Wells & Weber, 1991). As catecholamines are also responsible for vasoconstriction in the peripheral circulation, leading to increased resistance (Milligan & Wood, 1982), blood pressure is likely to rise. Wilson and Taylor (1993a) observed an increase in mean arterial blood pressure and heart rate after 20 h of lethal copper exposure in the rainbow trout (Fig. 2(j)). As a consequence of these physiological changes, cardiac work will be increased to the extent that death through circulatory failure seems probable. This likelihood of cardiac failure is increased by the additional stresses of reduced oxygen supply to the tissues, including the myocardium, and impaired acid–base regulation.

Attempts by the fish to compensate for insufficiencies in oxygen uptake at the gill during severe copper exposure may therefore compound the effects of ionoregulatory failure. In Wilson and Taylor's (1993a) study, heart rate, although greater than that in control fish, declined from an initial peak, despite the putative presence of catecholamines (Fig. 2(j)), indicating an inability to sustain a high heart rate, probably due to reduced myocardial contractility. The fish dies because its integrated responses to the physiological changes induced by the toxic environment culminate in its destruction. The breakdown appears to start with disruption of ionoregulation which in freshwater is vital to survival because the ionic gradients between the fish and its environment are very large.

Responses of seawater trout to copper

We have seen that acute exposure of freshwater trout to 4.9 mmol copper l^{-1} at pH 7.9 caused 100% mortality within 24 hours, as a result of severe ionoregulatory and respiratory disturbances. In contrast, trout acclimated to full-strength sea water (external [NaCl] three- to fourfold higher than blood [NaCl]) and exposed to 6.3 mmol copper l^{-1} at the same pH for 24 h, exhibited no significant change in plasma ion concentrations, no respiratory problems and survived with apparently undamaged gill lamellae (Wilson & Taylor, 1993*b*). It is clear that copper toxicity, and the physiological responses during acute exposure, are greatly reduced at high salinities. These results were unexpected, as our prediction was that an ionoregulatory inhibitor such as copper would cause a net increase in plasma ions in fish maintained in a hypersaline environment and indeed Stagg and Shuttleworth (1982) observed an elevation of the plasma ion concentration of flounder (*Platychthys flesus* L.) exposed to copper in seawater.

Thus, seawater(-adapted) teleosts can be less sensitive to copper than freshwater (adapted) species, as previously described by Eisler & Gardner (1973); Voyer (1975); Taylor, Maddock and Mance (1985). This may be a function of physicochemical factors such as higher calcium levels. Ambient [Ca^{2+}] in seawater is more than 26-fold higher than that in freshwater which would obviously provide more competition for external binding sites with copper, resulting in no ionoregulatory disturbances despite the large Na^+ and Cl^- gradients across the gills (Laurén & McDonald, 1985; Reid & McDonald, 1988). In addition, the high carbonate alkalinity of seawater and the presence of chelating agents, which can reduce the toxic forms of copper by complexation (Chakoumakos *et al.*, 1979; Miller & Mackay, 1980), may reduce copper availability. Alternatively, the lower toxicity could be related to physiological differences in the mechanisms of branchial salt transport and the location of the key transepithelial ion transporting enzymes in seawater (acclimated) fish. To test this hypothesis fully would require the use of isotopic unidirectional ion flux measurements and synthetic seawater media with controlled Ca^{2+} and Mg^{2+} levels to determine whether differences in [Ca^{2+}] and [Mg^{2+}] are, in fact, the determining factors in the differential responses to copper exposure seen in trout acclimated to freshwater and seawater. Longer-term exposure of fish to copper can result in uptake over the gills or via the gut and subsequent damage to internal organs such as the liver (Baker, 1969). Consequently, basolateral ion regulation may be affected by long-term exposure.

Sub-lethal effects of copper exposure

Measures of the relative toxicity of a substance to fish which rely on killing animals, such as the LC_{50}, used alone as a basis for judging the environmentally 'safe' concentrations of a pollutant are likely to lead to underestimation of its true consequence. The sub-lethal effects upon, for instance, the exercise performance, behaviour or reproductive capacity of individual organisms may have profound significance for the survival of the population as a whole. Howells, Howells and Alabaster (1983) found extinction of fish populations in a Welsh water system to occur at 50% of an experimentally determined LC_{50} for copper and zinc. Cairns (1966) stressed the need for caution in the application of results from acute toxicity tests and indicated the importance of measured changes in oxygen consumption and swimming capacity as sub-lethal measures of relative toxicity. Swimming performance is an indicator of the ability of the fish to feed, escape predation and maintain position in a current (Beamish, 1978) and is of particular significance to fish such as salmonids that migrate upstream to spawn.

It is somewhat surprising, therefore, to discover that there have been only two studies of the effect of copper upon swimming performance, one by Waiwood and Beamish (1978) and the other by the present authors (Beaumont, Butler & Taylor 1995a). In the former study, fingerling rainbow trout were exposed to a number of combinations of water hardness (30–360 mg l^{-1} EDTA), pH (6–8) and copper concentration (0–200 µg l^{-1} or 3.2 µmol l^{-1}). Critical swimming speed (U_{crit}; Brett, 1964) was determined after 0.5, 5, 10 and 30 days. The copper/acid concentrations chosen were not sub-lethal and mortality, particularly at the combination of lowest pH and hardness and highest copper concentration, was described as 'considerable'. The swimming performance of the survivors at day 5 was up to 35% lower than that of control fish, being most reduced in soft acidic water. Waiwood and Beamish (1978) also measured the oxygen consumption of copper exposed fish during exercise. At any given speed, oxygen consumption of copper exposed trout was greater than that of the control animals, but maximum uptake at U_{crit} was reduced. It was concluded that the former was the result of increased maintenance costs arising from inefficiencies of gas exchange at the damaged gill and from impaired osmoregulation. Gill damage was also proposed as the cause of reduced maximum uptake.

The study performed by the current authors has taken a slightly different approach, in that the copper concentrations used were entirely sub-lethal and that blood and tissue samples were taken from each fish in order to investigate whether disruption of swimming performance and gill ultrastructure could be correlated to changes to blood gas, acid–base and metabolite status. Adult brown trout (300–600 g) were acclimated to 5 °C in winter and 15 °C in summer in dechlorinated Birmingham tapwater for 2–4 weeks and then in artificial softwater (Ca^{2+} 50 µmol l^{-1}) for another 2–4 weeks. A catheter was inserted into

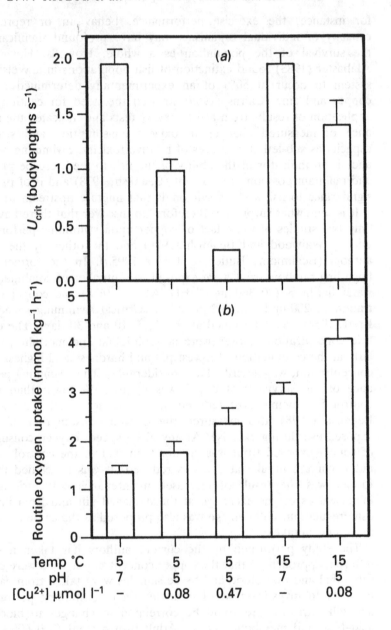

the dorsal aorta of each fish while under anaesthesia (MS222) to allow blood sampling with minimum disturbance. Trout were then exposed to a 96 h episode of copper at pH 5. The copper levels used had been previously determined by toxicity testing to be the sub-lethal copper concentrations (SLCC) for these experimental conditions, that is the highest dose that caused no mortality during the exposure period. The SLCC was 0.47 μmol l^{-1} Cu^{2+} at 5 °C and 0.08 μmol l^{-1} Cu^{2+} at 15 °C. In addition, winter trout were exposed to the summer SLCC (i.e. 0.08 μmol l^{-1} Cu^{2+} at 5 °C).

The effect of copper upon swimming performance was most marked in the winter trout exposed to 0.47 μmol l^{-1} Cu^{2+} (Fig. 3(a)). Only one of six fish tested swam steadily at the lowest test speed of 0.3 m s^{-1}. The others swam only for one or two minutes in brief bursts and glides. Those trout at 5 °C that displayed no ability for aerobic exercise, seemed to retain some capacity for anaerobic 'burst' swimming. This observation strongly suggests the effect of the copper/acid exposure to be upon aerobic exercise. Below 0.3 m s^{-1}, the trout were able to maintain position simply using their outstretched pectoral fins and the friction of the bottom of the flume (*cf* Randall, 1970). Sub-lethal copper exposure caused damage to the gill ultrastructure at all copper and temperature combinations (M.S.I. Mujallid, E.W. Taylor & P.J. Butler, in preparation), but the changes in the winter trout exposed to 0.47 μmol l^{-1} copper were particularly substantial, including hyperplasia of epithelial cells and proliferation of mucocytes in the secondary lamellae, necrosis of chloride cells and fusion of neighbouring secondary lamellae (Fig. 1(b)). Morphometric analysis showed the harmonic mean diffusion distance in these fish to have increased by over threefold from 3.62±0.42 μm in control fish to 11.54±1.63 μm. While there was no change in the arterial oxygen partial pressure (P_aO_2) of these trout at rest, there was a significant decline during exercise, indicating an underlying diffusional limitation in these fish.

Fig. 3. Effects on brown trout of exposure to sub-lethal levels of copper in soft, acid water (pH 5): (a) maximum sustainable swimming speed (U_{crit}) was about two body lengths s^{-1} in control fish (pH 7, no copper) at both 5 °C and 15 °C. Following exposure to 0.08 μmol l^{-1} copper U_{crit} fell to between 1 and 1.5 body lengths s^{-1}; exposure to 0.47 μmol l^{-1} at 5 °C reducing swimming speed to less than 0.33 body length s^{-1} (b) oxygen uptake increased with temperature and following exposure to copper. (From Beaumont *et al.*, 1995*a*.)

Exposure to 0.08 µmol l^{-1} copper had less drastic but still significant effects upon swimming performance, reducing U_{crit} by 25–50% at 15 and 5 °C (Fig. 3(a)). Ultrastructural changes at the gill were less severe, being characterized by large vacuoles in gill lamellae and structural deformations, including the curling over of the tips of some lamellae. There was no significant difference in the harmonic mean diffusion distance of trout exposed to this level of copper and that of control fish. Trout exposed to 0.08 µmol l^{-1} Cu^{2+} at pH 5 demonstrated no decline in P_aO_2 and in no group of copper exposed fish did the concentration of oxygen in arterial blood (C_aO_2) fall. In fact, at 15 °C, the oxygen content of copper/acid exposed trout rose significantly with exercise. This was achieved through an increase in haemoglobin concentration ([Hb]), most probably due to the release of erythrocytes from the spleen, since haematocrit was also elevated. Slight increases in haemoglobin concentration may also have been responsible for maintaining C_aO_2 at 5 °C.

Plasma lactate concentration was not elevated in fish exposed to sub-lethal levels of copper and in these trout, there is no evidence of a general systemic hypoxia or for a failure of branchial oxygen uptake. While there was no significant increase in plasma lactate in trout exposed to 0.47 µmol l^{-1} copper at 5 °C, red muscle lactate concentration of this group of fish was significantly elevated at rest from 4.7±0.7 µmol g^{-1} at neutral pH to 9.0±0.8 µmol g^{-1}. It has been suggested that haemoconcentration may lead to disruption of oxygen transport and a local tissue hypoxia (Randall & Brauner, 1991; Butler, Day & Namba, 1992). In addition to an increase in [Hb], the significant loss of plasma ions may lead to an osmotic loss of plasma water. In the fish exposed to 0.47 µmol l^{-1} Cu^{2+} at pH 5, there was a very significant rise in plasma protein which is indicative of an osmotic loss of plasma water. Both factors would elevate blood viscosity (Bushnell, Jones & Farrell, 1992) in which case effects upon the local circulation of blood to the exercising muscles may become significant.

Routine oxygen consumption rose in trout exposed to the SLCC at each acclimation temperature (Fig. 3(b)). Waiwood and Beamish (1978) also found copper and acid exposure to elevate the oxygen consumption of fish at rest. At 12 °C, exposure of fingerling rainbow trout to 0.4 µmol l^{-1} Cu^{2+} at pH 6 increased standard metabolism by 70%. In the same study, U_{crit} was decreased by 40% in comparison to that of control trout in copper free water at pH 7.8. In the study of Beaumont *et al.* (1995a), routine oxygen consumption rose by 38% in the summer fish (Fig. 3(b)). The U_{crit} of these trout was decreased by 26% (Fig. 3(a)), in almost the same proportion compared to the increase in oxygen consumption as found by Waiwood and Beamish (1978). However, it is unlikely that these relatively small increases in resting uptake would influence scope for exercise to such a great extent, when active metabolism can be as much as 12–15 times the resting level (Wood & Perry, 1985). Moreover, the winter trout exposed to 0.08 μmol l^{-1} Cu^{2+} at pH 5 did not have an increased metabolic rate at rest but displayed a decline in swimming performance similar to that of the brown trout exposed to this level of copper at 15 °C.

A number of pressure recordings made from cannulae inserted into the buccal cavity of some winter brown trout have indicated that respiratory frequency may increase by almost 40% following copper exposure at pH 5 (M.W. Beaumont, unpublished observations). As well as an increase in respiratory frequency, Sellers et al. (1975) also found ventilation volume to be greater following copper exposure. Jones and Schwarzfeld (1974) showed that, at rest, 10% of oxygen uptake is required to power the respiratory muscles. They also found the efficiency of these muscles to be low so that increasing ventilation becomes even more metabolically expensive. Thus, an element of the increased oxygen consumption of copper exposed fish at rest is likely to arise from elevated ventilation.

The accumulation of ammonia

Exposure of freshwater trout to acutely lethal levels of copper caused large and progressive increases in plasma ammonia levels (Fig. 2(h)). Seawater trout similarly experienced elevated ammonia levels when exposed to copper but they reached a plateau after 4 h at levels half that reached by 4 h and one-tenth of the value after 19 h in freshwater trout (Wilson & Taylor, 1993a,b). Ammonia accumulation during copper exposure has been observed previously. Laurén and McDonald (1985) exposed rainbow trout to a range of copper concentrations and, for example, 1.57 μmol l^{-1} Cu^{2+} caused plasma total ammonia concentration ($[T_{Amm}]$) to rise from 150 to 250–300 μmol l^{-1}. Wilson and Taylor (1993a) exposed freshwater rainbow trout to a higher level of 4.9 μmol l^{-1} Cu^{2+} and found plasma $[T_{Amm}]$ to rise from 63±11 to 1751±204 μmol l^{-1} in only 19 h (Fig. 2(h)). Exposure of brown trout to 0.47 μmol l^{-1} in acid freshwater (pH 5) for 96 h caused plasma ammonia to rise from 100 μmol l^{-1} in control fish (pH 7, no added copper) to 600 μmol l^{-1} (Beaumont et al., 1995a). In the same fish, total ammonia levels in red muscle rose from 17 mmol l^{-1} kg^{-1} to 37 mmol l^{-1}; values which illustrate that ammonia produced by tissues is 'trapped' in the relatively acidic intracellular compartment (J. Mair,

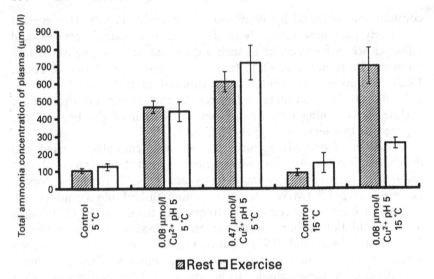

Fig. 4. Total plasma ammonia concentration (μmol l^{-1}) of brown trout acclimated to either 5 or 15 °C, exposed for 96 h to either control conditions (pH 7, no copper) or to the stated copper concentration at pH 5 and sampled either at rest or after exercise to U_{crit}. (From Beaumont *et al.*, 1995*b*.)

E.W. Taylor and P.J. Butler, in preparation). Exposure of trout to 0.08 μmol l^{-1} copper (pH 5) increased total plasma ammonia levels to 470 μmol l^{-1} at 5 °C and to 720 μmol l^{-1} at 15 °C (Fig. 4).

The mechanisms underlying this accumulation of copper are unclear. Copper exposure caused no increase on O_2 consumption in seawater rainbow trout (Wilson & Taylor, 1993*a*). If the normal relationship between $\dot{M}\,O_2$ and $\dot{M}\,NH_3$ (Brett & Zala 1975; Heisler, 1984) remained constant throughout exposure to copper, then the elevation of plasma [T_{Amm}] cannot be explained by any change in the rate of endogenous ammonia production. Indeed, exposure of common carp (*Cyprinus carpio*) to copper levels between 0.22 and 0.84 μmol l^{-1} had no effect on the rate of ammonia excretion into soft water at neutral pH (de Boeck *et al.*, 1995).

Although there was evidence of structural damage to the gills of trout exposed to sublethal levels of copper (Fig. 1(*b*)), there was no evidence of impairment to oxygen or carbon dioxide gas exchange, i.e. the fish were not hypoxic or hypercapnic. The branchial permeability coefficient of NH_3 is similar to that of CO_2 and much greater than that for O_2 (Cameron & Heisler, 1983). In addition, the acidity of the test water should enhance 'ammonia trapping', that is the conversion of excreted NH_3 to the impermeable NH_4^+ ions in the boundary layer outside the gills (Randall & Wright, 1989). It is possible that local ammonia concentrations are quite different to those of the bulk water. As ventilation rate increased in copper and acid-exposed trout (M.W. Beaumont, unpublished results; see also Sellers *et al.* 1975), the unstirred boundary layer is likely to be reduced, but may well contain ammonia in a higher concentration than that of the surrounding water. In addition the mucus layer may trap ammonia. Handy (1989) found that, even in control conditions, rainbow trout body mucus had a $[NH_4^+]$ almost four times that of the ambient water. The gill mucus layer, which expanded during copper/acid exposure (M.S. Mujallid, E.W. Taylor and P.J. Butler, in prep.), most likely acts in a similar manner. A decline in plasma $[T_{amm}]$ observed in the copper exposed trout at 15 °C when they were exercised (Beaumont *et al.*, 1995a, Fig. 4) could have arisen from a removal of the mucus layer by an increased water flow across the gill.

However, it is difficult to see how ammonia excretion could be affected by mucus without there being a similar effect upon respiratory gas exchange. The conditions of the experiments of Beaumont *et al.* (1995a) were seemingly ideal for the passive diffusion of NH_3. In contrast, there does appear to have been some ionoregulatory impairment of the copper/acid exposed trout, since both $[Na^+]$ and $[Cl^-]$ fell and it is known that copper exposure inhibits ATPases (Laurén & McDonald, 1985). These data indicate that plasma ammonia accumulation during copper exposure is caused by the inhibition of an active mechanism, which normally contributes significantly to the excretion of ammonia.

With passive diffusion of NH_4^+ ions dismissed as a route for excretion, at least in freshwater fish, due to the low permeability of most biological membranes to ionic compounds and to the barrier presented by tight junctions (Wright and Wood, 1985), the probable mechanisms of ammonia excretion are considered to be passive diffusion of free NH_3 and/or an active pathway, which may be either the direct exchange of NH_4^+ ions for Na^+ or a proton pump coupled to NH_3 excretion (Avella & Bornancin, 1989). As stated above, it seems unlikely that any changes in the gill NH_3 diffusion capacity occurred during copper exposure since no major changes in the other blood gas tensions P_aO_2 or P_aCO_2 were observed. This argues for an inhibition of apical Na^+/NH_4^+ exchange by copper as suggested by Laurén and McDonald (1985) in freshwater rainbow trout. The stabilization of $[T_{Amm}]$ after 4 h in seawater trout could then be explained if it is assumed that the rate of branchial NH_3 clearance by diffusion will again match the endogenous rate of NH_3 production once a new, larger P_{NH_3} gradient over the relatively undamaged gill lamellae is established. In freshwater, the gills are structurally damaged (Fig. 1) and ionoregulation breaks down after copper exposure (Fig. 2), so that apical Na^+/NH_4^+ exchange and NH_3 diffusion are likely to be impaired. Consequently, ammonia levels continue to rise rather than reaching a plateau.

Active exchange mechanisms have been proposed to account for the maintenance of ammonia excretion during episodes of high external ammonia (Cameron & Heisler, 1983; Cameron, 1986; Wilson & Taylor, 1992). Exposure to a T_{amm} of 1 mmol l^{-1} was non-toxic to trout in freshwater or seawater. The fish survived and were able to maintain plasma ammonia lower than ambient levels, despite the high permeability of the gill epithelium for NH_3 and its continuous production as a waste product of the fish's metabolism (Fig. 5(*a*)). Maintenance of a negative gradient for ammonia was accompanied by a net accumulation of non-respiratory protons (but no net pH changes) in freshwater, but a reduction in non-respiratory protons (and a subsequent alkalosis) in seawater (Fig. 5(*b*)).

These data were interpreted as indicating that the fish were able actively to excrete ammonia against a diffusion gradient, most likely as the ammonium ion $[NH_4^+]$, in exchange for sodium in seawater (leading to net efflux of protons) and in exchange for an influx of protons in freshwater (Wilson & Taylor, 1992). More recent data have suggested that, in freshwater at least, trout may not possess an active NH_4^+ exchange mechanism (Wilson *et al.*, 1994). Instead, acidification of the boundary layer at the external gill surface by respiratory CO_2 and/or H^+ may sufficiently reduce the local $[NH_3]$ to allow passive diffusion of NH_3 from blood to water, even when the $[NH_3]$ in the bulk water is much higher than that in the blood. In contrast, Playle and Wood (1989) estimated that, below approximately pH 5.1, the effect of ammonia excretion would predominate over the acidifying effect of CO_2 excretion and make the water more alkaline as it passes over the gill. This would have the effect of increasing the P_{NH_3} in the boundary layer and reducing the diffusion gradient. However, these authors measured a maximum change in expired water pH of only 0.7 units which would have an insignificant effect at the low water ammonia concentrations in the current experiments on copper toxicity. Currently, our understanding of the mechanisms for excretion of ammonia over the gills of fish is incomplete.

Ammonia and swimming performance

Regression analysis has shown there to be a significant linear relationship between U_{crit} and plasma $[NH_4^+]$ (Fig. 6). The coefficient of determi-

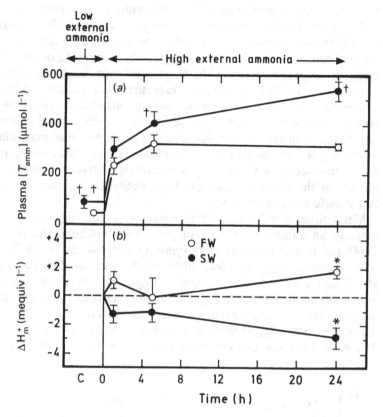

Fig. 5. (a) Plasma total ammonia (T_{amm}) in freshwater (FW) and seawater (SW) trout prior to, and during, 24 h of exposure to high external ammonia concentration (1000 μmol^{-1} T_{amm} at pH 7.9). External T_{amm} during the control period (C) was less than 30 $\mu mol\ l^{-1}$. † denotes values significantly different when compared with the corresponding value from FW trout ($P < 0.5$; Student's unpaired t-test). Values are means ±SEM ($n=6$). (b) The non-respiratory acid load for FW and SW trout during the control period (C) and during 24 h of exposure to high external ammonia concentration. * denotes a value significantly different from the control mean within the group ($P < 0.05$). (From Wilson & Taylor, 1992.)

nation (r^2) implies that almost 70% of the variation in swimming performance can be explained by the plasma [NH$_4^+$] (Beaumont, Butler & Taylor, 1995*b*). There is no evidence from this study that ammonia is a normal cause of fatigue in trout, as there was no ammonia accumulation in control animals at U_{crit}. Instead, what these data imply is that elevation in ammonia due to copper and acid exposure is the basis of reduced performance, i.e. ammonia causes fatigue to occur earlier in fish exposed to copper and low pH. A limited pilot experiment ($N=4$), in which ammonium bicarbonate was infused into the dorsal aorta of brown trout caused an average reduction in U_{crit} of 30%. These infusions took place over 24 h, shorter infusions had less effect and infusions up to 96 h in duration may therefore produce changes in swimming performance of more similar magnitude to those that occurred following 96 h of copper exposure. Ammonia and ammonium ions have a number of possible physiological effects that may influence swimming performance by affecting the metabolic status of the muscle or by interfering with central or peripheral nervous activity, transmission at the neuromuscular junction, excitation/contraction coupling and muscle electrophysiology.

Metabolism is influenced by ammonia at a number of stages. NH_4^+ ions are an allosteric activator of phosphofructokinase (Su & Storey, 1994) and inhibit pyruvate carboxylase (Zaleski & Bryla, 1977). Elevated ammonia may therefore increase the rate of flux through the glycolytic pathway depleting stored glycogen and, furthermore, disrupt its regeneration. Although U_{crit} is essentially aerobic, some recruitment of white muscle occurs even at sustainable speeds and particularly as the animal nears U_{crit} (Johnston & Moon, 1980; Butler & Day, 1994). The high ammonia concentration prior to exercise might, therefore,

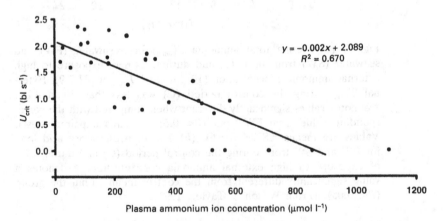

Fig. 6. The relationship between ammonium ion concentration in arterial plasma and swimming performance (U_{crit}) of individual brown trout acclimated to either 5 or 15 °C and exposed for 96 h to either control conditions (pH 7, no copper) or to copper concentrations of either 0.08 or 0.47 μmol l^{-1} at pH 5. (From Beaumont *et al.*, 1995b.)

have reduced swimming performance by impairing anaerobic capacity. The difficulty with this hypothesis, however, is to explain the results from the copper and acid exposed fish at 15 °C. Plasma $[T_{amm}]$ was lower in the exercised group than in the resting trout and it is the exercise $[NH_4^+]$ that fits the regression with swimming performance. If glycogen depletion was the cause of the reduced swimming capacity, perhaps it might be expected that the prior history of ammonia concentration is important rather than the level during exercise. However, a lowering of ammonia at the onset of exercise may allow sufficient gluconeogenesis to occur for some recovery of glycogen stores prior to the recruitment of the white muscle.

Ammonium ions also attenuate oxidative decarboxylation of pyruvate to lactate (McKhann and Tower, 1961) and, in the cat cerebral cortex, inhibit pyruvate dehydrogenase (Katunuma, Okada & Nishii, 1966). This latter enzyme is important for the conversion of pyruvate to acetyl coenzyme A, the link between glycolysis and the tricarboxylic acid cycle which occurs in the mitochondrion. Finally, due to an accelerated decrease in pyridine nucleotides, ammonium ions have an inhibitory role upon isocitrate dehydrogenase, a rate-limiting enzyme within the TCA cycle itself (Katunuma *et al.*, 1966). Indeed, Avillo *et al.* (1981) refer to unpublished data showing that high ammonia levels cause an impairment of the TCA cycle in rainbow trout. Elevated plasma ammonia could, therefore, have reduced swimming performance by slowing and even uncoupling oxidative phosphorylation, lowering the efficiency of aerobic metabolism.

Another area in which ammonia could be a factor is the neuromuscular coordination of exercise. Ammonium ions are able to substitute for K^+ in exchange mechanisms resulting in a depolarization of neurons (Binstock & Lecar, 1969). Sjøgaard (1991) has proposed that K^+ loss and subsequent depolarization is the cause of fatigue associated with low contraction forces in human skeletal muscle. The plasma $[K^+]$ of exercised trout that had been exposed to 0.08 μmol l^{-1} Cu^{2+} at pH 5 was not significantly different from that of control trout despite the difference in swimming performance; neither was plasma $[K^+]$ elevated in brown trout which were unable to swim following exposure to 0.47 μmol l^{-1} Cu^{2+} at pH 5. Unless potassium loss from muscle to the extracellular fluid is masked by fluxes from plasma to environment, this seems unlikely to be the cause of failure of swimming in these trout.

Ammonium ions may affect the neuromuscular system in other ways. The inhibition of glutaminase (O'Neill & O'Donovan, 1979) decreases glutamate, aspartate and GABA concentrations, all essential synaptic neurotransmitters. In the cat spinal cord, Raabe and Lin (1984) have shown that ammonia can decrease the hyperpolarizing action of post-synaptic inhibition. This was due to the inactivation of Cl^- extrusion from the neurons and occurred at ammonium concentrations much lower than those required to produce any other nervous effect and without any metabolic changes. The resulting disruption of synaptic transmission has been proposed as the cause of ammonia-induced convulsions and coma (Hillaby & Randall, 1979) but, at less lethal doses of ammonia, subtle disruption to co-ordination of exercise may lead to the loss of performance that has been observed in this study. The effect of ammonia upon nervous function would also be easily reversible and thus fit the observations of Beaumont *et al.* (1995*b*) on fish at 15 °C.

Elevation of plasma ammonia concentration is not a phenomenon related to copper exposure alone. For example, low pH (N. Day and P.J. Butler, in prep.), high pH (Lin & Randall, 1990) and aluminium (Booth *et al.*, 1988) elevate plasma ammonia, and these pollutants also decrease swimming performance (Ye & Randall, 1991; Butler *et al.*, 1992; Wilson & Wood, 1992). If a relationship between U_{crit} and ammonia can be confirmed, perhaps by direct manipulation of plasma ammonia through infusion of ammonium salts, then this single effect may prove, at least partly, to explain the reduced swimming performance of fish arising from their exposure to sublethal levels of a variety of environmental pollutants.

References

Alabaster, J.S. & Lloyd, R. (1980). *Water Quality Criteria for Freshwater Fish*. London, Boston: Butterworth Scientific. 360 pp.

Avella, M. & Bornancin, M. (1989). A new analysis of ammonia and sodium transport through the gills of the freshwater rainbow trout (*Salmo gairdneri*). *Journal of Experimental Biology*, **142**, 155–75.

Avillo, A., Margiocco, C., Melodia, F. & Mensi, P. (1981). Effects of ammonia on liver lysosomal functionality in *Salmo gairdneri* Richardson. *Journal of Experimental Zoology*, **218**, 321–6.

Baker, J.T.P. (1969). Histological and electron microscopical observations on copper poisoning in the winter flounder (*Pseudopleuronectes americanus*). *Journal of the Fisheries Research Board of Canada*, **26**(11), 2785–93.

Beamish, F.W.H. (1978). Swimming capacity. In *Fish Physiology*, vol. 7 (ed. W.S. Hoar and D.J. Randall), pp. 101–87. New York, London: Academic Press.

Beaumont, M.W., Butler, P.J. & Taylor, E.W. (1995*a*). Exposure of brown trout, *Salmo trutta*, to sub-lethal copper concentrations and its effect upon sustained swimming performance. *Aquatic Toxicology*, **33**, 45–63.

Beaumont, M.W., Butler, P.J. & Taylor, E.W. (1995*b*). Plasma ammonia concentration in brown trout (*Salmo trutta*) exposed to acidic water and sublethal copper concentrations and its relationship to decreased swimming performance. *Journal of Experimental Biology*, **198**, 2213–20.

Beckman, B.R. & Zaugg, W.S. (1988). Copper intoxication in chinook salmon (*Oncorhynchus tshawytscha*) induced by natural spring water: effects on gill Na^+, K^+-ATPase, haematocrit and plasma glucose. *Canadian Journal of Fisheries and Aquatic Science*, **45**, 1430–5.

Binstock, L. & Lecar, H. (1969). Ammonium ion currents in the squid giant axon. *Journal of General Physiology*, **53**, 342–61.

Booth, C.E., McDonald, D.G., Simons, B.P. & Wood, C.M. (1988). Effects of aluminium and low pH on net ion fluxes and ion balance in the brook trout (*Salvelinus fontinalis*). *Canadian Journal of Fisheries and Aquatic Science*, **45**, 1563–74.

Bowen, H.J.M. (1979). *Environmental Chemistry of the Elements*. London, New York: Academic Press.

Brett. J.R. & Zala, C.A. (1975). Daily pattern of nitrogen excretion and oxygen consumption of sockeye salmon (*Oncorhynchus nerka*) under controlled conditions. *Journal of the Fisheries Research Board of Canada*, **32**, 2479–86.

Bushnell, P.G., Jones, D.R. & Farrell, A.P. (1992). The Arterial System. In *Fish Physiology*, vol. 12 A. Hoar, W.S., Randall, D.J. and Farrell, A.P., eds, pp. 89–139. New York, London: Academic Press.

Butler, P.J. & Day, N. (1994). Acid water and white muscle recruitment in the brown trout (*Salmo trutta*). In *High Performance Fish*, *Proceedings of the International Fish Physiology Symposium*, MacKinlay, D.D., ed., pp. 312–17. Vancouver: Fish Physiology Association.

Butler, P.J., Day, N. & Namba, K. (1992). Interactive effects of seasonal temperature and low pH on resting oxygen uptake and swimming performance of adult brown trout (*Salmo trutta*). *Journal of Experimental Biology*, **165**, 195–212.

Cairns, J. (1966). Don't be half safe – the current revolution in bioassay techniques. Engineering Bulletin Purdue University Proceedings of the 21 Industrial Waste Conference, pp. 559–67.

Cameron, J.N. (1986). Responses to reversed NH_3 and NH_4^+ gradients in a teleost (*Ictalurus punctatus*), an elasmobranch (*Raja erinacea*) and a crustacean (*Callinectes sapidus*): Evidence for NH_4^+/H^+ exchange in the teleost and elasmobranch. *Journal of Experimental Zoology*, **239**, 183–95.

Cameron, J.N. & Heisler, N. (1983). Studies of ammonia in the rainbow trout: physico-chemical parameters, acid-base behaviour and respiratory clearance. *Journal of Experimental Biology*, **105**, 107–25.

Chakoumakos, C., Russo, R.C. & Thurston, R.V. (1979). Toxicity of copper to cutthroat trout (*Salmo clarki*) under different conditions of alkalinity, pH and hardness. *Environmental Science and Technology*, **13**, 213–19.

Christiansen, G.M., Mckim, J.M., Brungs, W.A. & Hunt, E.P. (1972). Changes in the blood of the brown bullhead (*Ictalurus nebulosus* Lesuer) following short and long term exposure to copper (II). *Toxicology and Applied Pharmacology*, **23**, 417–27.

Crespo, S. & Karnaky, K.L. (1983). Copper and zinc inhibit chloride transport across the opercular epithelium of sea-water adapted killifish, *Fundulus heteroclitus*. *Journal of Experimental Biology*, **102**, 337–41.

Culp, B.R. (1975). Aqueous complexation of copper with sewage and naturally occurring organics. PhD Thesis, Reported in Spear, P.A. & Pierce, R.C. (1979). Copper in the aquatic environment: chemistry, distribution and toxicology. National Research Council of Canada 16454.

Daye, P.G. & Garside, E.T. (1976). Histopathologic changes in surficial tissues of the brook trout, *Salvetinus fontinalis* (Mitchell), exposed to acute and chronic levels of pH. *Canadian Journal of Zoology*, **54**, 2140–55.

De Boeck, G., de Smet, H. & Blust, R. (1995). The effect of sublethal levels of copper on oxygen consumption and ammonia excretion in the common carp, *Cyprinus carpio*. *Aquatic Toxicology*, **32**, 127–41.

Eddy, F.B. & Fraser, J.E. (1982). Sialic acid and mucus production in rainbow trout (*Salmo gairdneri*) in response to zinc and seawater. *Comparative Biochemistry and Physiology*, **73C**, 357–9.

Eisler, R. & Gardner, G.R. (1973). Acute toxicity to an estuarine teleost of mixtures of cadmium, copper and zinc salts. *Journal of Fish Biology*, **5**, 131–42.

Ernst, R., Allen, H.E. & Mancy, K.H. (1975). Characterization of trace metals species and measurement of trace metal stability constants by electrochemical techniques. *Water Research*, **9**, 969–79.

Evans, D.H. (1987). The fish gill: site of action and model for toxic effects of environmental pollutants. *Environmental Health Perspectives*, **71**, 47–58.

Handy, R.D. (1989). The ionic composition of rainbow trout body mucus. *Comparative Biochemistry and Physiology*, **93A**, 571–5.

Heisler, N. (1984). Role of ion transfer processes in acid-base regulation with temperature changes in fish. *American Journal of Physiology*, **246**, R441–51.

Hillaby, B.A. & Randall, D.J. (1979). Acute ammonia toxicity and ammonia excretion in rainbow trout (*Salmo gairdneri*). *Journal of Fisheries Research Board of Canada*, **36**, 621–9.

Howarth, R.S. & Sprague, J.B. (1978). Copper lethality to rainbow trout in waters of various hardness and pH. *Water Research*, **12**, 455–62.

Howells, E.J., Howells, M.E. & Alabaster, J.S. (1983). A field investigation of water quality, fish and invertebrates in the Mawddach river system, Wales. *Journal of Fish Biology*, **22**, 447–69.

Hughes, G.M. (1976). Polluted fish respiratory physiology. In *Effects of Pollutants on Aquatic Organisms*, Lockwood, A.P.M., ed., pp. 163–83. London, New York: Cambridge University Press.

Irving, H. & Williams, R.J.P. (1953). The stability of transition metal complexes. *Journal of the Chemical Society*, **53**, 3192–210.

Johnston, I.A. & Moon, T.W. (1980). Exercise training in skeletal muscle of brook trout (*Salvelinus fontinalis*). *Journal of Experimental Biology*, **87**, 177–94.

Jones, D.R. & Schwarzfeld, T. (1974). The oxygen cost to the metabolism and efficiency of breathing in trout (*Salmo gairdneri*). *Respiration and Physiology*, **21**, 241–54.

Katunuma, N., Okada, M. & Nishii, Y. (1966). Regulation of the urea cycle and TCA cycle by ammonia. In *Advances in Enzyme Regulation*. vol. 4. Weber, G., ed., pp. 317–35. London: Pergamon Press.

Kirk, R.S. & Lewis, J.W. (1993). An evaluation of pollutant induced changes in the gills of rainbow trout using scanning electron microscopy. *Environmental Technology*, **14**, 577–85.

Laurén, D.J. & McDonald, D.G. (1985). Effects of copper on branchial ionoregulation in the rainbow trout *Salmo gairdneri* Richardson: modulation by water hardness and pH. *Journal of Comparative Physiology*, **155B**, 635–44.

Laurén, D.J. & McDonald, D.G. (1987). Acclimation to copper by rainbow trout *Salmo gairdneri*: biochemistry. *Canadian Journal of Fish Aquatic Science*, **44**, 105–11.

Lerman, A. & Childs, C.W. (1973). Metal organic complexes in natural waters: control of distribution by thermodynamic, physical and kinetic factors. In *Trace Metals and Metal–Organic Interactions in Natural Waters*, Singer, P.C., ed., pp. 201–36. Michigan: Ann Arbor.

Lewis, S.D. & Lewis, W.M. (1971). The effect of zinc and copper on the osmolarity of blood serum of the channel catfish, *Ictalurus punctatus* Rafinesque, and the golden shiner, *Notmigonus crysoleucas* Mitchill. *Transactions of the American Fisheries Society*, **100**, 639–43.

Lin, H. & Randall, D.J. (1991). Evidence for the presence of an electrogenic proton pump on the trout gill epithelium. *Journal of Experimental Biology*, **161**, 119–34.

Lorz, H.W. & McPherson, B.P. (1976). Effects of copper or zinc in fresh water on the adaptation to sea water and ATPase activity, and the effects of copper on migratory disposition of coho salmon (*Oncorhynchus kisutch*). *Journal of Fisheries Research Board of Canada*, **33**, 2023–30.

McDonald, D.G. (1983). The interaction of calcium and low pH on the physiology of the rainbow trout, *Salmo gairdneri*. 1 Branchial and renal net ion and H^+ fluxes. *Journal of Experimental Biology* **102**, 123–40.

McDonald, D.G. & Rogano, M.S. (1986). Physiological responses to stress in lake trout: effects of water hardness and genotype. *Transactions of the American Fisheries Society*, **122**, 1146–55.

McKhann, G.M. and Tower, D.B. (1961). Ammonia toxicity and cerebral metabolism. American Journal of Physiology **200**, 420–4.

McKim, J.M., Christiansen, G.M. & Hunt, E.P. (1970). Changes in the blood of brook trout (*Salvelinus fontinalis*) after short-term and long-term exposure to copper. *Journal of the Fisheries Research Board of Canada*, **27**, 1883–89.

Miles, H.M. (1971). Renal function in migrating adult coho salmon. *Comparative Biochemistry and Physiology*, **37A**, 787–826.

Miller, T.G. & MacKay,W.C. (1980). The effects of hardness, alkalinity and pH of test water on the toxicity of copper to rainbow trout (*Salmo gairdneri*). *Water Research*, **14**, 129–33.

Miller, T.G. & MacKay, W.C. (1982). The relationship of secreted mucus to copper and acid toxicity in rainbow trout. *Bulletin of Environmental Contamination Toxicology*, **28**, 68–74.

Milligan, C.L. & Wood, C.M. (1982). Disturbances in haematology, fluid volume, distribution and circulatory function associated with low environmental pH in the rainbow trout, *Salmo gairdneri*. *Journal of Experimental Biology*, **99**, 397–415.

Moore, J.W. & Ramamoothy, S. (1984). Heavy metals in natural waters. New York: Springer-Verlag.

Morel, F., McDuff, R.E. & Morgan, J.J. (1973). Interactions and chemostasis in aquatic chemical systems: role of pH, pE, solubility and complexation. In *Trace Metals and Metal–Organic Interactions in Natural Waters*. Singer, P.C., ed. pp. 157–200. Michigan: Ann Arbor.

Nieboer, E. & Richardson, D.H.S. (1980). The replacement of the nondescript term 'heavy metals' by a biologically and chemically significant classification of metal ions. *Environmental Pollution*, **1**, 3–26.

O'Neill, B.B. & O'Donovan, D.J. (1979). Cerebral energy metabolism in hyperammonaemia. *Biochemical Society Transactions*, **7**, 35–6.

Pagenkopf, G.K., Russo, R.C. & Thurston, R.V. (1974). Effect of complexation on toxicity of copper to fishes. *Journal of the Fisheries Research Board of Canada*, **31**, 462–5.

Playle, R.C. & Wood, C.M. (1989). Water chemistry changes in the gill microenvironment of rainbow trout: experimental observations and theory. *Journal of Comparative Physiology*, **159B**, 527–37.

Potts, W.T.W. & Fleming, W.R. (1971). The effect of environmental calcium and ovine prolactin on sodium balance in *Fundulus kansae*. *Journal of Experimental Biology*, **54**, 63–75.

Raabe, W. & Lin, S. (1984). Ammonia, post-synaptic inhibition and CNS energy state. *Brain Research*, **303**, 67–76.

Randall, D.J. (1970). Gas exchange in fish. In *Fish Physiology*, vol. 4. Hoar, W.S. & Randall, D.J., eds, pp. 253–92. New York, London: Academic Press.

Randall, D.J. & Brauner, C. (1991). Effects of environmental factors on exercise in fish. *Journal of Experimental Biology*, **160**, 113–26.

Randall, D.J. & Wright, P.A. (1989). The interaction between carbon dioxide and ammonia excretion and water pH in fish. *Canadian Journal of Zoology*, **67**, 2936–42.

Reid, S.D. & McDonald, D.G. (1988). The effects of cadmium, copper and low pH on calcium fluxes in rainbow trout, *Salmo gairdneri*,. *Canadian Journal of Fisheries and Aquatic Science*, **45**, 244–53.

Schreck, C.B. & Lorz, H.W. (1978). Stress response of coho salmon (*Oncorhyncus kisutch*) elicited by cadmium and copper and potential use of cortisol as an indicator of stress. *Journal of the Fisheries Research Board of Canada*, **35**, 1124–9.

Sellers, C.M., Heath, A.G. & Bass, M.L. (1975). The effect of sublethal concentrations of copper and zinc on ventilatory activity, blood oxygen and pH in rainbow trout (*Salmo gairdneri*). *Water Research*, **9** 401–8.

Shapiro, J. (1964). Effect of yellow organic acids on iron and other metals in water. *Journal of the American Water Works Association*, **56**, 1062–82.

Shaw, T.L. & Brown, V.M. (1974). The toxicity of some forms of copper to rainbow trout. *Water Research*, **8**, 377–82.

Shephard, K. & Simkiss, K. (1978). The effects of heavy metal ions on Ca^{2+}-ATPase extracted from fish gills. *Comparative Biochemistry and Physiology*, **61B**, 69–72.

Sjøgaard, G. (1991). Role of exercise-induced potassium fluxes underlying muscle fatigue: a brief review. *Canadian Journal of Physiology and Pharmacology*, **69**, 238–45.

Skidmore, J.F. & Tovell, P.W.A. (1972). Toxic effects of zinc sulphate on the gills of rainbow trout. *Water Research*, **6**, 217–30.

Soivio, A. & Nikinmaa, M. (1981). The swelling of erythrocytes in relation to the oxygen affinity of the blood of the rainbow trout, *Salmo gairdneri* Richardson. In *Stress in Fish*, Pickering, A.D., ed., pp. 103–19. London, New York: Academic Press.

Spear, P.A. & Pierce, R.C. (1979). Copper in the aquatic environment: chemistry, distribution and toxicology. National Research Council of Canada 16454.

Sposito, G. (1986). Distribution of potentially hazardous trace metals. In *Trace Metals in Biological Systems*, Vol 20, Sigel, H., ed., pp. 1–20. New York, Basel: Marcel Dekker Inc.

Stagg, R.M. & Shuttleworth, T.J. (1982). The effects of copper on ionic regulation by the gills of the seawater adapted flounder (*Platychthys flesus* L.). *Journal of Comparative Physiology*, **149**, 83–90.

Stiff, M.J. (1971). The chemical states of copper in polluted freshwater and a scheme of analysis to differentiate them. *Water Research*, **5**, 585–99.

Su, S.Y. & Storey, K.B. (1994). Regulation of rainbow trout white muscle phosphofructokinase during exercise. *International Journal of Biochemistry*, **26**, 519–28.

Taylor, D., Maddock, B.G. & Mance, G. (1985). The acute toxicity of nine 'grey list' metals (arsenic, boron, chromium, copper, lead, nickel, tin, vanadium and zinc) to two marine fish species: Dab (*Limanda limanda*) and grey mullet (*Chelon labrosus*). *Aquatic Toxicology*, **7**, 135–44.

Turnpenny, A.W.H., Sadler, K., Aston, R.J., Milner, A.G.P. & Lynam, S. (1987). The fish populations of some streams in Wales and Northern England in relation to acidity and associated factors. *Journal of Fish Biology*, **31**, 415–34.

Tuurala, H. & Soivio, A. (1982). Structural and circulatory changes in the secondary lamellae of *Salmo gairdneri* gills after sublethal exposures to dehydroabietic acid and zinc. *Aquatic Toxicology*, **2**, 21–9.

Ultsch, G.R. & Gros, G. (1979). Mucus as a diffusion barrier to oxygen: possible role in O_2 uptake at low pH in carp (*Cyprinus carpio*) gills. *Comparative Biochemistry and Physiology*, **62A**, 685–9.

Voyer, R.A. (1975). Effect of dissolved oxygen concentration on acute toxicity of cadmium to the mummichog, *Fundulus heteroclitus* L., at various salinities. *Transactions of the American Fisheries Society*, **104**, 129–34.

Waiwood, K.G. & Beamish, F.W.H. (1978). Effects of copper, pH and hardness on the critical swimming speed of rainbow trout (*Salmo gairdneri* Richardson). *Water Research*, **12**, 611–19.

Wells, R.M.G. & Weber, R.E. (1991). Is there an optimal haematocrit for rainbow trout, *Oncorhynchus mykiss* (Walbaum)? An interpretation of recent data based on blood viscosity measurements. *Journal of Fish Biology*, **38**, 53–65.

Westfall, B.A. (1945). Coagulation film anoxia in fishes. *Ecology*, **26**, 283–7.

Wilson, R.W. & Taylor, E.W. (1992). Transbranchial ammonia gradients and acid-base responses to high external ammonia concentration in rainbow trout (*Oncorhynchus mykiss*) acclimated to different salinities. *Journal of Experimental Biology*, **166**, 95–112.

Wilson, R.W. & Taylor, E.W. (1993*a*). The physiological responses of freshwater rainbow trout, *Oncorhynchus mykiss*, during acutely lethal copper exposure. *Journal of Comparative Physiology*, **163B**, 38–47.

Wilson, R.W. & Taylor, E.W. (1993*b*). Differential responses to copper in rainbow trout (*Oncorhynchus mykiss*) acclimated to sea water and brackish water. *Journal of Comparative Physiology*, **163B**, 239–46.

Wilson, R.W. & Wood, C.M. (1992). Swimming performance, whole body ions and gill Al accumulation during acclimation to sublethal aluminium in juvenile rainbow trout (*Oncorhynchus mykiss*). *Fish Physiology and Biochemistry*, **10**, 149–59.

Wilson, R.W., Wright, P.M., Munger, S. & Wood, C.M. (1994). Ammonia excretion in freshwater rainbow trout (*Oncorhynchus mykiss*) and the importance of gill boundary layer acidification: lack of evidence for Na^+/NH_4^+ exchange. *Journal of Experimental Biology*, **191**, 37–58.

Wood, C.M. (1989). The physiological problems of fish in acid waters. In *Acid Toxicity and Aquatic Animals*, Morris, R., Taylor, E.W., Brown, D.J.A. & Brown, J.A., eds., pp. 124–52. Cambridge, New York: Cambridge University Press.

Wood, C.M. & Perry, S.F. (1985). Respiratory, circulatory and metabolic adjustments to exercise in fish. In *Circulation, Respiration and Metabolism*, Gilles, R., ed., pp. 1–22. Heidelberg: Springer-Verlag.

Wright, P.A. & Wood, C.M. (1985). An analysis of branchial ammonia excretion in the freshwater rainbow trout: effects of environmental pH change and sodium uptake blockage. *Journal of Experimental Biology*, **114**, 329–53.

Yamamoto, K., Itazawa, Y. & Kobayashi, H. (1980). Supply of erythrocytes into the circulating blood from the spleen of exercised fish. *Comparative Biochemistry and Physiology*, **65A**, 5–11.

Ye, X. & Randall, D.J. (1991). The effect of water pH on swimming performance in rainbow trout (*Salmo gairdneri* R.). *Fish Physiology and Biochemistry*, **9**, 15–21.

Zaleski, J. & Bryla, J. (1977). Effects of oleate, palmitate and octanoate on gluconeogenesis in isolated rabbit liver cells. *Archives Biochimica et Biophysica*, **183**, 553–62.

J.A. BROWN and C.P. WARING

The physiological status of brown trout exposed to aluminium in acidic soft waters

Introduction

Low environmental pH and increased concentrations of various metals such as Cu, Mn, and Al interact in determining the toxicity of acidic soft waters to fish. Of the trace metals, Al has been highlighted as causing the greatest concern to date (Rosseland, Eldhurst & Staurnes, 1990), and numerous investigations have reported physiological disturbances in adult or juvenile fish, especially salmonid species, exposed to aluminium in soft acidic waters. Most of these studies have been concerned with the responses of the rainbow trout and brook trout (Neville, 1985; Playle, Goss & Wood, 1989; Witters *et al.*, 1990, 1991, 1992; Wood *et al.*, 1988*a,b*) while far fewer studies have centred on the brown trout, *Salmo trutta*, even though this is a major indigenous salmonid in European waters. The responses of brown trout to acidic waters in the absence of aluminium have been investigated in some depth by McWilliams & Potts (McWilliams, 1982, 1983), but the effects of aluminium in acidic waters are less clear though of greater environmental relevance. Some studies have suggested that aluminium-exposed brown trout show osmoregulatory and/or respiratory disturbances similar to those occurring in other salmonids (Muniz & Leivestad, 1980; Rosseland & Skogheim, 1984; Gagen & Sharpe, 1987; Battram, 1988; Dietrich & Schlatter, 1989*b*; Reader *et al.*, 1991; Sayer *et al.*, 1991) but in most of these studies aluminium concentrations were high and mortalities common. A knowledge of the effects of lower, sub-lethal concentrations of aluminium is particularly important in understanding acclimation and long-term survival as ecosystems recover. This review emphasizes some of our recent findings on the acute physiological responses of adult brown trout to both lethal and sub-lethal concentrations of aluminium in acidic soft waters, and the role which endocrine systems play, setting these against the backdrop provided by investigations of other species, especially those of other salmonid species.

Mechanism of aluminium toxicity

Physiological studies have clearly established that the gill is the main target organ of aluminium toxicity in acidic soft waters (Muniz & Leivestad, 1980; Neville, 1985; Witters et al., 1987; McDonald & Milligan, 1988; Wood et al., 1988a,b; Dietrich & Schlatter, 1989a; Playle & Wood 1989; Playle et al., 1989; Witters et al., 1990, 1991; McDonald & Wood 1993).

The chemistry of aluminium in aqueous solutions is complex (Driscoll & Schechner, 1990). In natural waters organic materials readily complex with aluminium reducing the presence of toxic species of labile aluminium and formation of polymers. Aluminium exists in solution as Al^{3+}, and as the charged hydroxide species, $Al(OH)_2^+$, $Al(OH)^{2+}$, $Al(OH)_4^-$ (Driscoll & Schechner, 1990). Aluminium in solution will form complexes, for example with fluoride, AlF^{2+}, AlF_2^+ and $Al(OH)F^+$ (Wilkinson, Campbell & Couture, 1990), and Al-hydroxides readily form polymers (Lyderson et al., 1990). The proportion of these various forms of aluminium varies dramatically with pH (Driscoll & Schechner, 1990). Water pH also influences aluminium solubility (which is minimal at around pH 6.5) and, if concentrations of aluminium exceed solubility for the particular pH, insoluble $Al(OH)_3$ may precipitate out. Thus, pH is intimately involved in determining aluminium chemistry and, perhaps not surprisingly, also influences branchial toxicity in fish. In most studies to date pH and aluminium content of the bulk water to which fish are exposed have been measured and often used to predict mathematically the proportion of the various aluminium species. However, it is the pH of the gill microclimate which matters and ultimately determines speciation, aluminium polymer formation and aluminium solubility as water passes across the gill surface (Playle & Wood, 1989).

The pH of the branchial microclimate is determined by branchial excretion of CO_2 and NH_3. The overall effect is a tendency to stabilize gill microclimate pH in the face of changes in water pH (Lin & Randall, 1990).

Carbon dioxide hydration to form bicarbonate tends to acidify neutral waters, but in water of less than pH 6, CO_2 hydration is considered to have little effect on branchial microclimate pH (Lin & Randall, 1990). At pH 5–6, conversion of NH_3 to ammonium ions is predicted to alkalinize the gill microclimate. As acidic bulk water passes over the gill the rising pH in the gill microclimate will both change the distribution of aluminium species and lower aluminium solubility which may, if aluminium concentrations are sufficient, cause precipitation of

Al(OH)$_3$ on the gills. Playle & Wood (1989, 1991) have investigated the alkalinization of acid water as it passes over the gill by placing a cannula near the gill filaments to sample water close to the gill membranes. In water of pH 4–5.8 containing 90–140 μg l^{-1} aluminium, the microenvironment of rainbow trout gills was 0.2–0.7 pH units more basic than inspired water pH, sufficient for solubility changes to cause precipitation of aluminium from the oversaturated solution onto the gill (Playle & Wood, 1989, 1990, 1991). The estimates of gill microclimate pH were admitted to be conservative and probably underestimate pH at the gill trailing edge (Playle & Wood, 1989). Theoretical calculations (Exley, Chappell & Birchall, 1991) suggested that the gill microclimate pH may be nearer to neutral. If this is true, then changes in speciation and reduced aluminium solubility in the gill microclimate will be even more significant.

Branchial toxicity has been suggested to result from the precipitation of insoluble Al(OH)$_3$ on the gill membranes and/or the interaction of gill ligands with positively charged species of aluminium and aluminium complexes, (Al^{3+}, Al(OH)$_2^+$, Al(OH)$^{2+}$, AlF^{2+}, AlF$_2^+$). The biochemistry of the gill ligands is poorly understood, but the microclimate, the protective mucus layer rich in sialic acid and the anionic binding sites of the apical boundary layer are undoubtedly of central importance in determining the binding of charged aluminium species (Reid, McDonald & Rhem, 1991; Wilkinson & Campbell, 1993). Rapid polymerization of aluminium hydroxide species either in solution or once bound to the gill membranes has been suggested to be important in determining aluminium toxicity (Lyderson *et al.*, 1990; Poleo, Lydersen & Muniz, 1991; Poleo & Muniz, 1993).

The interaction of charged aluminium species and complexes with the gill ligands, and precipitation of aluminium hydroxide will lead to accumulation of aluminium in gill tissue. A range of techniques, histochemistry, transmission electron microscopy and X-ray microanalysis have localized aluminium on and even within gill tissue (Karlsson-Norrgren *et al.*, 1986*a,b*; Youson & Neville 1987). Branchial accumulation of aluminium has been demonstrated in many species including rainbow trout (Karlsson-Norrgren *et al.*, 1986*a*; Youson & Neville, 1987; Neville & Campbell, 1988; Handy & Eddy, 1989; Playle & Wood, 1991), brook trout (Booth *et al.*, 1988), salmon (Wilkinson & Campbell, 1993), brown trout (Karlsson-Norrgren, 1986*a*; Dietrich & Schlatter, 1989*a*) and minnow (Norrgren, Wicklund-Glynn & Malborg, 1991). Our own studies showed accumulation of aluminium in the brown trout even when exposed to very low levels of aluminium (Fig. 1). Aluminium accumulation was found to be pH-dependent in some studies (Playle & Wood, 1989, 1991) though not in others (Norrgren *et al.*, 1991). This may reflect the protective function of the mucus which is readily sloughed off and may thus minimize accumulation of aluminium species bound to mucus, aluminium polymers and precipitated $Al(OH)_3$. Only one-tenth of aluminium extracted from water as it passed across the gills accumulated on the gills (Playle & Wood, 1991) which supports the notion of a protective sloughing of mucus with aluminium. No aluminium was localized in the chloride cells of aluminium-exposed rainbow trout (Goossenaerts *et al.*, 1988). However, electron microscopy localized putative aluminium particles both on the surface of and within chloride cells of the rainbow trout (Youson & Neville, 1987) so that intracellular toxic actions of aluminium may be important.

Given the effect of pH on aluminium speciation and solubility, it is perhaps not unexpected that the toxic actions of aluminium appear critically dependent upon water pH (Muniz & Leivestad 1980; Neville, 1985; Playle *et al.*, 1989; Playle & Wood, 1989). In the pH range 5.2–6.0, severe respiratory stress is apparent (Neville, 1985; Wood *et al.*, 1988*a*,*b*; Dietrich & Schlatter 1989*a*; Playle *et al.*, 1989; Witters, Van Puymbroeck & Vanderborght, 1991) and has been suggested to result from the irritant effect of the branchial binding of the hydroxide species prevalent at this pH ($Al(OH)_2^+$, $Al(OH)^{2+}$) and/or the effects of precipitated aluminium hydroxide. Gill irritation is believed to cause an increased discharge of mucus (Muniz & Leivestad, 1980; Rosseland & Skogheim, 1984; Handy & Eddy, 1989; Dietrich & Schlatter, 1989*a*; Mueller *et al.*, 1991), gill hyperplasia and lamellar epithelial thickening (manifested as an increased wet weight of gill tissue), fusion of neighbouring lamellae, oedema and lamellar distortion (Chevalier, Gauthier & Moreau, 1985; Karlsson-Norrgren *et al.*, 1986*a*; Tietge, Johnson & Bergman, 1988; Dietrich & Schlatter, 1989*a*; Ingersoll *et al.*, 1990; Mueller *et al.*, 1991; Reader *et al.*, 1991; Norrgren *et al.*, 1991). Increased mucus discharge together with oedema and tissue hyperplasia will increase the diffusion distance for O_2, and account for the reported decrease in arterial PO_2.

At a lower water pH of 5.2 or less, aluminium appears to result in an additional and usually dominant ionoregulatory disturbance. At lethal concentrations of aluminium a rapidly declining plasma Na^+ and Cl^- concentrations or sodium loss have been reported in many adult salmonid species including the rainbow trout, *Oncorhynchus mykiss* (Neville, 1985; Witters, 1986; Witters *et al.*, 1987, 1990; Playle *et al.*, 1989), the brook trout, *Salvelinus fontinalis* (Booth *et al.*, 1988; Wood *et al.*, 1988*a*; Walker, Wood & Bergman, 1991) and the brown trout,

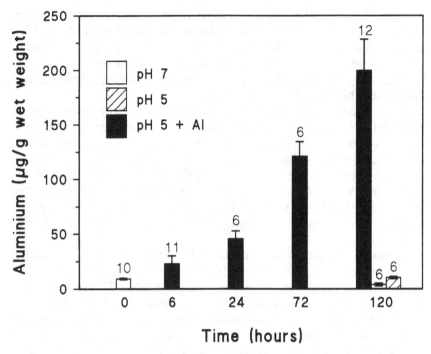

Fig. 1. Aluminium content of brown trout gills (μg g^{-1} wet weight) at time 0 h in soft synthetic water of pH 7, after 120 h confined exposure in soft water of pH 7, after 120 h exposure to soft water of pH 5, and after exposure to 12.5 μg l^{-1} aluminium in synthetic soft water of pH 5 for up to 120 h. Number of fish in each group are given above bars. Gills were rinsed in pH 7.1 soft water for 60 s, then digested in 5× vol 0.05 M H$_2$SO$_4$ for 8 h at 80 °C. After centrifug-ation, the supernatant was assayed for aluminium using the pyrocate-chol violet method. To allow for colorimetic interference by the tissue, gills of fish held in soft water of pH 7 were soaked in 10 mM EDTA and then used to prepare a gill digest; aliquots of the gill digest were added to the standards and blanks.

Salmo trutta (Battram, 1988; Sayer *et al.*, 1991; Reader *et al.*, 1991; Rosseland & Skogheim, 1984; Muniz & Leivestad, 1980). The ion loss due to aluminium exposure at water pHs of around 5.2 or below is believed to reflect the increasing proportion of Al^{3+} which appears to bind strongly to the gill membranes (Playle & Wood, 1991) competing with Ca^{2+} for binding sites. The cross-linking of calcium with gill ligands normally stabilizes the membranes and assures low paracellular electrolyte permeability (McDonald, 1983), so that displacement of

Ca^{2+} by Al^{3+} would manifest itself as opening and/or weakening of tight junctions and the observed profound increase in paracellular Na^+ and Cl^- loss (Battram, 1988; Booth et al., 1988; Witters et al., 1992). Fish may be capable of acclimating to sublethal aluminium exposure by increasing the gill binding affinity for Ca^{2+} thereby offsetting the competitive binding of Al^{3+} (Reid et al., 1991).

Recent studies indicate other ionic interactions may contribute to branchial ionoregulatory toxicity. Fluoro-aluminium species AlF^{2+} and AlF_2^+, common in the pH range 5–6 may complex with ligands at the gill surface (Wilkinson & Campbell, 1993; Wilkinson et al., 1993) and were reported to increase the ionoregulatory disturbance caused by aluminium (Wilkinson et al., 1990). Aluminium polymerization on the gill tissue rather than Al–F–gill–ligand was, however, suggested to be involved in ion loss in smoltifying Atlantic salmon (Poleo & Muniz, 1993).

Alongside the increased Na^+ and Cl^- efflux, aluminium exposure results in decreased ion influxes (Battram, 1988; Booth et al., 1988; Witters et al., 1992) which may reflect inhibition of branchial Na^+, K^+-ATPase activity (Staurnes, Sigholt & Reite, 1984). For example, in a recent investigation of parr-smolt transformation of Atlantic salmon a low concentration of aluminium 50 µg l^{-1} at pH 5, decreased Na^+,K^+-ATPase activity of smolts to the parr level, and thereby reduced seawater tolerance (Staurnes, Blix & Reite, 1993). This could have considerable impact on migratory success. However, in the wild, acclimation to sub-lethal levels of aluminium in acidic waters may be possible. Experiments involving a 10-week period of acclimation of brook trout to a sub-lethal pH (5.2) and sub-lethal aluminium (150 µg l^{-1}) indicated that sodium uptake was increased, perhaps reflecting an increased synthesis of the transport proteins responsible for sodium uptake (McDonald & Milligan, 1988). The decrease in chloride uptake by aluminium-exposed salmonids (Battram, 1988) is not well understood but may involve inhibition of $Cl^-–HCO_3^-$-ATPase.

In contrast to the work on rainbow trout and brook trout, our own recent studies of brown trout, exposed to lower concentrations of aluminium than those which have been generally used, suggest marked differences in the physiological responses from those reported to be typical of salmonids. In synthetic soft water of pH 5 (0.02 mM Ca^{2+}, 0.03 mM Na^+, 0.01 mM K^+, 0.04 mM Cl^-) aluminium at 50 µg l^{-1} and 25 µg l^{-1} caused relatively little or no ionoregulatory stress (Waring & Brown, 1995). Aluminium at 50 µg l^{-1} was lethal, while 33% of fish survived the 5-day period of exposure to 25 µg l^{-1} aluminium. A small transitory decline in plasma Na^+ and Cl^- concentrations occurred during exposure to a sublethal concentration of aluminium (12.5 µg l^{-1} at pH 5). The avoidance of ionoregulatory disturbances by these brown trout is intriguing. Ion fluxes were not measured in these brown trout and would clearly be of interest as a more sensitive index of ionoregulatory disturbance (Wood, 1992). At the moment it is not clear if an increase in ion efflux occurred, as is usually reported in aluminium-exposed salmonids in these conditions, followed by very rapid readjustment. Such adjustments would almost inevitably involve endocrine control. However, the properties of the gills of these brown trout may have protected them from aluminium-induced ion loss. Certainly, it is clear that brown trout populations acclimated long term to acidic waters can increase the branchial affinity for calcium ions (McWilliams, 1982, 1983). A high affinity for calcium ions could protect against aluminium displacement of calcium and resultant ion efflux.

Despite the lack of osmoregulatory imbalance in these aluminium-exposed brown trout, a profound respiratory stress occurred and was the cause of death at 50 µg l^{-1} and 25 µg l^{-1} aluminium. Even at the sublethal concentration of aluminium (12.5 µg l^{-1} at pH 5), total blood oxygen content (C_aO_2) declined, but within the 120 h exposure period recovery was apparent (Fig. 2). Exposure to an intermediate concentration of 25 µg l^{-1} Al at pH 5.0, conditions lethal for more than 50% fish, caused a decline in C_aO_2 to a stable but low level (Waring & Brown, 1995). The cause of the respiratory stress in these fish is uncertain. The aluminium concentrations are well below those generally occurring in naturally acidic waters (Stoner, Gee & Wade 1984) and demonstrate the harmful effect that even very low amounts of aluminium may have on sensitive salmonid species.

Recovery of blood oxygen and ionoregulatory status in brown trout sublethally stressed by aluminium exposure in soft acidic water indicates physiological adaptation in which hormonal factors are likely to play a significant role. The ionoregulatory and respiratory status of fish are influenced by a vast array of hormones: prolactin, arginine vasotocin, thyroid hormones, cortisol, angiotensin II, catecholamines, urotensins and cardiac peptides (Bern & Madsen, 1992). We have examined some aspects of the endocrine status of brown trout exposed to both lethal and sublethal concentrations of aluminium in order to gain insights into the role of hormones in control of the adaptive processes in fish exposed to aluminium in acidic waters. The hormones considered: prolactin, cortisol, catecholamines, and thyroid hormones might each be predicted, on the basis of their physiological actions, to potentially

122   J.A. BROWN AND C. P. WARING

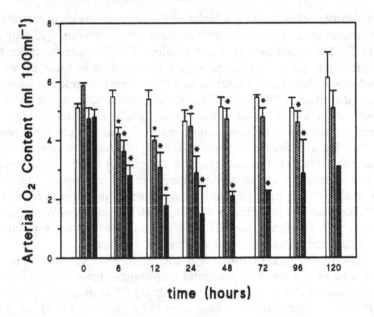

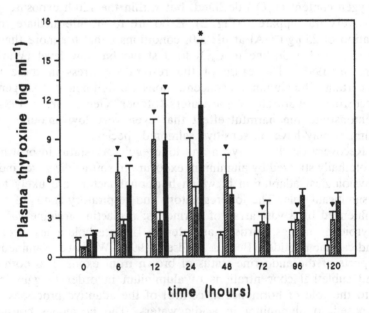

Fig. 2. Arterial oxygen content and plasma cortisol, thyroxine and triiodothyronine concentrations in trout exposed to water of pH 5.0 in the absence of aluminium and water of pH 5 containing aluminium

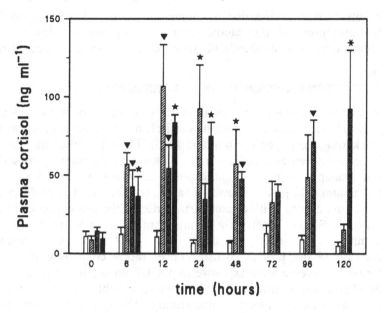

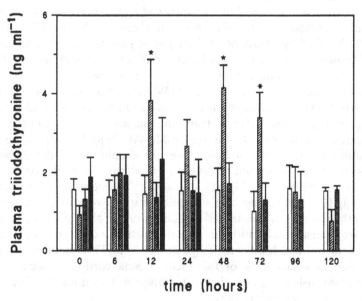

at 12.5 μg l^{-1}, 25 μg l^{-1} or 50 μg l^{-1} Al, asterisks: $P<0.05$, stars: $P<0.01$, triangles: $P<0.001$.

be involved in acclimation to aluminium. Others have suggested that the urotensins of the caudal neurosecretory system play a role in acclimation to aluminium/acid stress (Chevalier *et al.*, 1985, 1986).

Adrenocortical response to aluminium

A common response to a wide variety of stressors is the release of adrenocorticosteroids, especially cortisol, from the adrenocortical tissue (Pickering, Pottinger & Sumpter, 1986). Plasma cortisol has therefore been monitored as a general index of stress in aluminium-exposed fish. For example, rainbow trout exposed to water of pH 4.8 containing aluminium (112 µg l^{-1}) showed a rapid, pronounced and prolonged rise in plasma cortisol with a later further rise as the fish approached death (Goss & Wood, 1988). In a separate study, rainbow trout exposed to aluminium (597 µg l^{-1}) in acidic water (pH 4.7) showed a seven-fold increase in the plasma cortisol concentration compared to less than four-fold increase after exposure to pH 4.7 alone (Brown *et al.*, 1990a). Similarly, brook trout pre-acclimated to a sublethal acidic pH (5.2) and sublethal aluminium concentration (150 µg l^{-1}) and subsequently challenged with more acidic water (pH 4.8) and a higher concentration of aluminium (333 µg l^{-1}) increased plasma cortisol concentrations within a few minutes, but this returned to a basal concentration in 20 min, increasing again after 66 h in these conditions (Wood *et al.*, 1988b). We have examined the change in plasma cortisol concentrations of brown trout held in cages in an acidic softwater stream (yearly mean pH 5.3, Ca 0.05 mM, total Al 320 µg^{-1}), a stream which now has no natural population of brown trout (Whitehead & Brown, 1989). Plasma cortisol concentrations of fish in this stream were around seven-fold higher than those of fish from the adjacent nearby neutral stream (annual mean pH 6.9, Ca 0.07 mM, Al 70 µg l^{-1}). In more recent studies of the effects of lower levels of aluminium on the brown trout (at pH 5.0), significant elevations in plasma cortisol concentrations occurred within 6 h (Fig. 2). At a lethal concentration of aluminium (50 µg l^{-1}), circulating cortisol concentrations remained elevated until death while, at a sub-lethal concentration of aluminium (12.5 µg l^{-1}), more transient elevation in plasma cortisol occurred, with recovery evident from 72 h onwards (Fig. 2). At 25 µg l^{-1} aluminium, a concentration in which 67% of fish died, plasma cortisol concentrations of surviving fish remained elevated, implying a continued primary stress response (Donaldson, 1981).

Taken overall, the changes in plasma cortisol seem to provide a fair reflection of the stress status of salmonid fish and usually parallel changes in respiratory and/or ionoregulatory stress. The cortisol released as a direct response to aluminium-induced stress may have beneficial effects on carbohydrate metabolism (Wood, 1992; Brown, 1993). Cortisol appears to interact with other hormones, including the catecholamines, in control of hepatocyte function. The relative contributions of corticosteroid and adrenergic responses (see below) to the increased plasma glucose concentrations often observed in stressed fish (including fish stressed by aluminium in acidic waters) are still unclear. Cortisol appears to stimulate lipolysis and proteolysis, and have a gluconeogenic action (Lidman *et al.*, 1979; Leach & Taylor 1982; Sheridan, 1986), although administration of cortisol to unstressed rainbow trout, to achieve plasma concentrations typical of those occurring in chronically stressed fish failed to induce a hyperglycaemic response (Andersen *et al.*, 1991).

Perhaps more importantly, cortisol triggers the proliferation of chloride cells (Laurent & Perry, 1990), increasing branchial Na^+,K^+–ATPase and Ca^{2+}–ATPase activities (Flik & Perry, 1989; Madsen, 1990) and thus stimulating active influx of sodium chloride and calcium (Flik & Perry, 1989; Laurent & Perry, 1990). There is evidence for increased chloride cell densities in rainbow trout and brook trout during prolonged exposure to aluminium (Chevalier *et al.*, 1985; Karlsson-Norrgren *et al.*, 1986*a*,*b*; Tietge *et al.*, 1988). The ion transporting capacity of gills is dependent on the presence of mature chloride cells in contact with the external environment. It is significant, therefore, that perch from acidic (pH 4.4) aluminium-containing (50 µg l^{-1}) lakes and neutral lakes showed an increased presence of chloride cells with apical pits (Nikinma, Salama & Tuurula, 1990). Apical pits typify the chloride cells of seawater-adapted fish; these cells are crucial for ion secretion in seawater-adapted fish. In aluminium-stressed fish, chloride cell turnover may be altered to a greater extent than the number of chloride cells but only a few studies of acid/aluminium stressed fish have assessed turnover of these cells (Wendelaar Bonga *et al.*, 1990). Nevertheless, even at this stage of our understanding there is persuasive evidence that secretion of cortisol in aluminium-stressed fish may contribute to acclimation by aiding ionoregulatory recovery.

Catecholamines

In teleost fish, catecholamines, noradrenaline and adrenaline, are rapidly released into circulation from the richly innervated chromaffin cells of the anterior kidney in response to several forms of acute stress (Mazeaud & Mazeaud, 1981; Epple, Vogel & Nibbio, 1982; Ling &

Wells, 1985). Catecholamines may play a role in increasing plasma glucose during stress. Strong evidence for this role was provided in experiments where eels were injected with physiological doses of catecholamines (Epple, Hathaway & Nibbio, 1989). Studies in whole animals indicate that the plasma concentrations of catecholamines in stressed fish are sufficient to increase the active form of hepatic glycogen phosphorylase stimulating glycogenolysis and gluconeogenesis (Wright, Perry & Moon, 1989).

There is strong support for an essential role for catecholamines in the maintenance of the oxygen carrying capacity of the blood in the face of plasma acidosis which occurs in many forms of acute stress. Catecholamines stimulate the erythrocyte Na^+/H^+ exchange mechanism to maintain the intracellular pH when blood plasma pH is decreased (Primmett et al., 1986; Perry & Vermette, 1987). This, along with an increase in branchial permeability to oxygen, is believed to aid oxygen uptake (Perry, Daxboek & Dobson, 1985). Circulating catecholamines released during stress may also stimulate splenic contraction (Nilsson, 1983) releasing additional red blood cells into circulation, thus increasing the oxygen carrying capacity of the blood. However, although this effect may be beneficial for oxygen transport, it will also inevitably increase blood viscosity and may adversely increase blood pressure.

Although it is clear that respiratory stress often accompanies aluminium-induced stress, there have been relatively few measurements of plasma catecholamines in fish exposed to acidic soft waters containing aluminium. Plasma catecholamine concentrations of fish exposed to acidic soft waters in the absence of aluminium appear stable, except in very severe circumstances. For example, there was no apparent release of catecholamines in rainbow trout exposed to water of pH 4.0 for 5 h (Audet & Wood, 1993), in brown trout after 4 days at pH 4.5 (Butler & Day, 1993) or in carp, Cyprinus carpio, exposed to a gradual water acidification (Van Dijk et al., 1993). In the study of carp, the gradual acidification seems to have led to little ionoregulatory or respiratory stress, so the lack of catecholamine release is probably not surprising. In the other studies, analysis of plasma catecholamine concentration was restricted to a single time point and given the rapid release and metabolism of catecholamines this may have failed to detect episodes of catecholamine release. However, plasma adrenaline and noradrenaline concentrations of rainbow trout exposed to water of pH 4.0 for 72 h and sampled at several time points, also showed no significant increases except prior to death (Ye, Randall & He, 1991).

The occurrence of acidic soft waters without the presence of aluminium is environmentally rare, and the monitoring of plasma catecholamines in aluminium-stressed fish is therefore of greater interest.

Rainbow trout exposed for 46 h to 60 μg l⁻¹ aluminium in soft water of pH 5, a pH which in itself caused no physiological disturbance, showed a ten-fold increase in plasma adrenaline and noradrenaline concentrations (Witters *et al.*, 1991) which implies that catecholamines may play a role in aluminium-exposed fish. Plasma concentrations of both catecholamines were correlated to the depression of arterial PO_2 and decrease in arterial pH. A further study of rainbow trout exposed to 200 μg l⁻¹ aluminium suggested that rising catecholamine concentrations stimulated splenic contraction to increase the number of circulating erythrocytes (Witters *et al.*, 1990). In our own studies of the brown trout exposed to acidic soft water (pH 5.0) with a progressively increasing aluminium concentration, a transient increase in plasma noradrenaline concentration occurred after 6 h in aluminium (rising to 78 μg l⁻¹ filterable aluminium) followed by a marked increase in both plasma adrenaline and noradrenaline concentrations of surviving fish at 24 to 36 h when aluminium reached 230–257 μg l⁻¹ (Brown & Whitehead, 1995). These conditions were, however, lethal for more than 60% of the fish. At a much lower, sub-lethal and stable aluminium concentration (12.5 μg l⁻¹), noradrenaline concentrations were significantly elevated (although adrenaline concentrations were unchanged). Noradrenaline concentrations were significantly elevated relative to the initial concentration after 6 h and 12 h of aluminium exposure, recovered at 24 h and were significantly raised again at 48 h and 72 h. Intuitively, the release of catecholamines in response to respiratory stress might be expected to play a pivotal role in the survival of episodes of aluminium mobilization in acidic water.

Prolactin

Prolactin is a key osmoregulatory hormone in freshwater teleosts (Hirano, 1986), maintaining blood plasma Na^+ and Cl^- concentrations by actions on osmoregulatory tissues such as the gill, intestine, kidney and urinary bladder (Bern & Madsen, 1992). Prolactin decreases gill permeability to ions, in particular sodium, and osmotic permeability to water in freshwater fish (Clarke & Bern, 1980). There is also strong evidence of an important role in stimulating calcium uptake via the branchial Ca^{2+} pump (Flik, Fenwick & Wendelaar, 1989*a*). Prolactin could thus participate in osmotic and ionic regulation in response to any environmental stressor which has deleterious effects on ionoregulatory balance.

Despite the supposed beneficial actions of prolactin on the ionoregulatory status of freshwater teleosts, stress stimuli which are known to cause significant perturbations in plasma electrolyte concentrations have proven to have equivocal effects on plasma prolactin concentrations (Avella, Schreck & Prunet, 1991; Pottinger, Prunet & Pickering, 1992). Nevertheless, stimulation of prolactin release during acid- or aluminium-induced stress has been suggested to play a role in counteracting ionic disturbances in *Tilapia* sp (Flik *et al.*, 1989*b*; Wendelaar Bonga, Van der Meij & Flik, 1984*a*; Wendelaar Bonga *et al.*, 1984*b*; Wendelaar-Bonga, Balm & Flik, 1988). Exposure of acid-resistant tilapia to a severely acidic water pH 3.5–4.0 resulted in an initial decline in plasma osmolarity but this was followed by a full recovery within 10 days alongside an apparent increase in the activity of pituitary prolactin cells (Wendelaar Bonga *et al.*, 1984*a,b*). The ability of tilapia to respond to acidic water by synthesis and release of prolactin, and to restore plasma electrolytes, appears to be related to the rate of pH decline. Rapid acidification suppressed the prolactin response, while gradual reduction of water pH was accompanied by a stable plasma osmolarity (Wendelaar Bonga *et al.*, 1988). Long-term acclimation of tilapia to acidic water (3 months at pH 4.5) led to re-establishment of a positive sodium balance alongside an increase in prolactin cell activity (Flik *et al.*, 1989*b*).

In contrast to the increased synthesis and release of prolactin reported in tilapia, the prolactin response of salmonid species exposed to aluminium/acidic waters is less clear. Exposure of brook trout, *Salvelinus fontinalis*, to water of pH 4.5 resulted in an initial atrophy of pituitary prolactin cells with an apparent decrease in their biosynthetic activity (Notter *et al.*, 1976), followed by a compensatory increase. More recently, Fryer *et al.* (1988) suggested inhibited prolactin cell secretory activity occurred in brook trout exposed to aluminium (40 μg l^{-1} Al) at pH 4.5; over the 52-day study period, ionoregulatory disturbance in these fish recovered as prolactin secretory activity was restored.

Our own investigations of the prolactin responses of aluminium-stressed brown trout are the first to include measurements of circulating concentrations of prolactin. The wide variations in initial concentrations of plasma prolactin makes investigation of changes due to variations in water quality difficult. Irrespective of initial resting concentrations, however, exposure to 25 and 50 μg l^{-1} Al in soft water of pH 5.0 significantly depressed plasma prolactin concentrations at 12 h (Fig. 3) supporting the notion of decreasing prolactin cell activity in acid/aluminium stressed salmonids. This response is clearly maladaptive in freshwater fish which rely on prolactin to achieve ionoregulatory balance. It is therefore of particular interest that plasma prolactin concentrations were relatively stable in fish exposed to a sub-lethal concentration of aluminium (12.5 μg l^{-1}; pH 5.0; Fig. 3); this may have physiological importance in the restoration of the transient disturbance in ionoregulatory status noted in these fish.

Thyroid hormones

The thyroid gland of teleosts consists of isolated follicles found throughout the subpharyngeal and parapharangeal regions, around the ventral aorta and afferent branchial vessels, and in the head kidney. There is strong evidence that, in fish, the thyroid gland releases thyroxine (T_4) and that tri-iodothyronine (T_3), the physiologically active hormone, is primarily formed by peripheral conversion of T_4 to T_3, by hepatic and renal monodeiodinases (Eales, 1985; Leatherland *et al.*, 1990; Eales & Brown, 1993).

There have been several studies of the thyroid responses of the rainbow trout to acidic waters mainly by Brown and his colleagues in Winnipeg. After exposure of rainbow trout to pH 4.2 for 21 days there was no apparent effect on thyroid histology. Circulating T_4 was elevated but circulating T_3 was unchanged (Brown *et al.*, 1984). However, in a parallel time-course study T_3 and T_4 levels in control fish increased over the 21 days while there were no significant changes in acid-exposed fish, implying depression of thyroid function (Brown *et al.*, 1984). In a further study, exposure to water of pH 4.8 for 21 days did not significantly affect circulating concentrations of the thyroid hormones (Brown, Evans & Hara, 1986). More recent experiments on rainbow trout indicated that exposure to pH 4.7 did not significantly affect plasma concentrations of T_3, T_4 or their kinetics, but addition of aluminium (541 µg l^{-1}) increased T_4 degradation rate, while T_3 plasma appearance rates, T_3 plasma clearance rates and hepatic 5′monodeiodinase activity were depressed indicating depressed T_3 production (Brown *et al.*, 1990*a*). These studies are equivocal but do not suggest a major or consistent thyroid response in this species. However, this may reflect the high level of water calcium present in the Winnipeg tap water used in these experiments. High calcium alleviates the deleterious effects of acidic water on branchial function and thus reduces sodium and chloride loss (Muniz & Leivestad, 1980; McDonald, 1983; Mount, Hockett & Gern, 1988). The effects of water calcium on aluminium-induced ion losses are less certain: alleviation of ionoregulatory disturbances are reported in some cases (Muniz & Leivestad, 1980) but only at low pHs in other studies (Playle *et al.*, 1989). The lack of thyroid stimulation shown in rainbow trout exposed to acidic hard waters in the laboratory was also suggested in field studies of Atlantic salmon held in acidic Nova Scotian rivers, although here water calcium was low. Plasma T_3

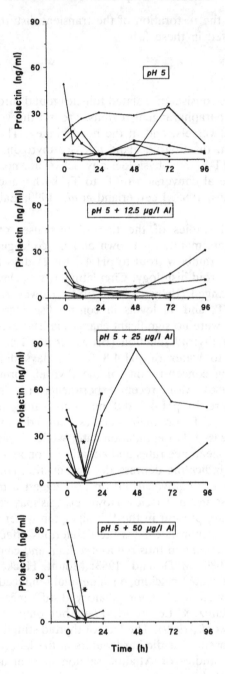

concentrations were reduced in fish held in the more acidic river (pH 4.7–5.2) (Brown *et al.*, 1990*b*). Although the studies on rainbow trout do not present an entirely consistent picture, taken together they do seem to give the impression of little or no change in plasma concentrations of T_4 and depressed plasma concentrations of T_3.

The reported thyroid responses of rainbow trout exposed to acidic/ aluminium-containing waters contrast markedly with the consistent response of brown trout noted in our own studies over the past 10 years. Brown trout exposed to either low pH alone or aluminium-containing acidic soft waters have repeatedly shown elevations in plasma thyroxine (T_4) concentrations (Edwards, Brown & Whitehead, 1987; Brown, Edwards & Whitehead, 1989; Whitehead & Brown, 1989, Fig. 2). However, plasma concentrations of triiodothyronine were usually stable suggesting a tight regulation of 5′-monodeiodinase activity. In our most recent studies, of the effect of exposure to low concentrations of aluminium, plasma T_4 concentrations increased five to ten-fold prior to death at a lethal concentration of aluminium (50 μg l^{-1}; pH 5.0) and increased 2–4-fold in survivors held in 25 μg l^{-1} aluminium at pH 5.0, although this response was not statistically significant. These responses do not seem to merely reflect very stressful or lethal conditions as exposure to a sub-lethal concentration of aluminium (12.5 μg l^{-1}; pH 5.0) similarly caused a significant increase in plasma T_4 concentrations which lasted throughout the 5-day exposure (Fig. 2). Plasma T_3 concentrations were unchanged in most experimental groups, but interestingly, were elevated transiently in fish exposed to the sub-lethal concentration of aluminium (12.5 μg l^{-1}; pH 5.0). This elevation of plasma T_3 concentrations tempts us to speculate that T_3 contributes to the observed amelioration of initial ionoregulatory disturbances shown only by these trout (Waring & Brown, 1995). Thyroid hormones have been shown to potentiate the cortisol-induced increase in branchial Na$^+$,K$^+$–ATPase activity in some species, such as the tilapia *Oreochromis mossambicus* (Dange, 1986) although this was not the case in rainbow trout (Madsen, 1990). A recent report showing that T_3 is required for the development of hypoosmoregulation in brown trout transferred to sea water (Leloup & Lebel, 1993) suggests that investi-

Fig. 3. Plasma prolactin concentrations in individual trout exposed to water of pH 5.0 in the absence of aluminium and water of pH 5 plus aluminium at 12.5 μg l^{-1}, 25 μg l^{-1} or 50 μg l^{-1}, asterisks $P<0.05$, stars $P<0.01$.

gation of the ionoregulatory role(s) of T_3 in the brown trout deserves further attention.

The caudal neurosecretory system (CNS)

The CNS secretes urotensins which participate in osmotic and ionic regulation of the body fluids of fish (Bern & Madsen, 1992).

Studies on the brook trout have suggested that the activity of the urophysis is increased in fish exposed to acidified and aluminium containing waters in the wild or in the short term in the laboratory (Chevalier et al., 1986; Hontela et al., 1989). Circulating levels of the urotensins have yet to be determined but persuasive morphological studies suggest increased synthetic activity of the neurosecretory cells and are supported by the decrease in urotensin II content of the urophyses. The implied increase in circulating urotensin II may assist in maintenance of the ionoregulatory status of fish exposed to acidic, aluminium containing waters.

Conclusions and future directions

Physiological disturbances occurring during exposure to sub-lethal levels of aluminium are of especial interest as these determine the critical environmental levels of aluminium for particular species. Progress has been made in determining the acclimatory processes and in forming hypotheses regarding the role that hormones such as cortisol, prolactin, the catecholamines and the urotensins play in these processes. However, the investigation of circulating concentrations of hormones alone stops short of providing an accurate picture of endocrine status – for this we need to investigate hormone kinetics.

It is clear that aluminium toxicity and the acclimation to aluminium centres on the fish gill. Branchial aluminium toxicity has been proposed to involve Al^{3+}, Al-hydroxides, Al-fluorides, insoluble $Al(OH)_3$ and aluminium polymerization. The toxicant action of these many forms of aluminium will depend both on water conditions, for example, pH, temperature, water calcium, and on the fish species in question. Even within a single species, the brown trout, profound differences in aluminium toxicity have become apparent; the commonly reported osmoregulatory disturbance was not of major concern to brown trout exposed to very low (sub-lethal) levels of aluminium, although respiratory distress was significant. To understand these differences, both within and between species, requires a fuller understanding of the gill microclimate which clearly plays a critical role in determining aluminium speciation, binding of positively charged species, polymerization and causing the precipitation of insoluble aluminium. The development of new approaches to study the gill microclimate and these interactions, as well as the wider use of established approaches, poses the next challenge – only then will we know the relative importance of the multiplicity of potential interactions at the gill surface and the changes which occur during acclimatory processes.

Acknowledgements

Studies on the brown trout by Brown, Waring, Whitehead and Edwards were supported by the Natural Environmental Research Council.

References

Andersen, D.E., Reid, S.D., Moon, T.W. & Perry, S.F. (1991). Metabolic effects associated with chronically elevated cortisol in rainbow trout (*Oncorhynchus mykiss*). *Canadian Journal of Fisheries and Aquatic Sciences*, **48**, 1811–17.

Audet, C. & Wood, C.M. (1993). Branchial morphological and endocrine responses of rainbow trout (*Oncorhynchus mykiss*) to a long-term sublethal acid exposure in which acclimation did not occur. *Canadian Journal of Fisheries and Aquatic Sciences*, **50**, 198–209.

Avella, M., Schreck, C.B. & Prunet, P. (1991). Plasma prolactin and cortisol concentration of stressed coho salmon, *Oncorhynchus kisutch*, in freshwater or salt water. *General and Comparative Endocrinology*, **81**, 21–7.

Battram, J.C. (1988). The effects of aluminium and low pH on chloride fluxes in brown trout, *Salmo trutta* L. *Journal of Fish Biology*, **32**, 937–47.

Bern, H.A. & Madsen, S.S. (1992). A selective survey of the endocrine system of the rainbow trout (*Oncorhynchus mykiss*) with emphasis on the hormonal regulation of ion balance. *Aquaculture*, **100**, 237–62.

Booth, C.E., McDonald, D.G., Simons, B.P. & Wood, C.M. (1988). The effects of aluminium and low pH on net ion fluxes and ion balance in the brook trout, *Salvelinus fontinalis. Canadian Journal of Fisheries and Aquatic Sciences*, **45**, 1563–74.

Brown, J.A. (1993). *Endocrine Responses to Environmental Pollutants*. Rankin, J.C. & Jensen, F.B. (eds.) pp. 276–96, Chapman & Hall, London.

Brown, J.A., Edwards, D. & Whitehead, C. (1989). Cortisol and thyroid hormone responses to acid stress in the brown trout, *Salmo trutta* L. *Journal of Fish Biology*, **35**, 73–84.

Brown, J.A. & Whitehead, C. (1995). Catecholamine release and interrenal response of brown trout, *Salmo trutta*, exposed to aluminium in acidic water. *Journal of Fish Biology* **46**, 524–35.

Brown, S.B., Eales, J.G., Evans, R.E. & Hara, T.J. (1984). Inter-renal, thyroidal, and carbohydrate responses of rainbow trout (*Salmo gairdneri*) to environmental acidification. *Canadian Journal of Fisheries and Aquatic Sciences*, 41, 36–45.

Brown, S.B., Evans, R.E. & Hara, T.J. (1986). Interrenal, thyroidal, carbohydrate and electrolyte responses in rainbow trout (*Salmo gairdneri*) during recovery from the effects of acidification. *Canadian Journal of Fisheries and Aquatic Sciences*, 43, 714–18.

Brown, S.B., MacLatchy, D.L., Hara, T.J. & Eales, J.G. (1990a). Effects of low ambient pH and aluminum on plasma kinetics of cortisol, T_3, and T_4 in rainbow trout (*Oncorhynchus mykiss*). *Canadian Journal of Fisheries and Aquatic Sciences*, 68, 1537–63.

Brown, S.B., Evans, R.E., Majewski, H.S., Sangalang, G.B. & Klaverkamp, J.F. (1990b). Responses of plasma electrolytes, thyroid hormones, and gill histology in Atlantic salmon (*Salmo salar*) to acid and limed river waters. *Canadian Journal of Fisheries and Aquatic Sciences*, 47, 2431–40.

Butler, P.J. & Day, N. (1993). The relationship between intracellular pH and swimming performance of brown trout exposed to neutral and sublethal pH. *Journal of Experimental Biology*, 176, 271–84.

Chevalier, G., Gauthier, L. & Moreau, G. (1985). Histopathological and electron microscopic studies of gills of brook trout *Salvelinus fontinalis* from acidified lakes. *Canadian Journal of Zoology*, 63, 2062–70.

Chevalier, G., Gauthier, L., Lin, R., Nishioka, R.S. & Bern, H.A. (1986). Effect of chronic exposure to an acidified environment on the urophysis of the brook trout *Salvelinus fontinalis*. *Experimental Biology*, 45, 291–9.

Clarke, W.C. & Bern, H.A. (1980). Comparative endocrinology of prolactin. In *Hormonal Proteins and Peptides* Vol VIII (ed. C.H. Li), pp. 105–97. Academic Press, New York.

Dange, A.D. (1986). Branchial Na^+/K^+-ATPase activity in freshwater and seawater acclimated tilapia *Oreochromis (Sarotherodon) mossambicus*: effects of cortisol and thyroxine. *General and Comparative Endocrinology*, 62, 341–3.

Dietrich, D. & Schlatter, Ch. (1989a). Aluminium toxicity to rainbow trout at low pH. *Aquatic Toxicology*, 15, 197–212.

Dietrich, D. & Schlatter, Ch. (1989b). Low levels of aluminium causing death of brown trout (*Salmo trutta fario*, L.) in a Swiss alpine lake. *Aquatic Sciences*, 51, 279–94.

Donaldson, E.M. (1981). The pituitary-interrenal axis as an indicator of stress in fish. *Stress and Fish*. Pickering, A.D. (ed.), Academic Press, New York.

Driscoll, C.T. & Schechner, W.D. (1990). The chemistry of aluminium in the environment. *Environmental Geochemistry and Health*, 12, 28–49.

Eales, J.G. (1985). The peripheral metabolism of thyroid hormones and regulation of thyroidal status in poikilotherms. *Canadian Journal of Zoology*, **63**, 1217–31.

Eales, J.G. & Brown, S.B. (1993). Measurement and regulation of thyroidal status in teleost fish. *Reviews in Fish Biology and Fisheries*, **3**, 299–347.

Edwards, D., Brown, J.A. & Whitehead, C. (1987). Endocrine and other physiological indicators of acid stress in the brown trout (*Salmo trutta*). *Annals of the Royal Society of Belgium*, **117** (Suppl. 1), 331–42.

Epple, A., Vogel, W.H. & Nibbio, A.J. (1982). Catecholamines in head kidney and body blood of eels and rats. *Comparative Biochemistry and Physiology*, **71C**, 115–18.

Epple, A., Hathaway, C.B. & Nibbio, B. (1989). Circulatory catecholamines in the eel: origins and functions. *Fish Physiology and Biochemistry*, **7**, 273–8.

Exley, C., Chappell, J.S. & Birchall, J.D. (1991). A mechanism for acute aluminium toxicity in fish. *Journal of Theoretical Biology*, **151**, 417–28.

Flik, G. & Perry, S.F. (1989). Cortisol stimulates whole body calcium uptake and the branchial calcium pump in freshwater trout. *Journal of Endocrinology*, **120**, 83–8.

Flik, G., Fenwick, J.C. & Wendelaar Bonga, S.E. (1989*a*). Calcitropic actions of prolactin in North American eel (*Anguilla rostrata* Le Seur). *American Journal of Physiology*, **257**, 74–9.

Flik, G., Van der Velden, J.A., Seegers, H.C.M., Kolar, Z. & Wendelaar Bonga, S.E. (1989*b*). Prolactin cell activity and sodium fluxes in tilapia *Oreochromis mossambicus* after long-term acclimation to acid water. *General and Comparative Endocrinology*, **75**, 39–45.

Fryer, J.N., Tam, W.H., Valentine, B. & Tikkala, R.E. (1988). Prolactin cell cytology, plasma electrolytes, and whole-body sodium efflux in acid-stressed brook trout (*Salvelinus fontinalis*). *Canadian Journal of Fisheries and Aquatic Sciences*, **45**, 1212–21.

Gagen, C.J. & Sharpe, W.E. (1987). Net sodium loss and mortality of three salmonid species exposed to a stream acidified by atmospheric deposition. *Bulletin of Environmental Contamination and Toxicology*, **39**, 7–14.

Goossenaerts, C., Van Grieken, R., Jacob, W., Witters, H. & Vanderborght, O. (1988). A microanalytical study of the gills of aluminium exposed rainbow trout (*Salmo gairdneri*). *International Journal of Environmental and Analytical Chemistry*, **34**, 227–37.

Goss, G.G. & Wood, C.M. (1988). The effects of acid and acid/aluminium exposure on circulating plasma cortisol levels and other blood parameters in the rainbow trout, *Salmo gairdneri*. *Journal of Fish Biology*, **32**, 63–76.

Handy, R.D. & Eddy, F.B. (1989). Surface absorption of aluminium by gill tissue and body mucus of rainbow trout, *Salmo gairdneri*, at the onset of episodic exposure. *Journal of Fish Biology*, **36**, 865–74.

Hirano, T. (1986). The spectrum of prolactin action in teleosts. In (Ralph, C.L. (ed.) *Comparative Endocrinology: Developments and Directions*. pp. 53–74. A.R. Liss, New York.

Hontela, A., Roy, Y., Coillie, R. van, Lederis, K. & Chevalier, G. (1989). Differential effect of low pH and aluminium on the caudal neurosecretory system of the brook trout, *Salvelinus fontinalis*. *Journal of Fish Biology*, **35**, 265–73.

Ingersoll, C.G., Gulley, D.D., Mount, D.R., Mueller, M.E., Fernandez, J.D., Hockett, J.R. & Bergman, H.L. (1990). Aluminium and acid toxicity to two strains of brook trout. *Canadian Journal of Fisheries and Aquatic Sciences*, **47**, 1641–8.

Karlsson-Norrgren, L., Dickson, W., Ljungberg, O. & Runn, P. (1986a). Acid water and aluminium exposure: gill lesions and aluminium accumulation in farmed brown trout, *Salmo trutta* L. *Journal of Fish Diseases*, **9**, 1–9.

Karlsson-Norrgren, L., Bjorklund, I., Ljungberg, O. & Runn, P. (1986b). Acid water and aluminium exposure: experimentally induced gill lesions in brown trout, *Salmo trutta* L. *Journal of Fish Diseases*, **9**, 11–25.

Laurent, P. & Perry, S.F. (1990). Effects of cortisol on gill chloride cell morphology and ionic uptake in the freshwater trout, *Salmo gairdneri*. *Cell and Tissue Research*, **259**, 429–42.

Leach, G.J. & Taylor, M.H. (1982). The effects of cortisol treatment on carbohydrate and protein metabolism in *Fundulus heteroclitus*. *General and Comparative Endocrinology*, **48**, 76–83.

Leatherland, J.F., Redding, P.K., Yong, A.N., Leatherland, A. & Lam, T.J. (1990). Hepatic 5'-monodeiodinase activity in teleosts *in vitro*: a survey of thirty three species. *Fish Physiology and Biochemistry*, **8**, 1–10.

Leloup, J. & Lebel, J.-M. (1993). Triiodothyronine is necessary for the actions of growth hormone in acclimation to seawater of brown trout (*Salmo trutta*) and rainbow trout (*Oncorhynchus mykiss*). *Fish Physiology and Biochemistry*, **11**, 165–73.

Lidman, U., Dave, G., Johansson-Sjobeck, M.-L., Larsson, A. & Lewander, K. (1979). Metabolic effects of cortisol in the European eel, *Anguilla* (L). *Comparative Biochemistry and Physiology*, **63A**, 339–44.

Lin, H. & Randall, D.J. (1990). The effect of varying water pH on the acidification of expired water in rainbow trout. *Journal of Experimental Biology*, **149**, 149–60.

Ling, N. & Wells, R.M.G. (1985). Plasma catecholamines and erythrocyte swelling following capture stress in a marine teleost fish. *Comparative Biochemistry and Physiology*, **82C**, 231–4.

Lyderson, E., Salbu, B., Poleo, A.B.S. & Muniz, I.P. (1990). The influences of temperature on aqueous aluminium chemistry. *Water, Air and Soil Pollution*, **51**, 203–15.

McDonald, D.G. (1983). The effects of H^+ upon the gills of freshwater fish. *Canadian Journal of Zoology*, **61**, 691–703.

McDonald, D.G., & Milligan, C.L. (1988). Sodium transport in the brook trout, *Salvelinus fontinalis*: effects of prolonged low pH exposure in the presence or absence of aluminium. *Canadian Journal of Fisheries and Aquatic Sciences*, **45**, 1606–13.

McDonald, D.G. & Wood, C.M. (1993). Branchial mechanism of acclimation to metals in freshwater fish. In *Fish Ecophysiology*. Rankin, J.C. & Jensen, F.B. (eds.) pp. 297–321, Chapman & Hall, London.

McWilliams, P.G. (1982). A comparison of physiological characteristics in normal and acid-exposed populations of the brown trout, *Salmo trutta*. *Comparative Biochemistry and Physiology*, **72A**, 515–22.

McWilliams, P.G. (1983). An investigation of the loss of bound calcium from the gills of the brown trout, *Salmo trutta*, in acid media. *Comparative Biochemistry and Physiology*, **74A**, 107–16.

Madsen, S.S. (1990). Effect of repetitive cortisol and thyroxine injections on chloride cell number and Na^+,K^+-ATPase activity in gills of freshwater acclimated rainbow trout, *Salmo gairdneri*. *Comparative Biochemistry and Physiology*, **95A**, 171–5.

Mazeaud, M.M. & Mazeaud, F. (1981). Adrenergic responses to stress in fish. In *Stress and Fish*. Pickering, A.D. (ed.) pp. 49–75. Academic Press, London.

Mount, D.R., Hockett, J.R. & Gern, W.A. (1988). Effect of long-term exposure to acid, aluminium and low calcium on adult brook trout, *Salvelinus fontinalis*. 2. Vitellogenesis and osmoregulation. *Canadian Journal of Fisheries and Aquatic Sciences* **45**, 1633–42.

Mueller, M.E., Sanchez, D.A., Bergman, H.L., McDonald, D.G., Rhem, R.G. & Wood, C.M. (1991). Nature and time course of acclimation to aluminium in juvenile brook trout (*Salvelinus fontinalis*). II. gill histology. *Canadian Journal of Fisheries and Aquatic Sciences*, **48**, 2016–27.

Muniz, I.R. and Leivestad, H. (1980). Toxic effects of aluminium on the brown trout, *Salmo trutta* L. In *Proceedings of International Conference on Ecological Impacts of Acid Precipitation*. Drablos, D. and Tolland, A. (eds.) pp. 320–1. Oslo-As, SNSF Project.

Neville, C.M. (1985). Physiological response of juvenile rainbow trout, *Salmo gairdneri*, to acid and aluminium – prediction of field responses from laboratory data. *Canadian Journal of Fisheries and Aquatic Sciences*, **42**, 2004–19.

Neville, C.M. & Campbell, P.G. (1988). Possible mechanism of aluminium toxicity in a dilute, acidic environment to fingerlings and older life stages of salmonids. *Water, Air and Soil Pollution*, **42**, 311–27.

Nikinma, M., Salama, A. & Tuurula, H. (1990). Respiratory effects of environmental acidification in perch (*Perca fluviatilis*) and rainbow trout (*Salmo gairdneri*). In *Acidification in Finland*. Kauppi, P. (ed.) Springer Verlag.

Nilsson, S. (1983). *Autonomic Nerve Function in the Vertebrates*. Springer Verlag, Berlin.

Norrgren, L., Wicklund-Glynn, A. & Malborg, O. (1991). Accumulation and effects of aluminium in the minnow (*Phoxinus phoxinus* L.) at different pH levels. *Journal of Fish Biology*, **39**, 833–47.

Notter, M.F.D., Mudge, J.E., Neff, W.H. & Anthony, A. (1976). Cytophotometric analysis of RNA changes in prolactin and stannius corpuscle cells of acid-stressed brook trout. *General and Comparative Endocrinology*, **30**, 273–84.

Perry, S.F., Daxboek, C. & Dobson, G.P. (1985). The effect of perfusion flow rate and adrenergic stimulation on oxygen transfer across the isolated saline-perfused head of rainbow trout (*Salmo gairdneri*). *Journal of Experimental Biology*, **116**, 251–69.

Perry, S.F. & Vermette, M.G. (1987). The effects of prolonged epinephrine infusion on the physiology of the rainbow trout *Salmo gairdneri*. I Blood respiratory acid base and ionic status. *Journal of Experimental Biology*, **128**, 235–53.

Pickering, A.D., Pottinger, T.G. & Sumpter, J.P. (1986). Independence of the pituitary-interrenal axis and melanotroph activity in the brown trout, *Salmo trutta* L., under conditions of environmental stress. *General and Comparative Endocrinology*, **64**, 206–11.

Playle, R.C. & Wood, C.M. (1989). Water pH and aluminum chemistry in the gill micro-environment of rainbow trout during acid and aluminum exposures. *Journal of Comparative Physiology*, **159B**, 539–50.

Playle, R.C. & Wood, C.M. (1990). Is precipitation of aluminium fast enough to explain aluminium deposition in fish gills? *Canadian Journal of Fisheries and Aquatic Sciences*, **47**, 1558–61.

Playle, R.C. & Wood, C.M. (1991). Mechanisms of aluminium extraction and accumulation at the gills of rainbow trout, *Oncorhynchus mykiss* (Walbaum), in acidic soft water. *Journal of Fish Biology*, **38**, 791–805.

Playle, R.C., Goss, G.G. & Wood, C.M. (1989). Physiological disturbances in rainbow trout (*Salmo gairdneri*) during acid and aluminum exposures in soft water of two calcium concentrations. *Canadian Journal of Zoology*, **67**, 314–24.

Poleo, A.B.S., Lydersen, E. & Muniz, I.P. (1991). The influence of temperature on aquatic aluminium chemistry and survival of Atlantic salmon (*Salmo salar* L) fingerlings. *Aquatic Toxicology*, **21**, 267–78.

Poleo, A.B.S. & Muniz, I.P. (1993). The effect of aluminium in soft water at low pH and different temperatures on mortality, ventilation frequency and water balance in smoltifying Atlantic salmon, *Salmo salar*. *Environmental Biology of Fishes*, **36**, 193–203.

Pottinger, T.G., Prunet, P. & Pickering, A.D. (1992). The effects of confinement stress on circulating prolactin levels in rainbow trout (*Oncorhynchus mykiss*) in fresh water. *General and Comparative Endocrinology*, **88**, 454–60.

Primmett, D.R.N., Randall, D.J., Mazeaud, M. & Boutilier, R.G. (1986). The role of catecholamines in erythrocyte pH regulation and oxygen transport in rainbow trout (*Salmo gairdneri*) during exercise. *Journal of Experimental Biology*, **122**, 139–48.

Reader, J.P., Dalziel, T.R.K., Morris, R., Sayer, M.D.J. & Dempsey, C.H. (1991). Episodic exposure to acid and aluminium in soft water: survival and recovery of brown trout, *Salmo trutta* L. *Journal of Fish Biology*, **39**, 181–96.

Reid, S.D., McDonald, D.G. & Rhem, R.R. (1991). Acclimation to sublethal aluminium: modification of metal-gill surface interaction of juvenile rainbow trout (*Oncorhynchus mykiss*). *Canadian Journal of Fisheries and Aquatic Sciences*, **48**, 1996–2005.

Rosseland, B.O. & Skogheim, O.K. (1984). A comparative study on salmonid fish species in acid aluminium-rich water. II. Physiological stress and mortality of one- and two-year-old fish. *Report Institute Freshwater Research, Drottingholm*, **61**, 186–94.

Rosseland, B.O., Eldhurst, T.D. & Staurnes, M. (1990). Environmental effects of aluminium. *Environment and Geochemistry of Health*, **12**, 17–27.

Sayer, M.D.J., Reader, J.P., Dalziel, T.R.K. & Morris, R. (1991). Mineral content and blood parameters of dying brown trout (*Salmo trutta* L.) exposed to acid and aluminium in soft water. *Comparative Biochemistry and Physiology*, **99C**, 345–8.

Sheridan, M.A. (1986). Effects of thyroxine, cortisol, growth hormone, and prolactin on lipid metabolism of coho salmon, *Oncorhynchus kisutch*, during smoltification. *General and Comparative Endocrinology*, **64**, 220–38.

Staurnes, M., Sigholt, T. & Reite, O.B. (1984). Reduced carbonic anhydrase and Na–K–ATPase activity in gills of salmonids exposed to aluminium-containing acid water. *Experientia*, **40**, 226–7.

Staurnes, M., Blix, P. & Reite, O.B. (1993). Effects of acid water and aluminium on parr smolt transformation and seawater tolerance in Atlantic salmon, *Salmo salar*. *Canadian Journal of Fisheries and Aquatic Sciences*, **50**, 1876–7.

Stoner, J.H., Gee, A.S. & Wade, K.R. (1984). The effect of acidification on the ecology of streams in the upper Tywi catchment in West Wales. *Environmental Pollution Series A*, **35**, 125–57.

Tietge, J.E., Johnson, R.D. & Bergman, H.L. (1988). Morphometric changes in gill secondary lamellae of brook trout (*Salvelinus fontinalis*) after long-term exposure to acid and aluminium. *Canadian Journal of Fisheries and Aquatic Sciences*, **45**, 1643–8.

Van Dijk, P.L.M., Van Den Thillart, G.E.E.J.M., Balm, P. & Wendelaar Bonga, S.E. (1993). The influence of gradual water acidification on the acid/base status and plasma hormone levels in carp. *Journal of Fish Biology*, **42**, 661–71.

Walker, R.L., Wood, C.M., Bergman, H.L. (1991). Effects of long-term exposure to sublethal concentrations of acid and aluminium on the ventilatory response to aluminium challenge in brook trout (*Salvelinus fontinalis*). *Canadian Journal of Fisheries and Aquatic Sciences*, **48**, 1989–95.

Waring, C.P. & Brown, J.A. (1995). Ionoregulatory and respiratory responses of brown trout, *Salmo trutta*, exposed to lethal and sublethal aluminium in acidic soft waters. *Fish Physiology and Biochemistry*, **13, 14**, 81–91.

Wendelaar Bonga, S.E., Balm, P.H.M. & Flik, G. (1988). Control of prolactin secretion in the teleost *Oreochromis mossambicus*: Effects of water acidification. *General and Comparative Endocrinology*, **72**, 1–12.

Wendelaar Bonga, S.E., Van Der Meij, J.C.A. & Flik, G. (1984a). Prolactin and acid stress in the teleost *Oreochromis* (formerly *Sartherodon*) *mossambicus*. *General and Comparative Endocrinology*, **55**, 323–32.

Wendelaar Bonga, S.E., Van Der Meij, J.C.A., Van Der Krabben, W.A.W.A. & Flik, G. (1984b). The effects of water acidification on prolactin cells and pars intermedia PAS-positive cells in the teleost fish *Oreochromis* (formerly *Sarotherodon*) *mossambicus*, and *Carassius auratus*. *Cell and Tissue Research*, **238**, 601–9.

Wendelaar Bonga, S.E., Flik, G., Balm, P.H.M. & van der Meij, J.C.A. (1990). The ultrastructure of chloride cells in the gills of the teleost *Oreochromis mossambicus* during exposure to acidified water. *Cell and Tissue Research*, **259**, 575–85.

Whitehead, C. & Brown, J.A. (1989). Endocrine responses of brown trout, *Salmo trutta* L., to acid, aluminium and lime dosing in a Welsh hill stream. *Journal of Fish Biology*, **35**, 59–71.

Wilkinson, K.J. & Campbell, P.G.C. (1993). Aluminium bioconcentration at the gill surface of juvenile Atlantic salmon in acidic media. *Environmental Toxicology and Chemistry*, **12**, 2083–95.

Wilkinson, K.J., Bertsch, P.M., Jagoe, C.H. & Campbell, P.G.C. (1993). Surface complexation of aluminum on isolated fish gill cells. *Environmental Science and Technology*, **27**, 1132–8.

Wilkinson, K.J., Campbell, P.G.C. & Couture, P. (1990). Effect of fluoride complexation on aluminium toxicity towards juvenile Atlantic salmon. *Canadian Journal of Fisheries and Aquatic Sciences*, **47**, 1446–52.

Witters, H.E. (1986). Acute acid exposure of rainbow trout, *Salmo gairdneri* Richardson: effects of aluminium and calcium on ion balance and haematology. *Aquatic Toxicology*, **8**, 197–210.

Witters, H.E., Vangenechten, J.H.D., Van Puymbroeck, S. & Vanderborght, O.L.J. (1987). Ionoregulatory and haematological responses of rainbow trout *Salmo gairdneri* Richardson to chronic acid and aluminium stress. *Annals of the Royal Society of Zoology*, Belgium **117** (suppl. 1), 411–20.

Witters, H.E., Van Puymbroeck, S., Van Den Sande, I. & Vanderborght, O.L.J. (1990). Haematological disturbances and osmotic shifts in rainbow trout, *Oncorhynchus mykiss* (Walbaum) under acid and aluminium exposure. *Journal of Comparative Physiology*, **160B**, 563–71.

Witters, H.E., Van Puymbroeck, S. & Vanderborght, O.L.J. (1991). Adrenergic response to physiological disturbances in rainbow trout, *Oncorhynchus mykiss*, exposed to aluminium at acid pH. *Canadian Journal of Fisheries and Aquatic Sciences*, **48**, 414–20.

Witters, H.E., Van Puymbroeck, S. & Vanderborght, O.L.J. (1992). Branchial and renal ion fluxes and transepithelial electrical potential differences in rainbow trout, *Oncorhynchus mykiss*: Effects of aluminium at low pH. *Environmental Biology of Fishes*, **34**, 197–206.

Wood, C.M. (1992). Flux measurements as indices of H^+ and metal effects on freshwater fish. *Aquatic Toxicology*, **22**, 239–64.

Wood, C.M., Playle, R.C., Simons, B.P., Goss, G.G. & McDonald, D.G. (1988a). Blood gases, acid–base status, ions, and haematology in adult brook trout (*Salvelinus fontinalis*) under acid/aluminum exposure. *Canadian Journal of Fisheries and Aquatic Sciences*, **45**, 1575–86.

Wood, C.M., Simons, B.P., Mount, D.R. & Bergman, H.L. (1988b). Physiological evidence of acclimation to acid/aluminium stress in brook trout (*Salvelinus fontinalis*). 2. Blood parameters by cannulation. *Canadian Journal of Fisheries and Aquatic Sciences*, **45**, 1597–605.

Wright, P.A., Perry, S.F. & Moon, T.W. (1989). Regulation of hepatic gluconeogenesis and glycogenolysis by catecholamines in rainbow trout during environmental hypoxia. *Journal of Experimental Biology*, **146**, 169–88.

Ye, X., Randall, D.J. & He, X. (1991). The effect of acid water on oxygen consumption, circulating catecholamines and blood ionic and acid–base status in rainbow trout (*Salmo gairdneri* Richardson). *Fish Physiology and Biochemistry*, **9**, 23–30.

Youson, J.H. & Neville, C.M. (1987). Deposition of aluminium in the gill epithelium of rainbow trout (*Salmo gairdneri* Richardson) subjected to sublethal concentrations of the metal. *Canadian Journal of Zoology*, **65**, 647–56.

R.W. WILSON

Physiological and metabolic costs of acclimation to chronic sub-lethal acid and aluminium exposure in rainbow trout

Introduction

Chronic exposure of freshwater fish to sub-lethal aluminium concentrations is frequently encountered in low pH soft water owing to the acid-induced leaching of Al from soils and sediments (Wright & Gjessing 1976; Cronan & Schofield, 1979; Dickson, 1980). It is now clear that, at pH values between 4.7 and 5.5, fish kills may be primarily due to the presence of Al rather than the H^+ concentration *per se* (Schofield & Trojnar 1980; Baker & Schofield 1982). Physiological studies have demonstrated that the gills are the primary target organ of acute exposure to both acid or acid combined with sub-lethal Al concentrations (for reviews see Wood, 1989, 1992; Wood & McDonald, 1987). Ionoregulatory toxicity is a key feature of acid exposure (with or without Al present) whereas, under moderately acid conditions (pH levels greater than about 5), sub-lethal Al exposure is compounded by pronounced respiratory toxicity (Neville, 1985; Wood *et al.*, 1988*a,c*; Walker, Wood & Bergman, 1988; Playle, Goss & Wood, 1989; Witters, Van Puymbroeck & Vanderbrought, 1991). The ionoregulatory and respiratory disturbances are associated with surface binding and precipitation of Al on the gills (Playle & Wood 1989, 1991; Reid, Rhem & McDonald, 1991; see also Brown & Waring, this volume). However, these disturbances are usually acute, limited to an initial 'shock' phase lasting 2–5 days, and followed by a slower period of acclimation (the development of increased resistance to Al), repair and various degrees of physiological recovery (McDonald & Wood, 1992). It is now apparent that not all processes can recover fully during chronic acid+Al exposure and that acclimation may bear certain physiological and metabolic costs (Wilson, Bergman & Wood, 1994*a,b*; Wilson, Wood & Houlihan, 1995). The present review will concentrate on recent studies which have examined the potential costs of chronic acclimation to sub-lethal acid (pH 5.2) and Al (30 µg l^{-1}) in juvenile rainbow trout. All the data shown are from laboratory studies where exposure conditions were matched as closely as possible (a synthetic softwater medium containing $[Ca^{2+}]$ = 25–32 µeq l^{-1}, $[Na^+]$ = 45–86 µeq l^{-1}, 15 °C) and relevant to the acid-threatened lakes and rivers of eastern North America and Europe.

Acclimation to chronic exposure to sub-lethal acid + Al

The term acclimation is used within the toxicological literature to define an increased resistance to high (usually lethal) levels of a toxicant which develops during prolonged sub-lethal exposure. Acclimation to Al in acidic softwaters has now been documented in a number of laboratory studies (Orr et al., 1986; McDonald et al., 1991; Wilson et al., 1994a) which may explain the continued presence of fish populations in acidified softwaters containing levels of Al in excess of the thresholds predicted by acute toxicity tests (Wright & Snekvik 1978; Schofield & Trojnar 1980; Kelso et al., 1986). The mechanism of acclimation is not precisely understood but is thought to revolve around a damage/repair process which involves biochemical, physiological and structural changes within the primary target organ, the gills (McDonald et al., 1991; Mueller et al., 1991; McDonald & Wood, 1992; Wilson et al., 1994a,b).

Acclimation usually occurs within the first 5–17 days of sub-lethal exposure (Orr et al., 1986; Wood et al., 1988a,c; McDonald et al., 1991; Wilson & Wood, 1992; Wilson et al., 1994a), and appears to coincide with a pronounced hypertrophy of gill mucous cells (Fig. 1). This is considered to be a compensatory rather than pathological response, which is directly related to acclimation (McDonald et al., 1991; Mueller et al., 1991; McDonald & Wood, 1992). The purpose of this mucocyte hyperplasia is presumably to accelerate the secretion of mucus that binds preferentially and carries away insoluble Al species from the sensitive gill surface (Playle & Wood, 1989, 1991; McDonald et al., 1991; Mueller et al., 1991; Wilson et al., 1994b). This clearance of precipitated Al may reduce the inflammatory response often associated with the respiratory component of Al toxicity (Playle & Wood, 1991; McDonald et al., 1991). However, this is not the sole mechanism of branchial acclimation. Additional responses which may be critical for the restoration of normal ionoregulatory function include changes in the affinity of gill surface ligands for Al (Reid et al., 1991; see pp. 148–9).

The overall effect of these branchial acclimatory changes is a reduction in the various forms of Al present at the gills. Total gill Al levels generally peak after 2–5 days' exposure to sub-lethal acid+Al (McDonald et al., 1991; Wilson & Wood, 1992; Wilson et al., 1994a)

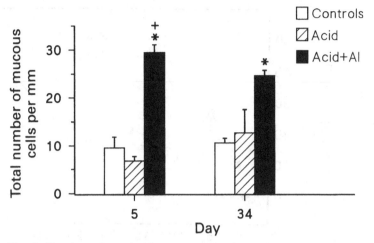

Fig. 1. Total number of gill mucous cells (on lamellae and filaments) expressed per mm of filament length, after 5 and 34 days exposure to either control softwater (pH 6.5; open bars), acid softwater (pH 5.2; hatched bars) or acid softwater plus sub-lethal Al (30 µg l^{-1}; solid bars). Means ± SEM are from 4 fish in each group. Asterisks indicate means significantly different from the 6.5/0 control group, and crosses indicate means significantly different from the 5.2/0 group ($P<0.05$).

and then fall coinciding with the development of acclimation. However, in two separate studies it has now been found that the accumulation of gill Al follows a biphasic pattern and that, during continued exposure (for up to 34 days), gill Al levels undergo a secondary more gradual increase and may even surpass the initial peak seen prior to acclimation (Fig. 2). Reduced toxicity is therefore not simply a function of a reduction in the total level of Al in the gills. What may be of more importance is the distribution of this Al at different sites within and on gills.

It is generally agreed that, at least initially, Al toxicity arises from the actions of bound and precipitated Al at the gill surface as described above (although a theoretical mechanism for intracellular toxicity has been suggested; Exley, Chappell & Birchall, 1991). From studies using histochemical, X-ray microanalysis, laser microprobe analysis and [26]Al-isotope techniques, it is apparent that within this initial phase (over the first few minutes to days) almost all the gill Al is found on the branchial surface (Goosenaaerts *et al.*, 1988; Oughton, Salbu & Bjornstad, 1992), and in particular within mucus-rich areas between gill lamellae (Norrgren, Wicklund Glynn & Malmborg, 1991). However,

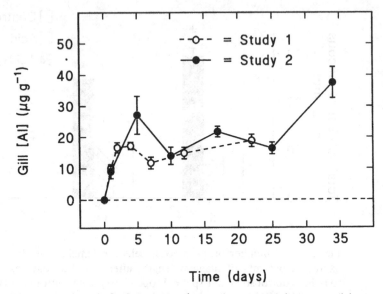

Fig. 2. Total gill [Al] (µg Al g^{-1} wet tissue weight) measured in two separate studies (Wilson & Wood, 1992, open circles; Wilson *et al.*, 1994*a*, filled circles) where juvenile rainbow trout were chronically exposed to acidified softwater (pH 5.2) plus sub-lethal Al (30 µg l^{-1}). Values are means ± SEM (*n*=8 and 10 for studies 1 and 2, respectively) and expressed relative to the control group (horizontal dotted line) held in pH 6.5 softwater. Trout exposed to acid alone (pH 5.2) had gill Al levels not significantly different from the control group. All values for the acid+Al group are significantly different from their respective control groups (*P*<0.05).

following prolonged exposure (from 1 week to 1 year) Al deposits can be found within the gill cells themselves (Karlsson-Norrgren *et al.*, 1986*a,b*; Youson & Neville, 1987; Evans, Brown & Hara, 1988; Norrgren *et al.*, 1991). It is therefore tempting to speculate that the biphasic pattern observed during analysis of *total* gill Al represents two separate processes: (i) a rapid accumulation on the branchial surface, and (ii) a second much slower internalization of Al within the gill cells themselves. The initial recovery of total gill Al levels between days 4 and 10 (Fig. 2) would correlate with the acclimatory changes in gill mucus secretion and gill binding affinity which serve to reduce *surface* Al and toxicity. Since internalized Al deposits in the gill are apparently membrane-bound precipitates and hence 'non-toxic' (Youson & Neville, 1987) the secondary increase in gill total Al (Fig. 2) may represent a slower development of cellular detoxification mechanisms which store and isolate the metal prior to exocytosis (McDonald & Wood, 1992). Thus, the internalization of Al and resultant secondary increase in total gill Al might actually represent a chronic response to additionally protect the gills from Al toxicity. This might also explain why some acclimated fish accumulate *more* total gill Al than control fish when challenged with an acutely lethal [Al] (rainbow trout; Wilson *et al.*, 1994*a*) whereas others accumulate less (brook trout; McDonald *et al.*, 1991; Reid *et al.*, 1991), although species differences should not be ruled out.

Specificity of acclimation

The specificity of acclimation to Al is poorly understood (that is, whether it will simultaneously increase a fish's resistance to other metals), but could add to our understanding of the acclimatory mechanism. Since the process of acclimation is probably designed to combat the mechanism of toxicity of a metal, it is possible that metals with similar modes of toxic action will elicit similar acclimatory responses and thus may exhibit crossover resistance (McDonald & Wood, 1992). For example, Zn and Cd both interfere with branchial control of Ca^{2+} balance, and acclimation to Cd has been shown to increase resistance to Zn and vice versa (Kito *et al.*, 1982; Duncan & Klaverkamp, 1983; Thomas *et al.*, 1985; Klaverkamp & Duncan, 1987). Copper has similar effects to Al in freshwater fish; both can impair respiratory gas exchange during acute exposure (Playle *et al.*, 1989; Wilson & Taylor, 1993) and both interfere with the active influx and passive efflux components of branchial ion transport (Laurén & McDonald, 1985, 1986; Booth *et al.*, 1988; Wood & McDonald, 1987). However, Wilson *et al.* (1994*a*) found no evidence for substantial crossover resistance to Cu in fish acclimated to sub-lethal acid+Al over a 34-day period. This implies that acclimation to Al involves changes that are specific to Al rather than the type of physiological toxicity it produces. However, although Cu and Al inhibit Na^+ uptake via a similar if not identical manner (inhibition of transport enzymes; Wood, 1992), they may stimulate passive Na^+ effluxes (J^{Na}_{out}) via subtly different mechanisms because J^{Na}_{out} is independent of the ambient $[Ca^{2+}]$ during Cu exposure (Laurén & McDonald, 1985, 1986), but is Ca^{2+}-dependent during Al exposure (Booth *et al.*, 1988; Wood *et al.*, 1988*a,c*; Playle *et al.*, 1989). This subtle difference between the mechanisms of ionoregulatory toxicities of Al and Cu may play a part in their lack of cross-over resistance. No other metals have yet been tested for cross-over resistance with Al. However, one recent study found that the acute toxicity of beryllium to perch (*Perca fluviatilis*) and roach (*Rutilus rutilus*) in soft acid water was analogous to that of aluminium (Jagoe *et al.*, 1993). The similar chemistry, solubility and aqueous speciation properties of beryllium (Vesely, Benes & Sevcik, 1989) compared to aluminium suggest that it may be a source of similar, as yet relatively unexplored, ecotoxicological problems for fish in soft acid waters. Given these observations, it would be interesting to test the cross-over resistance of Al with Be.

Acute and chronic effects of acid/Al on branchial ionoregulation

The mechanism of acclimation is intuitively linked to counteracting the mechanism of toxicity, and ultimately the recovery of physiological functions. In support of this hypothesis, it has been consistently observed that the development of acclimation to Al precedes the complete recovery of ionoregulatory status (whole body ion content). Another observation previously linking acclimation to physiological recovery was that fish exposed to acid alone, which does *not* induce acclimation, did not appear to recover ionoregulatory status but merely stabilized at a new reduced level (Audet, Munger & Wood, 1988; Audet & Wood, 1988; Wood *et al.*, 1988a; McDonald *et al.*, 1991; Wilson & Wood, 1992). However, Wilson *et al.* (1994a) have since confirmed that rainbow trout chronically exposed to acid alone (as well as acid+Al) can completely recover their whole body ion status given time, even though fish exposed to acid alone still failed to acclimate to low pH itself. In the case of exposure to acid in the absence of Al, increased resistance (acclimation) is *not* required for ionoregulatory recovery. On the other hand, some of the proposed mechanisms for acclimation to Al have an obvious role in the recovery of ion balance.

Ionoregulatory toxicity during acute Al exposure is manifested as a large net loss of Na^+ and Cl^- which results from both accelerated ionic effluxes and retarded ion uptake rates at the gills (Neville, 1985; Battram, 1988; Booth *et al.*, 1988; McDonald & Milligan, 1988; Wood *et al.*, 1988a; Witters, Van Puymbroeck & Vanderbroght, 1992). The increased ion efflux is associated with the disruption of the tight junctional permeability (i.e. paracellular diffusion pathway), leading to elevated diffusive losses of Na^+ and Cl^- from the blood to the very dilute external medium. Soluble aluminium is thought to bring this about by displacing Ca^{2+} from the intercellular junctions where it normally plays an important role in controlling the integrity and permeability of the branchial epithelium (Potts & Fleming, 1971;

Cuthbert & Maetz, 1972; McDonald & Rogano, 1986). It has now been shown that, during acclimation, the gill binding affinity for Al^{3+} is reduced relative to Ca^{2+} (Reid *et al.*, 1991). This would help regain control of paracellular ionic permeability and assist the recovery of ion efflux rates.

A reduction of active ion uptake during acute acid+Al exposure probably results from a direct inhibition of branchial enzymes involved in ion/acid–base transport such as the Na^+/K^+-ATPase and carbonic anhydrase (Staurnes, Sigholt & Reite, 1984). However, similar to ionic effluxes, the active uptake of Na^+ also recovers with time (McDonald & Milligan, 1988) and is the result of compensatory mechanisms within the gills. Although increased synthesis of transport proteins such as Na^+/K^+-ATPase has not been measured specifically, hyperplasia of gill chloride cells (associated with much of the gill's active ion transport capacity) is a prominent feature of long-term sub-lethal Al exposure (Tietge, Johnson & Bergman, 1988; Mueller *et al.*, 1991).

Acute and chronic effects of acid/Al on gill morphology and aerobic scope

Impaired respiratory gas exchange, as indicated by internal hypoxia, lactacidosis, and hyperventilation, is frequently observed in resting animals exposed to Al at low pH (Neville, 1985; Wood *et al.*, 1988*a*,*c*; Walker *et al.*, 1988; Playle *et al.*, 1989; Witters *et al.*, 1991). This respiratory toxicity is associated with gill histopathologies such as thickening of the lamellar epithelium and hyperplasia of the filamental epithelium (sometimes to the point of lamellar fusion) which can affect the blood–water diffusion distance and surface area available for gas exchange (Youson & Neville, 1987; Evans *et al.*, 1988; Tietge *et al.*, 1988; Mueller *et al.*, 1991; Audet & Wood, 1993). However, more subtle morphometric changes can be observed at sub-lethal levels of Al which are insufficient to impair resting aerobic metabolism (Wilson *et al.*, 1994*b*).

Typically, gill damage peaks during the first week of exposure and recovers thereafter (Mueller *et al.*, 1991). However, even when the initial gill damage is relatively mild, not all the gill morphometrics recover to their original state (Wilson *et al.*, 1994*b*). Fig. 3 is a schematic diagram of the gill lamellae which represents the changes in some gill morphometric parameters relevant to branchial gas exchange in rainbow trout after 5 and 34 days exposure to sublethal Al (30 µg l^{-1}) at pH 5.2. Although blood–water diffusion distance returns to the control level after 34 days, the lamellar height remains reduced and the filamental epithelium remains thickened. By estimating the lamellae perimeters, it was calculated that apparent lamellar surface area would be reduced by 22% even after 34 days exposure to acid+Al when, for example, ionoregulatory status is fully recovered (Wilson et al., 1994a,b). These longer lasting changes in gill morphology have been associated with branchial repair (McDonald & Wood, 1992) and may therefore be an unavoidable consequence of chronic exposure to acid+Al. Although the resting aerobic metabolism may be unaffected, or at least restored with time (Wood et al., 1988b; Walker, Wood & Bergman, 1991), the persistence of these structural abnormalities at the gills suggests that the *capacity* of the respiratory gas exchange system may remain compromised during prolonged exposure to Al. This only becomes apparent during increased aerobic activity such as sustained swimming.

Previous studies have shown that acute exposure to low environmental pH (in the absence of toxic metal concentrations) reduces swimming performance (Hargis, 1976; Waiwood & Beamish, 1978; Graham & Wood, 1981; Ye & Randall, 1991; Butler, Day & Namba, 1992). However, it has now been shown that exposure of rainbow trout to low pH combined with sub-lethal Al (30 µg l^{-1}) reduces the maximum aerobic swimming speed (U_{crit}; Brett, 1964) 1.5 to 2 times more than in trout exposed to acid alone, and in both media U_{crit} does not recover with time (Fig. 4; Wilson & Wood, 1992; Wilson et al., 1994b).

Aerobic swimming performance can be affected by changes at any level of the oxygen transport system and/or muscle contraction process. Randall and Brauner (1991) demonstrated that disturbances in the ion and water balance in muscle of coho salmon (*Oncorhynchus kisutch*) reduced U_{crit}, and suggested that this occurred via decreased muscle contractility. Given the severe ion losses encountered during the acute phase of acid/Al exposure a similar effect could be involved during these initial stages. However, since ionoregulatory recovery is complete within 25 days (Wilson et al., 1994a) it is unlikely that alterations at the level of the muscle contraction are important after this time.

In terms of oxygen transport and aerobic metabolism, Brett (1958) divided environmental stresses into two categories: 'limiting' stresses that reduce the maximum aerobic capacity (MO_2max), and 'loading' stresses that increase the cost of routine maintenance (MO_2basal). There is compelling evidence that the 1.5–2 fold greater reduction in U_{crit} in trout pre-exposed to acid+Al (compared to acid alone) is due to limiting factors related to the chronic morphological changes mentioned above. After 34 days' exposure to acid+Al MO_2max was reduced by 26%, similar to the 22% reduction in apparent lamellar surface area

Gill dimensions in Al-exposed fish compared
with control fish

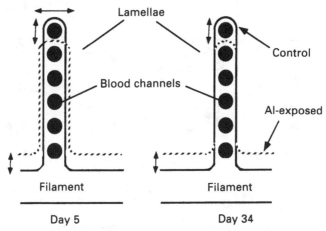

Day 5

1. 59% increase in filament thickness
2. 22% reduction in lamellar height
3. 30% increase in lamellar width
4. 33% increase in blood–water diffusion distance

Day 34

1. Filaments still thickened
2. Lamellar height still reduced
3. Lamellar width and blood–water diffusion distance recovered to control values

Fig. 3. A schematic diagram representing some of the gill morphometric changes measured in juvenile rainbow trout exposed to acidified softwater (pH 5.2) plus sub-lethal Al (30 μg l⁻¹) for 5 and 34 days. Gills of fish exposed to acid alone (pH 5.2) were not significantly different from those of the control fish. (See Wilson *et al.*, 1994*b* for further details.)

(see above). Although the blood–water diffusion distance had returned to control values by this time, increased mucus on the gills could act as an additional diffusion barrier to oxygen uptake (e.g. see Ultsch & Gros, 1979). In contrast, in the same study (Wilson *et al.*, 1994*b*) trout exposed to acid alone exhibited no significant changes in either gill morphometry or MO_2max compared to control fish (held at pH 6.5). The chronic reduction in U_{crit} caused solely by acidity (pH 5.2) is therefore not due to limitations in branchial oxygen uptake capacity. Instead, it can be seen from the regression plot of MO_2 versus swimming speed in Fig. 5 that chronic impairment of U_{crit} after 34 days exposure

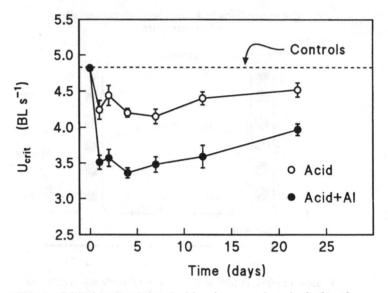

Fig. 4. Critical swimming velocities (expressed as body lengths per second) in juvenile rainbow trout at various intervals during 22 days pre-exposure to either control softwater (pH 6.5; dashed line), acid softwater (pH 5.2; open circles) or acid softwater plus sub-lethal Al (30 μg l^{-1}; solid circles). The dashed line represents the mean value for control fish swam in pH 6.5 softwater on days 0 and 24 (U_{crit} = 4.82 ± 0.11 BL s^{-1}, n = 30). Both other groups were swum in pH 5.2 water (with 30 μg l^{-1} added). Mean values ± SEM (n=8) are shown. (See Wilson & Wood, 1992 for further details.)

to pH 5.2 alone is the result of loading factors. These are seen as an elevation of the regression line above controls (i.e. the requirement for more oxygen at an equivalent subcritical swimming speed). Thus, acid-exposed trout reach the same maximum oxygen uptake capacity as control fish but at a lower speed. It has been speculated that increased ionoregulatory costs associated with acid exposure would contribute to this loading stress and lead to an increase in the metabolic rate required for homeostasis (Wilson *et al.*, 1994*b*). Such a loading stress would thereby impair U_{crit} by reducing the proportion of available oxygen that reaches the working aerobic muscle. Comparable loading factors must be present in trout exposed to acid+Al as the $\dot{M}O_2$/ swimming speed regression line is similarly elevated in these fish (Fig. 5). However, as stated before, the greater reduction in swimming capacity (U_{crit}) in trout exposed to acid+Al is the result of additional

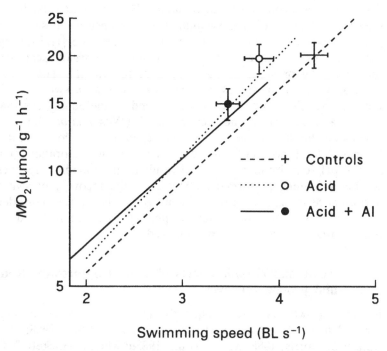

Fig. 5. Regression relationships between oxygen consumption (note log scale) and swimming speed (in BL s^{-1}) after exposing juvenile trout for 34 days to either control softwater (pH 6.5; dashed line), acid softwater (pH 5.2; dotted line) or acid softwater plus sub-lethal Al (30 μg l^{-1}; solid line). Group regression equations, where y = Log MO_2, x = speed, were: (i) $y = 0.239x + 0.256$ ($r = 0.962$) in the 6.5/0 group, (ii) $y = 0.267x + 0.236$ ($r = 0.836$) in the 5.2/0 group, (iii) $y = 0.225x + 0.358$ ($r = 0.874$) in the 5.2/Al group. The three slopes were not significantly different, but regression lines for both the acid and acid+Al groups were significantly elevated above the control group ($P<0.005$ and $P<0.05$, respectively) indicating increased maintenance costs at all sub-critical swimming speeds. The mean values for MO_2max and U_{crit} in the three groups have been plotted on the same axes.

morphological limitations to the branchial gas exchange capacity. Thus, reduced swimming performance in trout chronically exposed to acid+Al is due to a combination of both loading and limiting factors.

Perhaps the most interesting observation with respect to the loading stresses of acid exposure was the complete and rapid recovery of U_{crit} when trout were transferred back to circumneutral pH water (pH 6.5) after 36 days exposure to acid alone (Wilson *et al.*, 1994*b*). This suggests that their reduced swimming performance when swimming at pH 5.2 was *entirely* due to the presence of increased [H$^+$] during the swim test itself rather than any accumulated physiological damage caused by the previous 36 days exposure to low pH. The same was not true of trout pre-exposed to acid+Al that were swum in pH 6.5 water (containing no added Al) at the end of the 36-day experiment; MO_2max and U_{crit} showed no recovery at all (Wilson *et al.*, 1994*b*). The implication for fish in the wild is that, even after prolonged exposure to acid water (uncontaminated with Al), a rapid restoration of more neutral pH (e.g. by liming) should result in an immediate recovery of aerobic swimming performance. However, the recovery of swimming performance in acid+Al exposed trout probably requires considerably more time as normal branchial morphology must be restored before control MO_2max levels could be achieved.

Acute and chronic effects of acid/Al on growth, feeding, and protein turnover

Impaired growth has been frequently documented for laboratory fish exposed to sub-lethal acid+Al (Sadler & Lynam 1987, 1988; Reader, Dalziel & Morris, 1988; Mount *et al.*, 1988*a*; Mount, Hockett & Gern, 1988*b*; Ingersoll *et al.*, 1990*a,b*; Wilson & Wood, 1992; Wilson *et al.*, 1994*a*). What has not received so much attention is the cause of this reduced growth. Increased metabolic rate due to the demands of repair and acclimatory mechanisms are a likely drainage on the normal energy invested in growth. Indeed, increased metabolic rate during aerobic swimming in trout acclimated to Al is shown by the elevated regression line in Fig. 5. However, this analysis from laboratory swimming tests is rather artificial, and does not take into account a measure of the metabolic rate under routine, growing conditions. Unfortunately, such measurements under chronic acid+Al stress do not appear in the literature, so clarification of its role in reduced growth is currently unavailable. Even so, it appears that the role of appetite in reduced growth may far outweigh any changes in routine metabolism (Wilson *et al.*, 1994*a*; Wilson *et al.*, 1995).

It is now clear that during the initial 'shock' phase of acid+Al exposure, rainbow trout fed to satiation exhibit a dramatic reduction in appetite. This effect is immediate and can result in a 90% reduction in food intake over the first week of exposure (Fig. 6*a*). The cause of reduced appetite is unclear but may be related to hyperglycaemia-inducing hormonal changes; the raised plasma glucose in turn acting as a satiation signal to control feeding (Waiwood, Haya & Van Eeckhaute, 1992). This could also be the source of reduced feeding caused by other toxic metals (Cu, Lett, Farmer & Beamish, 1976; Drummond, Spoor & Olson, 1973; Zn, Farmer, Ashfield & Samant, 1979) and extremes of low pH in the absence of toxic metals (pH 4.2–4.7, Brown *et al.*, 1984; Tam *et al.*, 1988). However, feeding and growth is unaffected by exposure to pH 5.2 alone (Wilson *et al.*, 1994a, 1995).

Although initially pronounced, the knock-on effect this suppression of appetite has on growth rate is relatively short-lived, as normal feeding rates are resumed after about 10–12 days (Fig. 6(*a*)); Wilson *et al.*, 1994a, 1995). Specific growth rates measured in individually tagged animals were reduced by 73% during the first week of acid+Al exposure when appetite was at its lowest, but thereafter were indistinguishable from control animals at pH 6.5 (Fig. 6(*b*)). The chronic reductions in growth reported in earlier studies on fish exposed to sub-lethal acid+Al (Sadler & Lynam, 1987, 1988; Reader *et al.*, 1988; Mount *et al.*, 1988a,b; Ingersoll *et al.*, 1990a,b; Wilson and Wood, 1992) may therefore be largely if not entirely due to the initial effect that acid+Al exposure has on appetite. These temporary effects on feeding and growth are clearly attributable to the presence of Al as they are consistently absent in fish exposed to pH 5.2 alone (Wilson & Wood, 1992; Wilson *et al.*, 1994a, 1995; Fig. 6(*b*)).

As expected, changes in whole body protein synthesis (measured from the incorporation of ^{3}H-phenylalanine; Houlihan, McMillan & Laurent, 1986) follow a similar pattern to the specific growth rate (that is, reduced during the initial shock phase and unaffected thereafter; Wilson *et al.*, 1995). This is most likely a response to the decline in supply of dietary amino acids rather than a direct action of Al on whole body protein synthesis, since both whole body protein synthesis and growth recover to control levels once appetite and feeding rate has returned. This suggests that continued exposure to acid+Al does not compromise whole body protein synthesis or growth. However, a reduction in whole body translational efficiency (the amount of protein synthesized per unit of ribosomes) observed after 32 days suggests that Al may have a chronic effect on the mechanics of cellular protein synthesis (Wilson *et al.*, 1995), which appears to coincide with the slow accumulation of Al within the internal organs (Lee & Harvey, 1986; Karlsson-Norrgren *et al.*, 1986a; Booth *et al.*, 1988; Witters *et al.*, 1988). Thus, although a cellular mechanism has not been proposed, it is possible that chronic biochemical effects observed within internal tissues are caused directly by elevated tissue Al levels.

So, whole body protein synthesis (excluding gill and liver tissues) is unaffected during prolonged exposure to acid+Al. However, both acute and chronic changes in gill protein turnover have been observed (Wilson

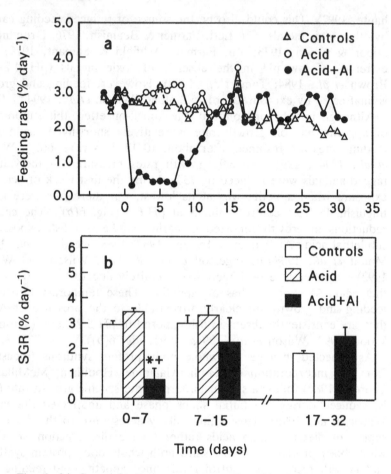

Fig. 6. (a) Group feeding rates expressed as % of the average wet body weight eaten per day in juvenile rainbow trout exposed to either control softwater (pH 6.5; open triangles and dashed line), acid softwater (pH 5.2; open circles) or acid softwater plus sub-lethal Al (30 μg l^{-1}; solid circles) for 32 days (approximately 100 fish per tank at the start, mean weight approximately 20 g). (b) Weight specific growth rates (%/day) in the same three groups as above (open bars = controls; hatched bars = acid-exposed group; solid bars = group exposed to acid+Al) during three stages of a 32-day exposure period. Asterisks indicate significantly different from the control group, and crosses indicate significantly different from the acid-exposed group ($P<0.05$). Values are means ± SEM ($n=10$).

et al., 1995). The rate of protein synthesis in gill tissue varies according to the feeding state and whole body protein synthesis in fish (Houlihan *et al.*, 1988; McMillan & Houlihan, 1988; Houlihan, 1991). Since acid+Al had such a pronounced effect on feeding, it is appropriate to express gill protein synthesis as the ratio of gill : whole body protein synthesis rate in this case (Fig. 7). This shows that, after 7 days, exposure to acid+Al 'relative' gill protein synthesis was almost double that of the control group. This coincides with the peak of gill damage during exposure to acid+Al (e.g. Mueller *et al.*, 1991; Wilson *et al.*, 1994*b*) and is presumably a correlate of increased cellular turnover. Gill protein degradation was even more elevated than protein synthesis, indicating that breakdown of damaged gill tissue predominates over the synthesis of new cellular material during this early 'damage' phase of acid+Al exposure. The intensification of branchial protein turnover whilst protein metabolism in the remaining whole body is depressed enhances the view that acute Al toxicity is directed principally at the gills.

Since gill protein synthesis (but not degradation) was still increased after 32 days of exposure, there is evidence that continued exposure to acid+Al also produces *chronic* elevations in protein synthesis within the gills, which infer some increased metabolic cost. This could be associated with branchial processes for maintaining acclimation (e.g. increased mucus production), increased cost of combating ionoregulatory problems (e.g. increased production of transport proteins), or simply elevated cell turnover due to the continuation of Al-induced gill damage. Whatever the reason for this increased protein turnover, the additional metabolic cost will probably be small relative to the fishes overall energy budget (the increase in gill protein synthesis only represents about 2.3% of the total protein synthesized by the whole body per day). One perplexing observation from these metabolism studies is that rainbow trout show a substantially *higher* gross food conversion efficiency during both the damage and recovery phases of chronic exposure to acid+Al compared to both control and acid-exposed trout (Wilson *et al.*, 1994*a*, 1995). This conflicts with the assumption that chronic exposure to metals should act as a loading factor to routine metabolism. However, although measurements of routine metabolic rate have not been made, reduced spontaneous activity has been quantitatively documented in cutthroat trout (Woodward *et al.*, 1989) and brook trout (Cleveland *et al.*, 1986, 1989) exposed to sub-lethal acid+Al. If reduced activity is accompanied by a lower overall routine metabolic rate, it could result in a greater percentage of the dietary protein being retained for growth. This

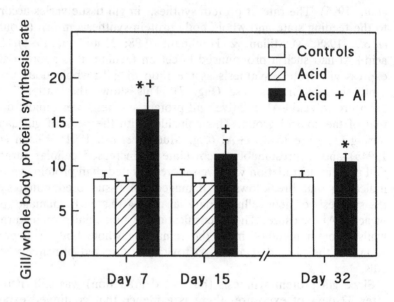

Fig. 7. Fractional rates of protein synthesis in the gill tissue of juvenile rainbow trout exposed for 32 days to either control softwater (pH 6.5; open bars), acid softwater (pH 5.2; hatched bars) or acid softwater plus sub-lethal Al (30 μg l^{-1}; solid bars). Rates are expressed as the ratio of gill to whole body protein synthesis (both in % day^{-1}) to take into account the variability in feeding between groups (see text). Asterisks indicate significantly different from the control group, and crosses indicate significantly different from the acid-exposed group for the same exposure day ($P<0.05$). Values are means ± SEM ($n=10$).

might normally be considered an advantage. However, enhanced food conversion efficiency may occur in the laboratory when the food supply is not limited, but the ultimate effect of decreased appetite and routine activity in the wild would surely be a reduction in an animal's overall fitness to feed, avoid predation, and reproduce (Little & Finger, 1990).

Conclusions

The physiological and metabolic effects of acid+Al exposure can be divided into two phases: an acute 'damage' phase lasting 4–7 days, and a chronic 'repair/recovery' phase following on from this. Acclimation (increased resistance) to Al develops prior to physiological recovery. Ionoregulation can recover fully given time (at least 25 days) but gill morphology does not completely recover. Aerobic scope and swimming performance are therefore chronically compromised although both limiting and loading factors contribute to this. Growth is temporarily reduced, and is almost entirely due to a dramatic loss of appetite during the first week of exposure. Following this acute phase feeding, growth and whole body protein synthesis rates return to normal. However, rates of protein synthesis in the gills are both acutely and chronically elevated implying a persistent cost of branchial acclimation to Al, although this cost is small compared to the animals total energy budget. Finally, throughout acid+Al exposure, fish are much more efficient at converting food into whole body growth, a paradox which may be an artefact of laboratory conditions (feeding to satiation). Future studies might seek to investigate this paradox and relate this observation to the overall fitness of fish exposed to acid+Al in the wild.

Acknowledgements

The data reported here was from several studies performed at McMaster University and the University of Wyoming, and was supported by a NSERC strategic grant in Environmental Quality to C.M. Wood (Department of Biology, McMaster University, Canada). I would like to especially thank Dr C.M. Wood for his continuing contributions, Dr H.L. Bergman for the Wyoming collaboration, and Professor D.F. Houlihan for assistance with the protein synthesis measurements.

References

Audet, C. & Wood, C.M. (1988). Do rainbow trout acclimate to low pH? *Canadian Journal of Fisheries and Aquatic Sciences*, **45**, 1399–405.

Audet, C. & Wood, C.M. (1993). Branchial morphological and endocrine responses of rainbow trout (*Oncorhynchus mykiss*) during a long term sublethal acid exposure in which acclimation did not occur. *Canadian Journal of Fisheries and Aquatic Sciences*, **50**, 198–209.

Audet, C., Munger, R.S. & Wood, C.M. (1988). Long-term sublethal acid exposure in rainbow trout (*Salmo gairdneri*) in soft water: effects on ion exchanges and blood chemistry. *Canadian Journal of Fisheries and Aquatic Sciences*, **45**, 1387–98.

Baker, J.P. & Schofield, C.L. (1982). Aluminum toxicity to fish in acidic waters. *Water, Air and Soil Pollution*, **18**, 289–309.

Battram, J.C. (1988). The effects of aluminium and low pH on chloride fluxes in the brown trout, *Salmo trutta* L. *Journal of Fish Biology*, **32**, 937–47.

Beamish, F.W.H., Howlett, J.C. & Medland, T.E. (1989). Impact of diet on metabolism and swimming performance in juvenile lake trout, *Salvelinus namaycush*. *Canadian Journal of Fisheries and Aquatic Sciences*, **46**, 384–8.

Booth, C.E., McDonald, D.G., Simons, B.P. & Wood, C.M. (1988). Effects of aluminum and low pH on net ion fluxes and ion balance in the brook trout (*Salvelinus fontinalis*). *Canadian Journal of Fisheries and Aquatic Sciences*, **45**, 1563–74.

Brett, J.R. (1958). Implications and assessments of environmental stress. In *The Investigation of Fish-power Problems*. Larkin, P.A. (ed.) pp. 69–93. Institute of Fisheries, University of BC.

Brett, J.R. (1964). The respiratory metabolism and swimming performance of young sockeye salmon. *Journal of Fisheries Research Board Canada*, **21**, 1183–226.

Brown, S.B., Eales, J.G., Evans, R.E. & Hara, T.J. (1984). Interrenal, thyroidal, and carbohydrate responses of rainbow trout (*Salmo gairdneri*) to environmental acidification. *Canadian Journal of Fisheries and Aquatic Sciences*, **41**, 36–45.

Butler, P.J., Day, N. & Namba, K. (1992). Interaction of seasonal temperature and low pH on resting oxygen uptake and swimming performance of adult brown trout, *Salmo trutta. Journal of Experimental Biology*, **165**, 195–212.

Calow, P. (1991). Physiological costs of combating chemical toxicants: ecological implications. *Comparative Biochemistry and Physiology*, **100C**, 3–6.

Cleveland, L., Little, E.E., Hamilton, S.J., Buckler, D.R. & Hunn, J.B. (1986). Interactive toxicity of aluminum and acidity to early life stages of brook trout. *Transactions of the American Fisheries Society*, **115**, 610–20.

Cleveland, L., Little, E.E., Wiedmeyer, R.H. & Buckler, D.R. (1989). Chronic no-observed effect concentrations of aluminum for brook trout exposed in low-calcium, dilute acidic water. In *Environmental Chemistry and Toxicology of Aluminum*. Lewis, T.E. (ed.) pp. 229–45. Lewis Publishers, Inc., Chelsea, MI.

Cronan, C.S. & Schofield, C.L. (1979). Aluminum leaching response to acid precipitation: effects of high elevation watersheds in the North-east. *Science* (Washington, DC), **204**, 306.

Cuthbert, A.W. & Maetz, J. (1972). The effects of calcium and magnesium on sodium fluxes through the gills of *Carassius auratus* L. *Journal of Physiology*, **221**, 633–43.

Dickson, W. (1980). Properties of acidified waters. pp. 75–83. In Drablos, D. and Tollan, A. (ed.) *Proceedings of the International Conference on Ecological Impact of Acid Precipitation*, SNCF Project, Norway.

Drummond, R.A., Spoor, W.A. & Olson, G.F. (1973). Some short-term indicators of sublethal effects of copper on brook trout (*Salvelinus fontinalis*). *Journal of the Fisheries Research Board of Canada*, **30**, 698–701.

Duncan, D.A. & Klaverkamp, J.F. (1983). Tolerance and resistance to cadmium in white suckers (*Catostomus commersoni*) previously exposed to cadmium, mercury, zinc or selenium. *Canadian Journal of Fisheries and Aquatic Sciences*, **40**, 128–38.

Evans, R.E., Brown, S.B. & Hara, T.J. (1988). The effects of aluminum and acid on the gill morphology in rainbow trout, *Salmo gairdneri*. *Environmental Biology of Fishes*, **22**, 299–311.

Exley, C., Chappell, J.S. & Birchall, J.D. (1991). A mechanism for acute aluminium toxicity in fish. *Journal of Theoretical Biology*, **151**, 417–28.

Farmer, G.J., Ashfield, D. & Samant, H.S. (1979). Effects of zinc on juvenile Atlantic salmon *Salmo salar*: acute toxicity, food intake, growth and bioaccumulation. *Environmental Pollution*, **19**, 109–17.

Goossenaerts, C., Van Grieken, R., Jacob, W., Witters, H. & Vander-borght, O. (1988). A microanalytical study of the gills of aluminium exposed rainbow trout (*Salmo gairdneri*). *International Journal of Environmental Analytical Chemistry*, **34**, 227–37.

Graham, M.S. & Wood, C.M. (1981). Toxicity of environmental acid to the rainbow trout: interactions of water hardness, acid type, and exercise. *Canadian Journal of Zoology*, **59**, 1518–26.

Hargis, J.R. (1976). Ventilation and metabolic rate of young rainbow trout (*Salmo gairdneri*) exposed to sublethal environmental pH. *Journal of Experimental Zoology*, **196**, 39–44.

Houlihan, D.F., McMillan, D.N. & Laurent, P. (1986). Growth rates, protein synthesis and protein degradation rates in rainbow trout: effects of body size. *Physiology and Zoology*, **59**(4), 482–93.

Houlihan, D.F., Hall, S.J., Gray, C. & Noble, B.S. (1988). Growth rates and protein turnover in Atlantic cod, *Gadus morhua*. *Canadian Journal of Fisheries and Aquatic Sciences*, **43**, 951–64.

Houlihan, D.F. (1991). Protein turnover in ectotherms and its relationships to energetics. pp. 1–43. In Gilles, R. (ed.) *Advances in Comparative and Environmental Physiology*. Springer-Verlag, Berlin Heidelberg.

Ingersoll, C.G., Mount, D.R., Gulley, D.D., Fernandez, J.D., LaPoint, T.W. & Bergman, H.L. (1990a). Effects of pH, aluminum, and calcium on survival and growth of eggs and fry of brook trout (*Salvelinus fontinalis*). *Canadian Journal of Fisheries and Aquatic Sciences*, **47**, 1580–92.

Ingersoll, C.G., Gulley, D.D., Mount, D.R., Mueller, M.E., Fernandez, J.D., Hockett, J.R. & Bergman, H.L.(1990b). Aluminum and acid toxicity to two strains of brook trout (*Salvelinus fontinalis*). *Canadian Journal of Fisheries and Aquatic Sciences*, **47**, 1641–8.

Jagoe, C.H., Matey, V.E., Haines, T.A. & Komov, V.T. (1993). Effect of beryllium on fish in acid water is analogous to aluminium toxicity. *Aquatic Toxicology*, **24**, 241–56.

Karlsson-Norrgren, L., Dickson, W., Ljungberg, O. & Runn, P. (1986*a*). Acid water and aluminum exposure: gill lesions and aluminum accumulation in farmed brown trout *Salmo trutta* L. *Journal of Fish Diseases*, **9**, 1–8.

Karlsson-Norrgren, L., Bjorklund, I., Ljungberg, O. and Runn, P. (1986*b*). Acid water and aluminum exposure: Experimentally induced gill lesions in brown trout, *Salmo trutta* L. *Journal of Fish Diseases*, **9**, 11–25.

Kelso, J.R.M., Minns, C.K., Gray, J.E. & Jones, M.L. (1986). Acidification of surface waters in eastern Canada and its relationship to aquatic biota. *Canadian Special Publication Fisheries and Aquatic Sciences*, **87**, 42 pp.

Kito, H., Kazawa, T., Ose, Y., Sato, T. & Ishikawa, T. (1982). Protection by metallothionein against cadmium toxicity. *Comparative Biochemistry and Physiology*, **73**, 135–9.

Klaverkamp, J.F. & Duncan, D.A. (1987). Acclimation to cadmium toxicity by white suckers: cadmium binding capacity and metal distribution in gill and liver cytosol. *Environmental Toxicological Chemistry*, **6**, 275–89.

Laurén, D.J. & McDonald, D.G. (1985). Effects of copper on branchial ionoregulation in the rainbow trout, *Salmo gairdneri* Richardson. Modulation by water hardness and pH. *Journal of Comparative Physiology*, **155B**, 635–44.

Laurén, D.J. & McDonald, D.G. (1986). Influence of water hardness, pH, and alkalinity on the mechanisms of copper toxicity in juvenile rainbow trout, *Salmo gairdneri*. *Canadian Journal of Fisheries and Aquatic Sciences*, **43**, 1488–96.

Lee, C. & Harvey, H.H. (1986). Localization of aluminum in tissues of fish. *Water, Air and Soil Pollution*, **30**, 649–55.

Lett, P.F., Farmer,G.J. & Beamish, F.W.H. (1976). Effect of copper on some aspects of the bioenergetics of rainbow trout (*Salmo gairdneri*). *Journal of Fisheries Research Board Canada*, **33**, 1335–42.

Little, E.E. & Finger, S.E. (1990). Swimming behaviour as an indicator of sublethal toxicity in fish. *Environmental and Toxicological Chemistry*, **9**, 13–19.

McDonald, D.G. & Milligan, C.L. (1988). Sodium transport in the brook trout, *Salvelinus fontinalis*: effects of prolonged low pH exposure in the presence and absence of aluminum. *Canadian Journal of Fisheries and Aquatic Sciences*, **45**, 1606–16.

McDonald, D.G. & Rogano, M.S. (1986). Branchial mechanisms of ion and acid-base regulation in the freshwater rainbow trout, *Salmo gairdneri*. *Canadian Journal of Zoology*, **66**, 2699–708.

McDonald, D.G., Wood, C.M., Rhem, R.G., Mueller, M.E., Mount, D.R. & Bergman, H.L. (1991). Nature and time course of acclimation to aluminum in juvenile brook trout (*Salvelinus fontinalis*). 1. Physiology. *Canadian Journal of Fisheries and Aquatic Sciences*, **48**, 2006–15.

McDonald, D.G. & Wood, C.M. (1992). Branchial mechanisms of acclimation to metals in freshwater fish, pp. 295–319. In *Fish Ecophysiology*. Rankin, C. & Jensen, F.B. (ed.) Chapman and Hall, England.

McMillan, D.N. & Houlihan, D.F. (1988). The effect of refeeding on tissue protein synthesis in rainbow trout. *Physiology and Zoology*, **61**(5), 429–41.

Mount, D.R., Ingersoll, C.G., Gulley, D.D., Fernandez, J.D., LaPoint, T.W. & Bergman, H.L. (1988*a*). Effect of long-term exposure to acid, aluminum, and low calcium on adult brook trout (*Salvelinus fontinalis*). 1. survival, growth, fecundity, and progeny survival. *Canadian Journal of Fisheries and Aquatic Sciences*, **45**, 1623–32.

Mount, D.R., Hockett, J.R. & Gern, W.A. (1988*b*). Effect of long-term exposure to acid, aluminum, and low calcium on adult brook trout (*Salvelinus fontinalis*). 2. vitellogenesis and osmoregulation. *Canadian Journal of Fisheries and Aquatic Sciences*, **45**, 1623–32.

Mueller, M.E., Sanchez, D.A., Bergman, H.L., McDonald, D.G., Rhem, R.G. & Wood, C.M. (1991). Nature and time course of acclimation to aluminum in juvenile brook trout (*Salvelinus fontinalis*). 2. Histology. *Canadian Journal of Fisheries and Aquatic Sciences*, **48**, 2016–27.

Neville, C. (1985). Physiological response of juvenile rainbow trout, *Salmo gairdneri*, to acid and aluminum – prediction of field responses from laboratory data. *Canadian Journal of Fisheries and Aquatic Sciences*, **42**, 2009–19.

Norrgren, L., Wicklund Glynn, A. & Malmborg, O. (1991). Accumulation and effects of aluminium in the minnow (*Phoxinus phoxinus* L.) at different pH levels. *Journal of Fish Biology*, **39**, 833–47.

Orr, P.L., Bradley, R.W., Sprague, J.B. & Hutchinson, N.J. (1986). Acclimation-induced change in toxicity of aluminum to rainbow trout (*Salmo gairdneri*). *Canadian Journal of Fisheries and Aquatic Sciences*, **43**, 243–6.

Oughton, D.H., Salbu, B. & Bjornstad, E. (1992). Use of an aluminium-26 tracer to study the deposition of aluminium species on fish gills following mixing of limed and acidic waters. *Analyst*, **117**, 619–21.

Playle, R.C., Goss, G.G. & Wood, C.M. (1989). Physiological disturbances in rainbow trout (*Salmo gairdneri*) during acid and aluminum exposures in soft water of two calcium concentrations. *Canadian Journal of Zoology*, **67**, 314–24.

Playle, R.C. & Wood, C.M. (1989). Water pH and aluminum chemistry in the gill micro-environment of rainbow trout during acid and aluminum exposure. *Journal of Comparative Physiology*, **159B**; 539–50.

Playle, R.C. & Wood, C.M. (1991). Mechanisms of aluminium extraction and accumulation at the gills of rainbow trout, *Oncorhynchus mykiss* (Walbaum), in acidic soft water. *Journal of Fish Biology*, **38**, 791–805.

Potts, W.T.W. & Fleming, W.R. (1971). The effects of environmental calcium and ovine prolactin on sodium balance in *Fundulus kansae*. *Journal of Experimental Biology*, **55**, 63–76.

Randall, D. & Brauner, C. (1991). Effects of environmental factors on exercise in fish. *Journal of Experimental Biology*, **160**, 113–26.

Reader, J.P., Dalziel, T.R.K. & Morris, R. (1988). Growth, mineral uptake and skeletal calcium deposition in brown trout, *Salmo trutta* L., yolk-sac fry exposed to aluminium and manganese in soft acid water. *Journal of Fish Biology*, **32**, 607–24.

Reid, S.D., Rhem, R.G. & McDonald, D.G. (1991). Acclimation to sublethal aluminum: modifications of metal-gill surface interactions of juvenile rainbow trout (*Oncorhynchus mykiss*). *Canadian Journal of Fisheries and Aquatic Sciences*, **48**, 1995–2004.

Sadler, K. & Lynam, S. (1987). Some effects on the growth of brown trout from exposure to aluminium at different pH levels. *Journal of Fish Biology*, **31**, 209–19.

Sadler, K. & Lynam, S. (1988). The influence of calcium on aluminium-induced changes in the growth rate and mortality of brown trout, *Salmo trutta* L. *Journal of Fish Biology*, **33**, 171–9.

Schofield, C.L. & Trojnar, J.R. (1980). Aluminum toxicity to brook trout (*Salvelinus fontinalis*) in acidified waters. In *Polluted Rain*. Toribara, T.Y., Miller, M.W. & Morrow, P.E. (ed.) pp. 341–63. Plenum Press, New York.

Staurnes, M., Sigholt, T. & Reite, O.B. (1984). Reduced carbonic anhydrase and Na^+–K^+-ATPase activity in gills of salmonids exposed to aluminium-containing acid water. *Experientia*, **40**, 226–7.

Tam, W.H., Fryer, J.N., Ali, I., Dallaire, M.R. & Valentine, B. (1988). Growth inhibition, gluconeogenesis, and morphometric studies of the pituitary and interrenal cells of acid-stressed brook trout (*Salvelinus fontinalis*). *Canadian Journal of Fisheries and Aquatic Sciences*, **45**, 1197–211.

Thomas, D.G., Brown, M.W., Shurben, D., del G. Solbe, J.F., Cryer, A. & Kay, J. (1985). A comparison of the sequestration of cadmium and zinc in the tissue of rainbow trout (*Salmo gairdneri*) following exposure to the metals singly or in combination. *Comparative Biochemistry and Physiology*, **82C**, 55–62.

Tietge, J., Johnson, R. & Bergman, H.L. (1988). Morphometric changes in gill secondary lamellae of brook trout (*Salvelinus fontinalis*) after long-term exposure to acid and aluminum. *Canadian Journal of Fisheries and Aquatic Sciences*, **45**, 1643–8.

Ultsch, G.R. & Gros, G. (1979). Mucus as a diffusion barrier to oxygen: possible role in O_2 uptake at low pH in carp (*Cyprinus carpio*) gills. *Comparative Biochemistry and Physiology*, **62A**, 685–9.

Vesely, J., Benes, P. & Sevcik, K. (1989). Occurrence and speciation of beryllium in acidified freshwaters. *Water Research*, **23**, 711–17.

Waiwood, K.G. & Beamish, F.W.H. (1978). Effects of copper pH and hardness on the critical swimming performance of rainbow trout (*Salmo gairdneri* Richardson). *Water Research*, **12**, 611–19.

Waiwood, B.A., Haya, K. & Van Eeckhaute, L. (1992). Energy metabolism of hatchery-reared juvenile salmon (*Salmo salar*) exposed to low pH. *Comparative Biochemistry and Physiology*, **101C**, 49–56.

Walker, R.L., Wood, C.M. & Bergman, H.L. (1988). Effects of low pH and aluminum on ventilation in the brook trout, *Salvelinus fontinalis*. *Canadian Journal of Fisheries and Aquatic Sciences*, **45**, 1614–22.

Walker, R.L., Wood, C.M. & Bergman, H.L. (1991). Effects of longterm pre-exposure to sublethal concentrations of acid and aluminum on the ventilatory response to aluminum challenge in brook trout (*Salvelinus fontinalis*). *Canadian Journal of Fisheries and Aquatic Sciences*, **48**, 1989–95.

Wilson, R.W. & Wood, C.M. (1992). Swimming performance, whole body ions, and gill Al accumulation during acclimation to sublethal aluminium in juvenile rainbow trout (*Oncorhynchus mykiss*). *Fish Physiology and Biochemistry*, **10**, 149–59.

Wilson, R.W. & Taylor, E.W. (1993). The physiological responses of freshwater rainbow trout, *Oncorhynchus mykiss*, during acutely lethal copper exposure. *Journal of Comparative Physiology*, **163B**, 38–47.

Wilson, R.W., Bergman, H.L. & Wood, C.M. (1994a). Metabolic costs and physiological consequences of acclimation to aluminum in juvenile rainbow trout (*Oncorhynchus mykiss*). 1: Acclimation specificity, resting physiology, feeding, and growth. *Canadian Journal of Fisheries and Aquatic Sciences*, **51**, 527–35.

Wilson, R.W., Bergman, H.L. & Wood, C.M. (1994b). Metabolic costs and physiological consequences of acclimation to aluminum in juvenile rainbow trout (*Oncorhynchus mykiss*). 2: Gill morphology, swimming performance, and aerobic scope. *Canadian Journal of Fisheries and Aquatic Sciences*, **51**, 536–44.

Wilson, R.W., Wood, C.M. & Houlihan, D.F. (1995). Growth and protein turnover during acclimation to acid/aluminum in juvenile rainbow trout (*Oncorhynchus mykiss*). *Canadian Journal of Fisheries and Aquatic Sciences* (in press).

166   R.W. WILSON

Witters, H.E., Vangenechten, J.H.D., Van Puymbroeck, S. & Vand-
erbroght, O.L.J. (1988). Internal or external toxicity of aluminium
in fish exposed to acid water. *Commission of the European Com-
munities Report Europe*, 965–70.
Witters, H.E., Van Puymbroeck, S. & Vanderbrought, O.L.J. (1991).
Adrenergic response to physiological disturbances in rainbow trout,
Oncorhynchus mykiss, exposed to aluminum at acid pH. *Canadian
Journal of Fisheries and Aquatic Sciences*, **48**, 414–20.
Witters, H.E., Van Puymbroeck, S. & Vanderbroght, O.L.J. (1992).
Branchial and renal ion fluxes and transepithelial electrical potential
differences in rainbow trout, *Oncorhynchus mykiss*: effects of alu-
minium at low pH. *Environmental Biology of Fishes*, **34**, 197–206.
Wood, C.M. (1989). The physiological problems of fish in acid waters.
In *Acid Toxicity and Aquatic Animals*. Morris, R., Taylor, E.W.,
Brown, D.J.A. & Brown, J.A. (eds.) SEB Seminar Series, **34**, 125–
52. Cambridge University Press, Cambridge, UK.
Wood, C.M. (1992). Flux measurements of indices of H^+ and metal
effects on freshwater fish. *Aquatic Toxicology*, **22**, 239–64.
Wood, C.M. & McDonald, D.G. (1987). The physiology of acid/
aluminium stress in trout. *Annales de la Société Royale Zoologique
de Belgique*, **117**, 399–410.
Wood, C.M., McDonald, D.G., Booth, C.E., Simons, B.P., Ingersoll,
C.G. & Bergman, H.L. (1988a). Physiological evidence of acclim-
ation to acid/aluminum stress in adult brook trout, (*Salvelinus
fontinalis*). 1. Blood composition and net sodium fluxes. *Canadian
Journal of Fisheries and Aquatic Sciences*, **45**, 1587–96.
Wood, C.M., Simons, B.P., Mount, D.R. & Bergman, H.L. (1988b).
Physiological evidence of acclimation to acid/aluminum stress in
adult brook trout, (*Salvelinus fontinalis*). 2. Blood parameters by
chronic cannulation. *Canadian Journal of Fisheries and Aquatic
Sciences*, **45**, 1597–605.
Wood, C.M., Playle, R.C., Simons, B.P., Goss, G.G. & McDonald,
D.G. (1988c). Blood gases, acid-base status, ions, and hematology
in adult brook trout, (*Salvelinus fontinalis*) under acid/aluminum
exposure. *Canadian Journal of Fisheries and Aquatic Sciences*, **45**,
1575–86.
Woodward, D.F., Farag, A.M., Mueller, M.E., Little, E.E. & Vert-
ucci, F.A. (1989). Sensitivity of endemic snake river cutthroat trout
to acidity and elevated aluminum. *Transactions of the American
Fisheries Society*, **118**, 630–43.
Wright, R.F. & Gjessing, E.T. (1976). Acid precipitation: changes in
the chemical composition of lakes. *Ambio*, **5** 219–23.
Wright, R.F. & Snekvik, E. (1978). Chemistry and fish populations
in 700 lakes in southernmost Norway. *Verhandlungen der Inter-
nationalen Vereinigung für Limnologie*, **20**, 765–75.

Ye, X. & Randall, D.J. (1991). The effect of water pH on swimming performance in rainbow trout (*Salmo gairdneri*, Richardson). *Fish Physiology and Biochemistry*, **9**, 15–21.

Youson, J. & Neville, C. (1987). Deposition of aluminum in the gill epithelium of rainbow trout (*Salmo gairdneri* Richardson) subjected to sublethal concentrations of the metal. *Canadian Journal of Zoology*, **65**, 647–56.

F.B. JENSEN

Physiological effects of nitrite in teleosts and crustaceans

Introduction

Nitrite is known to be toxic to both vertebrates and invertebrates. In humans and other terrestrial animals, prime concern has been on pathophysiological consequences of dietary nitrite intake. Aquatic animals take up nitrite directly from the ambient water, which typically causes higher internal nitrite concentrations than in terrestrial animals. Since nitrite can build up transiently in many aquatic habitats either naturally or as result of anthropogenic activities, the sub-lethal and lethal effects of nitrite attracts both ecophysiological and ecotoxicological interest. One prime toxic action of nitrite is that it oxidizes haemoglobin to methaemoglobin. Fish can, however, accommodate relatively high levels of unfunctional haemoglobin without mortality. During nitrite exposure, mortality may be associated with both high and only moderately elevated methaemoglobin levels (e.g. Margiocco *et al.*, 1983; Eddy & Williams, 1987). This inconsistency between the degree of methaemoglobinaemia and mortality suggests that additional effects must be involved in the toxicity of nitrite. A similar argument applies to aquatic crustaceans, whose respiratory pigment, haemocyanin, is less affected by nitrite than haemoglobin. An extensive literature is available on the toxicity of nitrite and factors affecting it (for review see Lewis & Morris, 1986). During the last decade, insight into physiological effects has also improved. Such knowledge is essential in order to understand the mechanisms of nitrite toxicity. The present review focuses on physiological effects of nitrite in fishes and crustaceans.

Nitrite in aquatic habitats

Nitrite is an intermediate product in bacterial nitrification and denitrification processes. An imbalance in either of these processes can lead to elevated ambient nitrite concentrations. The concentration of nitrite in unpolluted waters is typically very low (in the μM range) but elevated concentrations are regularly found in hypoxic lakes and ponds and in the oxygen minimum zones of the oceans (Anderson *et al.*, 1982; Eddy & Williams, 1987). Occasionally, very high levels (several mM) develop in lake shores with decaying organic matter (Eddy & Williams, 1987). Pollution with nitrogenous wastes (sewage effluents, fertilizers, etc.) can elevate [NO_2^-] in various aquatic habitats. In fish farms and recirculated aquaculture systems, high nitrite concentrations can occur by means of incomplete oxidation of ammonia excreted by the animals. Bacterial oxidation of ammonia to nitrate involves two groups of bacteria: *Nitrosomonas*, which oxidizes ammonia to nitrite, and *Nitrobacter*, which oxidizes nitrite to nitrate. Imbalance in this nitrification process often develops in newly started biological filters, with water nitrite concentrations of up to 1 mM as a consequence.

Nitrite uptake and accumulation

When exposed to nitrite, most freshwater fishes accumulate nitrite in their blood plasma to concentrations far above the environmental concentration (Bath & Eddy, 1980; Eddy, Kunzlik & Bath, 1983; Margiocco *et al.*, 1983; Palacheck & Tomasso, 1984; Jensen, Andersen & Heisler, 1987) (Fig. 1(*a*)). The fact that nitrite is taken up against a concentration gradient suggests that the uptake mechanism is active. Nitrite appears to enter the fish via the same route as chloride by being a competitive inhibitor of the active chloride uptake mechanism in the gills (Williams & Eddy, 1986). This uptake mechanism rationalizes why an elevation of environmental chloride levels protects against nitrite accumulation and toxicity (Perrone & Meade, 1977; Bath & Eddy, 1980) and why species with relative high Cl^- influx rates (such as rainbow trout, perch and pike) are more sensitive to nitrite than species with low uptake rates (carp, tench and eel) (Williams & Eddy, 1986).

In addition to interspecific difference in nitrite tolerance, some species show intraspecific differences. Thus, nitrite-exposed rainbow trout fall into two distinct groups. Some specimens accumulate nitrite, and die, more rapidly than others, even if the animals originate from the same stock of fish (Margiocco *et al.*, 1983; J. Stormer & F.B. Jensen, unpublished observations) (Fig. 1(*b*)). This phenomenon may relate to intraspecific differences in uptake rates and/or sensitivity to nitrite.

In freshwater crayfish, nitrite is also taken up via the active branchial chloride uptake mechanism, resulting in very high extracellular nitrite concentration during nitrite exposure (Gutzmer & Tomasso, 1985; Jensen, 1990*b*; Harris & Coley, 1991). In contrast to fish gills, crustacean gills are covered by a thin layer of the chitinous exoskeleton.

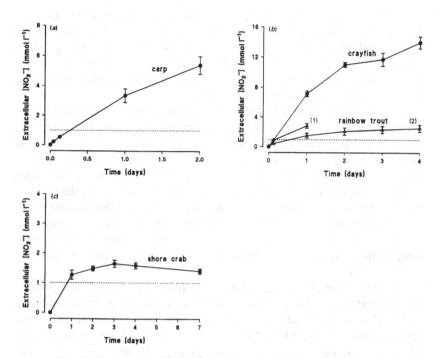

Fig. 1. Extracellular $[NO_2^-]$ as a function of time during exposure to 1 mM ambient nitrite at 15–16 °C. (a) carp (*Cyprinus carpio*), in freshwater with $[Cl^-]$ = 0.3 mM. (b) rainbow trout (*Oncorhynchus mykiss*) and crayfish (*Astacus astacus*), in freshwater with $[Cl^-]$ = 1.3 mM. (c) shore crab (*Carcinus maenas*), in brackish water with $[Cl^-]$ = 235 mM. Rainbow trout were divided into two groups: (1) those dying between day 1 and 2, and (2) those surviving to day 4. The dotted lines depict the environmental nitrite level. Note the different axis scales. Data (means ± SE) from: (a), carp (N=7), Jensen *et al.* (1987); (b) rainbow trout (1: N=8; 2: N=5), (J. Stormer & F.B. Jensen, unpublished), crayfish (N=8), Jeberg & Jensen (1994); (c) shore crab (N=8) (M. Jeberg & F.B. Jensen, unpublished observations.)

Nitrite and chloride must permeate this cuticle before reaching the uptake sites. The crayfish gill cuticle is, however, highly permeable to both Cl^- and NO_2^- (Aubert & Lignon, 1986), whereby the cuticle does not constitute a barrier to the passage of any of these ions.

Nitrite is accumulated at a higher rate and to higher extracellular concentrations in crayfish than in fishes (Jensen, 1990*b*; Harris & Coley, 1991; Jeberg & Jensen, 1994) (Fig. 1(*b*)). This may reflect that crayfish have larger active Cl^- uptake rates in order to maintain chloride homeostasis. The high nitrite uptake may also originate in a higher Cl^-/Cl^- self-exchange in crayfish than in fish (Wheatly, 1989), which changes to NO_2^-(in)/Cl^-(out) exchange when nitrite is present in the water. Alternatively, the affinity of nitrite for the uptake mechanism may be higher in crayfish than in fish.

Several environmental factors affect nitrite accumulation and toxicity. Among these, the water chloride concentration undoubtedly is the most important. Elevated ambient chloride concentrations inhibit the uptake of nitrite and protect against its toxic effects in both fishes and crayfish (Perrone & Meade, 1977; Bath & Eddy, 1980; Gutzmer & Tomasso, 1985). Other environmental ions, such as bicarbonate and calcium, also provide some protection (Lewis & Morris, 1986). Temperature may be a further important factor, but this variable has received only little attention. In the crayfish *Astacus astacus*, an increase in acclimation temperature significantly increases nitrite accumulation (Jeberg & Jensen, 1994).

The high concentration of chloride and other ions in seawater predicts that seawater animals are less sensitive to nitrite than freshwater animals. This is confirmed by relatively high LC_{50} values in seawater species (Eddy *et al.*, 1983; Ary & Poirrier, 1989). Significant amounts of nitrite do, however, enter seawater fish (Eddy *et al.*, 1983) and marine crustaceans such as shore crabs (Jensen, 1990*b*; Fig. 1(*c*)) and shrimps (Chen & Chen, 1992). Shore crabs in seawater with a salinity of 15‰ are hyperosmoregulating. The route of nitrite entry may, however, be mainly via diffusion rather than via active transport. Haemolymph nitrite concentrations come to equilibrium at values only slightly above the seawater nitrite concentration (Fig. 1(*c*)).

Dietary intake of nitrite is a possible route of nitrite uptake, even though it evidently will be secondary to the branchial uptake in freshwater organisms. Predators are, however, in a potential danger of ingesting large amounts of nitrite when feeding on nitrite-loaded fish or crayfish. The significance of ingested nitrite remains to be studied in aquatic animals. Uptake of nitrite across the intestine may similarly be a relevant topic of investigation in marine fishes, given their high drinking rates necessitated by osmoregulation.

Nitrite entry into fish red cells

Nitrite-induced oxidation of red cell haemoglobin to methaemoglobin implicitly demands that nitrite accumulated in plasma crosses the red cell membrane. It may be initially assumed that the transport could be mediated by the membrane bound anion exchanger (Band 3 protein). Experimental evidence does, however, not support this hypothesis. Nitrite entry into carp red cells is unaffected by inhibition of the anion exchanger with DIDS (Jensen, 1990*a*). Also, the plasma NO_2^- concentration is much lower than the plasma Cl^- and HCO_3^- concentrations, whereby competition for the external anion site of Band 3 does not favour nitrite entry via this route (Jensen, 1992).

The influx of nitrite into carp red cells is strongly oxygenation dependent. At physiological pH, nitrite rapidly enters deoxygenated red cells whereas it hardly enters oxygenated red cells (Jensen, 1990*a*; 1992) (Fig. 2). When pH is lowered, nitrite starts to enter oxygenated red cells but at a lower rate than into deoxygenated cells (Jensen, 1992). The increased nitrite influx with reduction in pH and haemoglobin oxygenation is positively correlated with parallel increases in red cell membrane potential, suggesting that entry is via conductive transport

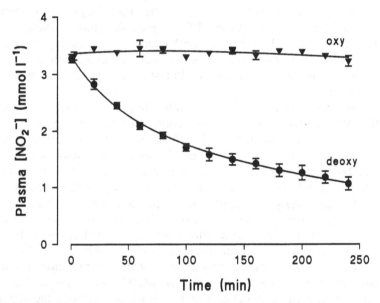

Fig. 2. Time-dependent changes in plasma nitrite concentration in oxygenated and deoxygenated carp blood at pH 8.14. Nitrite was added to the blood at time zero. A subsequent decrease in the plasma nitrite concentration reflects an influx of nitrite to the red blood cells. The curves represent non-linear curve fits of the data to the biexponental equation: plasma $[NO_2^-] = Pe^{-at} + Qe^{-bt}$. (Modified from Jensen, 1992.)

(Jensen, 1992). Mathematical modelling based on conductive transport and subsequent intracellular removal of some of the nitrite suggests that an extracellular nitrite load should lead to a biexponential decay of extracellular $[NO_2^-]$, as is indeed the case (Jensen, 1992; Fig. 2). Also, the initial nitrite influx into deoxygenated red cells at physiological pH does not show saturation kinetics but increases linearly with plasma $[NO_2^-]$ up to at least 20 mM (L. Pilgaard & F.B. Jensen, unpublished observations). All available evidence thus supports the idea that the main route of entry into the red cells is via conductive transport.

Effects on oxygen transport in fishes

Methaemoglobin formation

Entry of nitrite into the red cells is associated with oxidation of the haem iron atom $(Fe(II) \rightarrow Fe(III))$, whereby functional haemoglobin (Hb) is converted into methaemoglobin (metHb), that cannot transport O_2. Formation of metHb is one of the most frequently documented effects of nitrite (cf. Kiese, 1974; Lewis & Morris, 1986; Eddy & Williams, 1987). The degree of methaemoglobinaemia depends on those factors that determine the degree of NO_2^- accumulation (e.g. species, exposure time and environment) but also on several cellular and molecular factors.

The haem groups of Hb are embedded in hydrophobic pockets of the molecule which provides protection against oxidation (Perutz, 1978; Jaffé, 1981). The small rate of autoxidation that does occur in the red cell is effectively countered by metHb reductase systems. Of these, the NADH metHb reductase is the most important in both mammals (Jaffé, 1981) and fish (Freemann, Beitinger & Huey, 1983). Blood metHb levels accordingly are kept low in the absence of nitrite or other oxidizing agents. In the presence of nitrite, the rate of metHb formation increases, until a balance develops between oxidation and reduction at an elevated metHb level (Jensen, 1990a). When the NO_2^- load increases, as is the case during *in vivo* nitrite accumulation, the metHb concentration increases consecutively. MetHb levels above 80% of the total Hb can easily develop, which drastically decreases the arterial O_2 content and leads to tissue O_2 shortage (Jensen *et al.*, 1987).

The mechanism by which nitrite oxidizes oxygenated Hb is complex. The kinetics of the reaction is characterized by an initial lag phase followed by an autocatalytic increase in reaction rate in both mammalian (Kiese, 1974) and fish (Jensen, 1990a, 1993) haemoglobins. The oxidation process proceeds via a series of intermediate reactions involving free radicals, with the haem groups being oxidized by the radical NO_2^- during the autocatalytic phase (Kosaka & Tyuma, 1987). In mammalian Hbs, the reaction is slower and via a different mechanism in the absence than in the presence of oxygen (Kiese, 1974). In rainbow trout haemolysates, a lowering of HbO_2 saturation is sufficient to significantly slow down nitrite-induced metHb formation and to alter the reaction kinetics (Jensen, 1993). The oxidation of trout Hb by nitrite is also slowed down by increased concentrations of red cell organic phosphates (ATP, GTP), inorganic ions, NADH and catecholamines (Jensen, 1993). Nitrite-induced Hb oxidation will therefore be influenced by a large series of organismic, cellular and molecular factors in the *in vivo* situation.

Red cell volume and oxygen affinity

Methaemoglobin formation is traditionally believed to increase the oxygen affinity of haem groups that remain functional. The reason for this is that metHb tends to assume an R- or 'oxy'-like quaternary structure, which has a higher O_2 affinity than the T or 'deoxy' structure (Perutz, 1978). An increased O_2 affinity has been reported in nitrite-treated mammalian blood (Darling & Roughton, 1942). It is thus of considerable interest that the opposite response, i.e. a large reduction in blood O_2 affinity, is seen in nitrite-exposed carp (Jensen *et al.*, 1987; Jensen, 1990*a*).

This reduction in O_2 affinity is related to a profound shrinkage of the red cells. Nitrite-induced methaemoglobinaemia in carp is associated with a net KCl efflux from the red cell, which is followed by osmotically obliged water, whereby cell volume decreases (Jensen, 1990*a*, 1992). Shrinkage of the red cells increases the intracellular concentrations of haemoglobin and organic phosphates (ATP and GTP), leading to an increased complex binding between organic phosphate and Hb that lowers O_2 affinity (Jensen *et al.*, 1987; Jensen, 1990*a*). A relative decrease in red cell pH, and perhaps direct binding of nitrite or nitrate (that is produced in the oxidation reaction) to the Hb, may also contribute to decrease O_2 affinity (Jensen, 1990*a*).

Via red cell shrinkage, the intracellular haemoglobin concentrations can increase to values that surpass the solubility limit, and it is possible that intracellular Hb crystals are formed (Jensen *et al.*, 1987). Degeneration of red cell structural and functional properties may cause an increased removal of red cells from the circulation and contribute to the decrease in total blood Hb seen in nitrite-exposed fish (Brown & McLeay, 1975; Jensen *et al.*, 1987).

The large reduction in arterial O_2 concentration that develops in nitrite-exposed carp is primarily due to methaemoglobin formation, but the decrease in O_2 affinity contributes significantly as illustrated in Fig. 3. Following 48 h of nitrite exposure the arterial O_2 concentration is extremely low in carp (Jensen *et al.*, 1987) (Fig. 3). It is evident, that the tolerance of the fish to additional stresses such as exercise or hypoxia must be greatly reduced. Indeed, swimming performance is

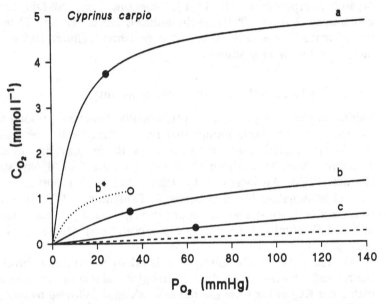

Fig. 3. Concentration versus tension O_2 equilibrium curves of blood from carp in normoxic water (curve a, 4.9% metHb) and after 24 h (curve b, 64.6% metHb) and 48 h (curve c, 83.3% metHb) of 1 mM nitrite exposure. Arterial points are shown by filled circles. Apart from increasing the metHb level, nitrite exposure is associated with a decreased blood O_2 affinity. Curve b* is a theoretical curve with the same metHb level as for curve b, but with an unchanged O_2 affinity as in curve a. The vertical distance between the filled circle on curve a and the open circle on curve b* gives the decrease in arterial O_2 content caused by the elevated metHb level after 24 h, whereas the vertical distance between the open circle on curve b* and the filled circle on curve b gives the additional decrease in arterial O_2 content caused by the O_2 affinity decrease. The dashed line at the bottom is the curve for physically dissolved O_2. (Modified from Jensen *et al.*, 1993.)

impaired by high metHb levels in chinook salmon (Brauner, Val & Randall, 1993).

Blood oxygen transport in fishes benefits from catecholamine-stimulated red cell Na^+/H^+ exchange in various stress situations (Nikinmaa, 1990). The β-adrenergic response causes red cell swelling and elevates red cell pH, thus increasing HbO_2 affinity in hypoxia or protecting it from a decrease during extracellular acidoses. Nitrite-induced metHb formation strongly reduces catecholamine-stimulated Na^+/H^+ exchange across the red cell membrane in rainbow trout (Nikinmaa & Jensen, 1992). This inhibition of the red cell β-adrenergic response may further compromise blood O_2 transport in nitrite-exposed fish.

Ventilation, blood flow and tissue pigments

Whereas effects of nitrite on blood O_2 transport are well documented, only limited information is available on possible effects at other steps in the oxygen transport cascade. The arterial oxygen tension is increased and the CO_2 tension is slightly decreased during severe methaemoglobi-naemia in carp, which indicates a moderate increase in ventilation (Jensen *et al.*, 1987). The hyperventilation is much smaller than when the blood O_2 content is decreased by environmental hypoxia (Williams, Glass & Heisler, 1992).

The nitrite-induced reduction in blood O_2 capacitance in principle can be counterbalanced by an increased cardiac output (cf. Jensen, Nikinmaa & Weber, 1993) but cardiac output has not been assessed in nitrite-exposed fish. Nitrite induces vascular smooth muscle relax-ation, thus causing vasodilation. This response is mediated via stimula-tion of guanylate cyclase and subsequent accumulation of cGMP in the cells (e.g. Laustiola *et al.*, 1991). The consequences of nitrite-induced vasodilation for blood flow and blood pressure in fish are unknown.

In the tissues, nitrite can oxidize intracellular myoglobin to metmyo-globin (Doeller & Wittenberg, 1991), thus inactivating myoglobin and its role in cellular O_2 transport. Mitochondrial cytochrome oxidase is also inhibited by nitrite (Paitian, Markossian & Nalbandyan, 1985).

Effects on oxygen transport in crustaceans

In crustaceans, each O_2 binding site of the respiratory pigment, haemo-cyanin, contains two copper atoms that change oxidation state upon bind-ing and release of O_2. Formation of methaemocyanin by nitrite occurs predominantly at low pH and in the presence of a large excess of nitrite and appears to be unimportant at physiological pH (Gondko, Serafin & Mazur, 1985; Tahon *et al.*, 1988; Jensen, 1990*b*). Thus, nitrite does not disrupt reversible oxygen binding of haemocyanin to the same extent as is the case with haemoglobin. Nitrite-induced disturbances in O_2 transport are therefore more modest in crustaceans than in fish.

The concentration of haemocyanin in the haemolymph decreases in nitrite-exposed crayfish (Jensen, 1990*b*). The concentration of pigment is, however, generally low in crustaceans (Taylor, 1982), whereby they have relatively high circulatory requirements (ml haemolymph pumped by the heart per ml O_2 consumed). Oxygen affinity of crayfish haemolymph is not influenced by short-term *in vitro* treatment with 10 mM sodium nitrite (Jensen, 1990*b*). Changes in haemolymph oxygen affinity during *in vivo* nitrite exposure have not been investigated. Alterations in the *in vivo* oxygen affinity are possible, since haemocyanin oxygen affinity is influenced by a large series of inorganic (e.g. H^+, Ca^+) and organic (e.g. lactate, urate, dopamine) compounds (Truchot & Lallier, 1992), some of which may change in concentration during nitrite exposure.

Acid–base balance

During the first many hours of nitrite exposure, the arterial pH is slightly elevated in carp as result of an increase in plasma $[HCO_3^-]$ and a small decrease in $P\text{co}_2$ (Jensen *et al.*, 1987). Tissue lactic acid production, resulting from limitations in tissue O_2 supply, results in a progressive increase in plasma lactate. An extracellular lactacidosis is, however, not induced in carp. It appears that lactic acid is produced at a rate that does not exceed the capacity of the branchial acid–base regulatory ion transfer mechanisms to mediate the transfer of excess protons to the water (Jensen *et al.*, 1987).

A small respiratory alkalosis is observed in the crayfish *Astacus astacus* during sub-lethal nitrite exposure (Jensen, 1990*b*). A more severe nitrite accumulation in the crayfish *Pacifastacus leniusculus*, causes an increase in haemolymph $P\text{co}_2$ and a decrease in pH (Harris & Coley, 1991). In neither of the crayfish species does haemolymph lactate increase, supporting the view that tissue O_2 delivery is not severely perturbed.

Effects on electrolyte balance

Anions

The requirement for electroneutrality in electrolyte solutions means that the accumulation of nitrite is bound to have consequences for the extracellular electrolyte balance. The rise in $[NO_2^-]$ must be balanced either by a parallel rise in cation concentration (which would increase osmolarity) or by a parallel decrease in other anions. The latter seems to be the case. Osmolarity as well as the sum of anions and the sum of cations stays constant during nitrite exposure in carp (Jensen *et al.*, 1987).

Depletion of extracellular chloride is a characteristic effect in nitrite-exposed fish and crustaceans (Jensen *et al.*, 1987; Jensen, 1990*b*; Harris & Coley, 1991). The active branchial Cl^- uptake is partially converted to NO_2^- uptake, whereas the passive Cl^- efflux persists, leading to a net Cl^- loss at the gills. The decrease in extracellular $[Cl^-]$ is, however, larger than the rise in $[NO_2^-]$. In carp, a linear relationship with a slope of -1 exists between plasma $[Cl^-]$ and the sum of $[NO_2^-]$, [lactate] and $[HCO_3^-]$, demonstrating stoichiometrical balance between the decrease in $[Cl^-]$ and increases in $[NO_2^-]$ and [lactate] and changes in $[HCO_3^-]$ (Jensen *et al.*, 1987). The additional decrease in $[Cl^-]$ associated with the rise in plasma lactate may be a result of acid–base regulatory HCO_3^-/Cl^- exchange across the gills, where the HCO_3^- taken up buffers the protons from lactic acid and becomes eliminated as molecular CO_2 (cf. Jensen *et al.*, 1987). In crayfish, lactate levels remain low but haemolymph $[Cl^-]$ nevertheless decreases much more than $[NO_2^-]$ increases (Jensen, 1990*b*; Harris & Coley, 1991). Thus, in *Astacus*, when haemolymph $[NO_2^-]$ is increased by 10 mM, haemolymph $[Cl^-]$ decreases by some 40 mM (Jensen, 1990*b*). When osmolarity and cation changes are also taken into account, it is evident that some unknown anion(s) must increase in concentration in nitrite-exposed crayfish. The identity of this (these) anion(s) is presently under investigation. It appears to be primarily nitrate that increases in concentration (F.B. Jensen, in preparation).

The number of gill chloride cells increases in nitrite-exposed rainbow trout (Gaino, Arillo & Mensi, 1984; Williams & Eddy, 1988). An increased chloride cell density may enhance the chloride uptake capacity in response to the chloride depletion but will presumably at the same time be maladaptive by increasing nitrite uptake. Cortisol is known to cause proliferation of chloride cells on the gill (Perry & Laurent, 1993). The increased cortisol levels in nitrite-exposed fish (Tomasso, Davis & Simco, 1981) could therefore be a trigger of the elevated number of chloride cells. A further reported effect is that nitrite inhibits carbonic anhydrase in trout gills, which may reduce Cl^- uptake by limiting the formation of HCO_3^- for the branchial Cl^-/HCO_3^- exchange mechanism (Gaino *et al.*, 1984).

Cations

Recent studies have shown that nitrite causes large disturbances in potassium balance. In studies on nitrite-induced extracellular K^+

changes, it is important to sample blood from undisturbed fish via chronic indwelling blood catheters. Other sampling methods (e.g. netting the animal with subsequent puncture of caudal vessels) may be associated with fish struggling, that provokes K^+ release from the muscles and blurs nitrite-induced effects. In carp with indwelling dorsal aorta catheters, nitrite exposure elicits a large elevation of extracellular K^+ values (Jensen *et al.*, 1987). A nitrite-induced increase in plasma $[K^+]$ is also seen in cannulated rainbow trout (M. Nikinmaa & F.B. Jensen, unpublished). These findings suggest that K^+ is released from intracellular compartments.

The elevation of extracellular $[K^+]$ in nitrite-exposed carp is paralleled by a decrease in the red cell K^+ content (Fig. 4). Potassium release from the red cells contributes significantly to the plasma $[K^+]$ increase, but release of K^+ from other intracellular compartments is also involved (Jensen, 1990*a*). In crayfish, muscle tissue K^+ is depleted during nitrite exposure, suggesting that K^+ release from intracellular compartments constitutes a general response to nitrite (Jensen, 1990*b*).

The mechanism underlying the K^+ efflux is known for carp red cells. Nitrite-influx to the cells converts a large fraction of the Hb molecules to metHb and stimulates K^+-Cl^- cotransport out of the cells (Jensen, 1992). An additional, but minor, leukotriene-sensitive route also seems to be involved (Jensen, 1992). This route could be via separate, conductive K^+ and Cl^- transport. The loss of K^+ from crayfish muscle may similarly be accompanied by Cl^- loss (Jensen, 1990*b*), but the precise mechanism is not yet known.

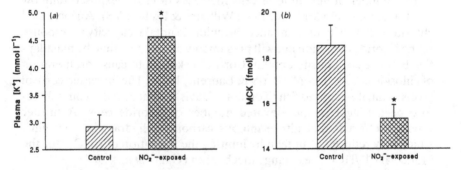

Fig. 4. (a) Plasma potassium concentration and (b) mean red blood cell potassium content in carp before (control) and during nitrite exposure (plasma nitrite elevated to 3 mM). (Data from Jensen, 1990*a*.)

Potassium loss from crayfish muscle is not associated with elevation of extracellular [K$^+$] as seen in fish (Jensen, 1990*b*). This may be due to (i) an expansion of the extracellular volume, (ii) K$^+$ efflux to the environment, or (iii) K$^+$ uptake in other tissues. Tissue [K$^+$] in the midgut gland of crayfish does not decrease but rather tends to increase slightly during nitrite exposure (Jeberg & Jensen, 1994). This suggests that the K$^+$ efflux mechanism that is stimulated in muscle tissue is absent (or less affected) in the midgut gland, and that some of the K$^+$ released from muscle may be taken up by the midgut gland (Jeberg & Jensen, 1994).

The disturbances of potassium balance could be significant in the toxicity of nitrite by perturbing a number of vital physiological functions (Jensen, 1990*b*). Depletion of intracellular K$^+$ and elevation of extracellular K$^+$ levels will affect membrane potentials, and may interfere with neurotransmission, skeletal muscle contractions, and heart function. The cardiotoxic effect of severe extracellular hyperkalaemia (e.g. Clausen, 1986) may cause heart failure and death in nitrite-exposed animals. Depletion of intracellular K$^+$ levels could perturb metabolism, given the influence of changes in intracellular K$^+$ on enzymatic reactions (Yancey *et al.*, 1982).

In contrast to the major changes in potassium balance, other cations are either unchanged or show only small changes. The plasma K$^+$ increase in carp is countered by a decrease in [Na$^+$], which, on the basis of the high background concentration of Na$^+$, will be physiologically unimportant. Other extracellular cations remain constant in concentration (Jensen *et al.*, 1987). In crayfish, extracellular Na$^+$ decreases more than in fish but the decrease is much smaller than the decrease in chloride (Jensen, 1990*b*; Harris & Coley, 1991). The decrease in sodium may be due to an expansion of the extracellular volume (Jensen, 1990*b*) but a reduced branchial Na$^+$ uptake is also a possibility (Harris & Coley, 1991). Extracellular and intracellular concentrations of divalent cations are unchanged in *Astacus astacus* during sub-lethal nitrite exposure (Jensen, 1990*b*) but during severe nitrite accumulation in *Pacifastacus leniusculus* haemolymph Ca^{2+} increases, which may result from buffering of the extracellular acidosis observed in this species by exoskeletal CaCO$_3$ (Harris & Coley, 1991).

N-nitroso compounds

Nitrite can be converted to HNO$_2$ and subsequently to N$_2$O$_3$ under acid conditions. This (and other) nitrogen oxide(s) can react with nitrogenous compounds (e.g. amines) to yield *N*-nitroso compounds (e.g. *N*-nitrosamines). Many *N*-nitroso compounds are mutagenic and carcinogenic: *N*-nitrosamines are metabolized, and nitrosamides spontaneously decompose, to highly reactive carbonium ions (e.g. Hotchkiss *et al.*, 1992) that can alkylate DNA and cause base substitutions during DNA replication (e.g. Bartsch & Montesano, 1984).

Formation of *N*-nitroso compounds has received much attention in connection with dietary nitrite and nitrate intake in mammals. To what extent endogenous nitrosation occurs in various body compartments of nitrite-exposed fish or crustaceans is largely unknown. The potential role of *N*-nitroso compounds in the toxicity of nitrite in aquatic animals warrants investigation.

Liver damage and effects on the immune system

Nitrite is bioaccumulated not only in plasma but also in various tissues such as gill, liver, brain and muscle (Margiocco *et al.*, 1983). Entry of nitrite into liver cells of rainbow trout causes mitochondrial damage, whereas brain mitochondria are less affected (Arillo *et al.*, 1984). Damage of liver mitochondria may be related to tissue O_2 shortage and to a direct toxic action of nitrite or one of its metabolic derivatives. Structural damage to the liver could interfere with a large series of essential metabolic processes, impairing the metabolic integrity of the organism.

Sub-lethal nitrite exposure increases the susceptibility of channel catfish to bacterial diseases (Hanson & Grizzle, 1985). This finding suggests that nitrite represses the immune system.

Recovery from nitrite intoxication

When fish are transferred to nitrite-free water after a period of nitrite exposure, plasma nitrite levels decline. In rainbow trout (Eddy *et al.*, 1983) and sea bass (Scarano & Saroglia, 1984) plasma nitrite and blood methaemoglobin levels return to control values after about 24 h of recovery. In carp, nitrite off-loading is considerably slower than the preceding nitrite loading, and nitrite and metHb levels remain elevated even 2–3 days after return to clean water (Jensen *et al.*, 1987; Williams *et al.*, 1992). In crayfish, haemolymph clearance of nitrite is also slower than uptake as is the restoration of haemolymph Cl^- levels (Harris & Coley, 1991).

Even though nitrite and metHb recovers towards low control values, other post-exposure effects may develop. In sea bass, recovery in metHb levels is succeeded by a drop in total Hb concentration lasting up to three weeks (Scarano & Saroglia, 1984).

Concluding remarks

Many different physiological changes are induced in nitrite-exposed animals. Nitrite toxicity may accordingly result from the combined action of several physiological disturbances, the relative importance of which may vary with species, nitrite load, length of exposure, physiological condition and other factors. Further study is needed in order to obtain an integrated understanding of the role played by individual physiological effects.

Acknowledgement

Supported by the Danish Natural Science Research Council (11–9659–1).

References

Anderson, J.J., Okubo, A., Robbins, A.S. & Richards, F.A. (1982). A model for nitrite and nitrate distributions in oceanic oxygen minimum zones. *Deep-Sea Research*, **29**, 1113–40.

Arillo, A., Gaino, E., Margiocco, C., Mensi, P. & Schenone, G. (1984). Biochemical and ultrastructural effects of nitrite in rainbow trout: liver hypoxia as the root of the acute toxicity mechanism. *Environmental Research*, **34**, 135–54.

Ary, R.D., Jr. & Poirrier, M.A. (1989). Acute toxicity of nitrite to the blue crab (*Callinectes sapidus*). *Progressive Fish-Culturist*, **51**, 69–72.

Aubert, A. & Lignon, J.M. (1986). Cuticular anionic selectivity in the gill lamina of the crayfish. In *Comparative Physiology of Environmental Adaptations – Abstracts*, p. 13. European Society for Comparative Physiology and Biochemistry 8th Conference, Strasbourg.

Bartsch, H. & Montesano, R. (1984). Relevance of nitrosamines to human cancer. *Carcinogenesis*, **5**, 1381–93.

Bath, R.N. & Eddy, F.B. (1980). Transport of nitrite across fish gills. *Journal of Experimental Zoology*, **214**, 119–21.

Brauner, C.J., Val, A.L. & Randall, D.J. (1993). The effect of graded methaemoglobin levels on the swimming performance in chinook salmon (*Oncorhynchus tshawytscha*). *Journal of Experimental Biology*, **185**, 121–35.

Brown, D.A. & McLeay, D.J. (1975). Effect of nitrite on methemoglobin and total hemoglobin of juvenile rainbow trout. *Progressive Fish-Culturist*, **37**, 36–8.

Chen, J.-C. & Chen, S.F. (1992). Accumulation of nitrite in the haemolymph of *Penaeus monodon* exposed to ambient nitrite. *Comparative Biochemistry and Physiology*, **103C**, 477–81.

184   F.B. JENSEN

Clausen, T. (1986). Regulation of active Na^+-K^+ transport in skeletal muscle. *Physiological Reviews*, **66**, 542–80.

Darling, R.C. & Roughton, F.J.W. (1942). The effect of methemoglobin on the equilibrium between oxygen and hemoglobin. *American Journal of Physiology*, **137**, 56–68.

Doeller, J.E. & Wittenberg, B.A. (1991). Myoglobin function and energy metabolism of isolated cardiac myocytes: effect of sodium nitrite. *American Journal of Physiology*, **261**, H53–62.

Eddy, F.B., Kunzlik, P.A. & Bath, R.N. (1983). Uptake and loss of nitrite from the blood of rainbow trout, *Salmo gairdneri* Richardson, and Atlantic salmon, *Salmo salar* L. in fresh water and in dilute sea water. *Journal of Fish Biology*, **23**, 105–16.

Eddy, F.B. & Williams, E.M. (1987). Nitrite and freshwater fish. *Chemistry and Ecology*, **3**, 1–38.

Freeman, L., Beitinger, T.L. & Huey, D.W. (1983). Methemoglobin reductase activity in phylogenetically diverse piscine species. *Comparative Biochemistry and Physiology*, **75B**, 27–30.

Gaino, E., Arillo, A. & Mensi, P. (1984). Involvement of the gill chloride cells of trout under acute nitrite intoxication. *Comparative Biochemistry and Physiology*, **77A**, 611–17.

Gondko, R., Serafin, E. & Mazur, J. (1985). The kinetics of the reaction of thiocyanate and nitrite ions with *Orconectes limosus* oxyhemocyanin. *Zeitschrift für Naturforschung*, **40c**, 677–81.

Gutzmer, M.P. & Tomasso, J.R. (1985). Nitrite toxicity to the crayfish *Procambarus clarkii*. *Bulletin of Environmental Contamination and Toxicology*, **34**, 369–76.

Hanson, L.A. & Grizzle, J.M. (1985). Nitrite-induced predisposition of channel catfish to bacterial diseases. *Progressive Fish-Culturist*, **47**, 98–101.

Harris, R.R. & Coley, S. (1991). The effects of nitrite on chloride regulation in the crayfish *Pacifastacus leniusculus* Dana (Crustacea: Decapoda). *Journal of Comparative Physiology*, **161B**, 199–206.

Hotchkiss, J.H., Helser, M.A., Maragos, C.M. & Weng, Y.M. (1992). Nitrate, nitrite, and *N*-nitroso compounds. In *Food Safety Assessment*. Finley, J.W., Robinson, S.F. & Armstrong, D.J. (eds.). *ACS Symposium Series*, vol. **484**, pp. 400–18.

Jaffé, E.R. (1981). Methaemoglobinaemia. *Clinics in Haematology*, **10**, 99–122.

Jeberg, M.V. & Jensen, F.B. (1994). Extracellular and intracellular ionic changes in crayfish *Astacus astacus* exposed to nitrite at two acclimation temperatures. *Aquatic Toxicology*, **29**, 65–72.

Jensen, F.B. (1990a). Nitrite and red cell function in carp: control factors for nitrite entry, membrane potassium ion permeation, oxygen affinity and methaemoglobin formation. *Journal of Experimental Biology*, **152**, 149–66.

Jensen, F.B. (1990*b*). Sublethal physiological changes in freshwater crayfish, *Astacus astacus*, exposed to nitrite: haemolymph and muscle tissue electrolyte status, and haemolymph acid–base balance and gas transport. *Aquatic Toxicology*, **18**, 51–60.

Jensen, F.B. (1992). Influence of haemoglobin conformation, nitrite and eicosanoids on K^+ transport across the carp red blood cell membrane. *Journal of Experimental Biology*, **171**, 349–71.

Jensen, F.B. (1993). Influence of nucleoside triphosphates, inorganic salts, NADH, catecholamines, and oxygen saturation on nitrite-induced oxidation of rainbow trout haemoglobin. *Fish Physiology and Biochemistry*, **12**, 111–17.

Jensen, F.B., Andersen, N.A. & Heisler, N. (1987). Effects of nitrite exposure on blood respiratory properties, acid–base and electrolyte regulation in the carp (*Cyprinus carpio*). *Journal of Comparative Physiology*, **157B**, 533–41.

Jensen, F.B., Nikinmaa, M. & Weber, R.E. (1993). Environmental perturbations of oxygen transport in teleost fishes: causes, consequences and compensations. In *Fish Ecophysiology*. Rankin, J.C. & Jensen, F.B. (eds.), pp. 161–79. Chapman & Hall, London.

Kiese, M. (1974). *Methemoglobinemia: A Comprehensive Treatise*. CRC Press, Cleveland.

Kosaka, H. & Tyuma, I. (1987). Mechanism of autocatalytic oxidation of oxyhemoglobin by nitrite. *Environmental Health Perspectives*, **73**, 147–51.

Laustiola, K.E., Vuorinen, P., Pörsti, I., Metsä-Ketelä, T., Manninen, V. & Vapaatalo, H. (1991). Exogenous GTP enhances the effects of sodium nitrite on cyclic GMP accumulation, vascular smooth muscle relaxation and platelet aggregation. *Pharmacology and Toxicology*, **68**, 60–3.

Lewis, W.M. & Morris, D.P. (1986). Toxicity of nitrite to fish: a review. *Transactions of the American Fisheries Society*, **115**, 183–95.

Margiocco, C., Arillo, A., Mensi, P. & Schenone, G. (1983). Nitrite bioaccumulation in *Salmo gairdneri* Rich. and hematological consequences. *Aquatic Toxicology*, **3**, 261–70.

Nikinmaa, M. (1990). *Vertebrate Red Blood Cells. Adaptation of Function to Respiratory Requirements*. Springer–Verlag, Berlin.

Nikinmaa, M. & Jensen, F.B. (1992). Inhibition of adrenergic proton extrusion in rainbow trout red cells by nitrite-induced methaemoglobinaemia. *Journal of Comparative Physiology*, **162B**, 424–9.

Paitian, N.A., Markossian, K.A. & Nalbandyan, R.M. (1985). The effect of nitrite on cytochrome oxidase. *Biochemical and Biophysical Research Communications*, **133**, 1104–11.

Palachek, R.M. & Tomasso, J.R. (1984). Toxicity of nitrite to channel catfish (*Ictalurus punctatus*), tilapia (*Tilapia aurea*), and largemouth bass (*Micropterus salmoides*): evidence for a nitrite exclusion mechanism. *Canadian Journal of Fisheries and Aquatic Sciences*, **41**, 1739–44.

Perrone, S.J. & Meade, T.L. (1977). Protective effect of chloride on nitrite toxicity to coho salmon (*Oncorhynchus kisutch*). *Journal of the Fisheries Research Board of Canada*, **34**, 486–92.

Perry, S.F. & Laurent, P. (1993). Environmental effects on fish gill structure and function. In *Fish Ecophysiology*. Rankin, J.C. & Jensen, F.B. (eds.), pp. 231–64. Chapman & Hall, London.

Perutz, M.F. (1978). Hemoglobin structure and respiratory transport. *Scientific American*, **239**, 68–86.

Scarano, G. & Saroglia, M.G. (1984). Recovery of fish from functional and haemolytic anaemia after brief exposure to a lethal concentration of nitrite. *Aquaculture*, **43**, 421–6.

Tahon, J.-P., Van Hoof, D., Vinckier, C., Witters, R., De Ley, M. & Lontie, R. (1988). The reaction of nitrite with the haemocyanin of *Astacus leptodactylus*. *Biochemical Journal*, **249**, 891–6.

Taylor, E.W. (1982). Control and co-ordination of ventilation and circulation in crustaceans: responses to hypoxia and exercise. *Journal of Experimental Biology*, **100**, 289–319.

Tomasso, J.R., Davis, K.B. & Simco, B.A. (1981). Plasma corticosteroid dynamics in channel catfish (*Ictalurus punctatus*) exposed to ammonia and nitrite. *Canadian Journal of Fisheries and Aquatic Sciences*, **38**, 1106–12.

Truchot, J.-P. & Lallier, F.H. (1992). Modulation of the oxygencarrying function of hemocyanin in crustaceans. *News in Physiological Sciences*, **7**, 49–52.

Wheatly, M.G. (1989). Physiological responses of the crayfish *Pacifastacus leniusculus* to environmental hyperoxia. I. Extracellular acid–base and electrolyte status and transbranchial exchange. *Journal of Experimental Biology*, **143**, 33–51.

Williams, E.M. & Eddy, F.B. (1986). Chloride uptake in freshwater teleosts and its relationship to nitrite uptake and toxicity. *Journal of Comparative Physiology*, **156B**, 867–72.

Williams, E.M. & Eddy, F.B. (1988). Anion transport, chloride cell number and nitrite-induced methaemoglobinaemia in rainbow trout (*Salmo gairdneri*) and carp (*Cyprinus carpio*). *Aquatic Toxicology*, **13**, 29–42.

Williams, E.M., Glass, M.L. & Heisler, N. (1992). Blood oxygen tension and content in carp, *Cyprinus carpio* L., during hypoxia and methaemoglobinaemia. *Aquaculture and Fisheries Management*, **23**, 679–90.

Yancey, P.H., Clark, M.E., Hand, S.C., Bowlus, R.D. & Somero, G.N. (1982). Living with water stress: evolution of osmolyte systems. *Science*, **217**, 1214–22.

P.-E. OLSSON

Metallothioneins in fish: induction and use in environmental monitoring

General background

Heavy metals cause an abundance of effects on all living animals. In fish there is a multitude of information on different alterations that are caused by cadmium, copper, zinc and mercury (Sorensen, 1991). Heavy metals affect every level of organization from the society (behaviour) to the organism (reproductive) and sub-cellular level (enzymatic and composition changes). The effects can be general to all of the metals or highly metal specific. Since the organism can not destroy the metals by metabolic degradation, the only way in which an organism can protect itself from heavy metal poisoning is by decreasing the rate of uptake, by binding the metals to a ligand that will hinder the metal from disrupting normal physiological processes or increase the rate of excretion.

Metallothionein with its exceptionally high content of thiol groups is extremely effective in binding heavy metal ions. Metallothioneins are widely distributed in nature and are generally considered to be involved in zinc and copper homeostasis, heavy metal detoxification and possibly as a free radical scavenger (Hamer, 1986). As well as being able to bind group IB and IIB heavy metals, metallothionein is also induced by addition of these metals (Zafarullah, Olsson & Gedamu, 1989b). Studies on fish and fish cell lines have further shown that the hepatic metallothionein levels are elevated by glucocorticoids, noradrenalin and progesterone (Hyllner et al., 1989; Olsson et al., 1990a, George et al., 1992). Metallothionein has further been shown to be endogenously regulated during different developmental life stages (Olsson et al., 1990b). An understanding of the complex and highly controlled regulation of metallothionein in fish is necessary if the system is to be used in environmental monitoring. Therefore, the main focus of the present chapter is to describe the regulation and function of the major heavy metal binding protein, metallothionein, in fish and how it can be utilized in environmental studies.

Structure and metal binding properties

Following the first isolation of mammalian metallothioneins, it took almost 20 years until these proteins were indicated to be present in teleost species (Marafante *et al.*, 1976). Following the identification of metallothionein in goldfish, the first partial amino acid sequence of a teleost metallothionein was obtained from plaice, showing that metallo-thionein had an N-terminal end that was distinct from mammalian metallothioneins, although the protein sequence was conserved (Overnell, Berger & Wilson, 1981). During the following years, much attention has focused on the use of metallothionein as an indicator of heavy metal pollution. In 1987, the first cDNA sequences of fish metallothioneins were obtained from rainbow trout (*Oncorhynchus mykiss*) (Bonham, Zafarullah & Gemadu, 1987). Since then, several cDNA sequences have been obtained as well as some partial amino acid sequences (Kay *et al.*, 1987; Leaver & George, 1989; Chan, Davidson & Fletcher, 1989; Kille, Kay & Sweeney, 1991; Hylland, 1992; Kille *et al.*, unpublished results, Olsson *et al.*, unpublished results).

The sequences obtained on fish metallothionein show that all fish species so far studied have the cysteines conserved in number and position. When compared to mammals, there is one exception, the cysteine at position 55 in teleosts is found at position 57 in mammals. Except for the rainbow trout MT-A, which consist of 61 amino acids, all the fish metallothioneins contain 60 amino acids. The extra amino acid in rainbow trout MT-A is due to an insertion of an alanine at position 31. This position is at the junction between the second and third intron and constitutes the border between the two domains of metallothionein.

To date, eight different metallothionein amino acid or mRNA sequences have been obtained from different fish species. The sequences are derived from four different orders of fish, the salmoniformes (rainbow trout and pike), the pleuronectiformes (winter flounder and plaice), the cypriniformes (stone loach and goldfish) and the gadiformes (cod). Comparison of the amino acid sequences of these fish show that species belonging to the same order show high sequence homology (Fig. 1). The plaice and winter flounder metallothioneins differ only in one amino acid, a serine at position 57 in the plaice is altered into a threonine in winter flounder. Rainbow trout metallothionein A and B and pike metallothionein all differ from each other in three amino acids. The stone loach and goldfish metallothioneins also differ in three amino acids. When comparing the sequences across the orders, the conservation is much lower with anywhere from 8 to 16 amino acids differing between the metallothioneins. The same conservation in sequence is seen at the nucleotide level where the pleuronectiformes differ in three nucleotides, the salmoniformes in 14 to 16 nucleotides and the cypriniformes in 20 nucleotides. Between orders, the difference is about 40 nucleotides out of 183 or 186.

Upon binding metals, the thionein backbone folds into two metal binding domains. These are the N-terminal β-domain and the C-terminal α-domain. The binding of metals to these domains results in the formation of two metal clusters of which cluster A, that is located in the α-domain of the protein shows a preferential incorporation of zinc and cadmium while cluster B preferentially fills up with copper atoms. The metal binding properties of the two domains opens up the possibility to construct metal specific chelator peptides in order to protect the host animal/plant from specific metal exposures.

Another structural feature that is important to consider in metallothioneins is the N-terminal region. It has been postulated that antibodies to metallothionein are directed primarily towards the N-terminal portion of the protein. The N-terminal amino acid sequence leading up to the first cystein varies between mammals, avians and teleost. Avians have a MDPQDC sequence, mammals have a MDPNC sequence, and teleosts have a MDPC sequence (Andrews, 1990; Wei & Andrews, 1988; Karin & Richards, 1984; Bonham *et al.*, 1987). These variations may explain the low cross-reactivity of antibodies raised towards animals of one phylum when tested against metallothionein from another phylum. When different species of teleosts are compared to each other, the first amino acid difference is seen at position five where an E is

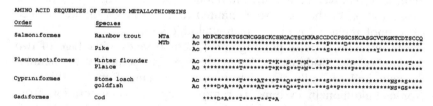

Fig. 1. Comparison of the amino acid sequences of metallothionein from different fish species. Ac, Acetylated N-terminal amino acid; *, conserved amino acid; -, deleted residue. Amino acids that differ from the rainbow trout metallothionein-A sequence are shown. All sequences except the partial cod amino acid sequence are derived from nucleotide sequences. References: Bonham *et al.*, 1987; Leaver & George, 1989; Chan *et al.*, 1989; Kille *et al.*, 1991; Hylland, 1992; Kille *et al.*, unpublished observations.)

replaced by a D in goldfish, stone loach and cod. The next alteration is at position seven where an A in cod and goldfish has replaced an S in the other sequenced teleost metallothionein. Whether this affects the antigenicity of metallothionein has not been tested, but it has been shown that antibodies raised against cod and perch metallothioneins demonstrate distinct cross-reactivity patterns with different fish species (Hylland, 1992).

Methods of detecting metallothionein

Salient features of metallothioneins are their low molecular weight, heat stability, the abundance of cysteines in the protein and the thereby conveyed capability to sequester heavy metal ions. These features have been used in designing special methods for the detection of metallothioneins. Thus, besides the common techniques utilizing chromatography and antibody-derived methods for identification of metallothionein, other special methods have been designed. These include different metal saturation methods (Piotrowski, Bolanowska & Sapota, 1973; Onosaka & Cherian, 1982) and differential pulse polarography (Olafson & Olsson, 1991).

The metal saturation method is dependent on three main features. First, the high affinity of metallothionein for Cd or Hg. Secondly, that metallothionein is soluble after heat treatment or in 5% TCA. Thirdly, the ability of heat or TCA to remove non-metallothionein-bound Cd and Hg. The metal saturation method is easy to perform and allows processing of large numbers of samples in a short time. It is, however, an indirect method and there is a risk of high background due to inefficient removal of non-metallothionein bound ^{203}Hg or ^{109}Cd. The Cd-saturation method is further limited to Cd- and Zn-thioneins since Cd cannot replace Cu or Hg in metallothionein.

The differential pulse polarographic method takes advantage of two features of metallothionein. These are the heat solubility of metallothionein and secondly the high cysteine content of metallothionein. Non-metallothionein cysteine is removed by heat denaturation from the specimens to be tested, and the cysteine content of the sample is then measured by differential pulse polarography. The presence of glutathione does not interfere with the method due to the different linkage of amino acids. Differential pulse polarography is a semi-direct method to detect metallothionein. Since it detects the cysteine moiety of metallothionein, it has an advantage over antibody reactions in that it is not dependent on the primary or secondary structure of the protein. Thus differential pulse polarography is a good choice for measuring metallothionein levels in field samples of species where the structure of metallothionein is not known. The limitation of this method is mainly that each assay takes between 5 and 10 minutes, allowing the determination of only 5–10 samples an hour.

Metallothionein transcription can be measured in most teleost species using heterologous fish metallothionein cDNA probes under low stringency hybridization. The increased amount of sequence information of metallothionein from different teleost families facilitates the use of specific oligonucleotide derived probes for detection of metallothionein from a number of closely related fish species.

Metallothionein gene promoters in fish

To date, four metallothionein genes have been characterized in teleosts. These include two rainbow trout genes, the stone loach gene and the pike gene (Zafarullah, Bonham & Gedamu, 1988; Kille *et al.*, 1993; Murphy *et al.*, 1990; Kling & Olsson, 1993). Common to all the fish metallothionein promoter regions is the presence of several metal responsive elements (MREs). These MREs show a high degree of conservation when compared to mammalian MREs. At first, two MREs were identified within 120 bp upstream of the TATA box in both rainbow trout metallothionein-A and metallothionein-B (Zafarullah *et al.*, 1988; Murphy *et al.*, 1990). This was consistent with the localization of mammalian MREs. However, further characterization of more distal parts of metallothionein promoters revealed the presence of additional MREs located at −576 to −760 in rainbow trout metallothionein-A, −524 to −710 in stone loach metallothionein and −611 to −880 in pike metallothionein (Kling & Olsson, 1993; Kille, Stephens & Kay, 1991). The MREs have been shown to be the site of binding for metal transcription factors (MTF). To date, three possible MTFs have been isolated, these are the yeast ACE1, the mouse MTF1 and the mouse MafY (Thiele, 1988; Radtke *et al.*, 1993; Xu, 1993).

In addition to the MREs, both the rainbow trout and the pike metallothionein genes contain putative AP1 sites in the distal portion of the promoters (Kling & Olsson, 1993; Kille *et al.*, unpublished observations). There is a striking homology between the distal portion, containing two AP1 sites, of the promoter regions of the rainbow trout metallothionein-A gene and the pike metallothionein gene. In the rainbow trout metallothionein-B gene, there are two CAAT boxes and putative *cis*-elements for adenovirus E1A and CREB (Gedamu & Zafarullah, 1993). There are also one glucocorticoid responsive element (GRE) half site as well as a region with 85% similarity to a GRE in the rainbow trout MT-A gene (Kling & Olsson, 1993). Unlike the human metallothionein genes, no GC boxes or SP1 sites have yet been identified in the fish metallothionein genes.

Induced metallothionein gene transcription

Several groups have shown that administration of metals such as cadmium, copper, zinc or mercury result in induction of metallothionein synthesis (Olsson & Hogstrand, 1987; Zafarullah, Olsson & Gedamu, 1989a; Norey et al., 1990; Gagne, Marion & Denizeau, 1990b). While the binding affinity for metals to thionein is highest for mercury followed by copper, cadmium and zinc in decreasing order, this is not reflected by the ability of these metals to induce metallothionein synthesis. Thus, zinc is the most potent inducer of metallothionein in most studied vertebrate species, closely followed by cadmium and mercury, while copper is often a poor inducer. The inducibility of metallothionein by these different metals in vivo may also be a reflection of other systems that the specific metal species interact with. Studies on tissue culture have shown that the inducibility of metallothionein is temperature dependent in fish (Hyllner et al., 1989). Induction of metallothionein mRNA in RTH-149 (a rainbow trout hepatoma cell line) cells at 20 °C results in detectable transcription and translation within about 3–6 hours after metal treatment (Price-Haughey, Bonham & Gedamu, 1986; Zafarullah et al., 1989a). Primary cultures of rainbow trout hepatocytes grown at 15 °C showed detectable metallothionein protein levels 12 hours after cadmium or mercury treatment (Gagne et al., 1990a,b). One experiment has been performed to test the temperature dependency of the system, and this study showed that rainbow trout hepatocytes grown at 9 °C responded to zinc treatment with increased metallothionein protein levels after 4 days, while a decrease down to 6 °C delayed the response to 8 days (Hyllner et al., 1989).

The half-life of the metallothionein mRNA has been determined at 20 °C in RTG-2 cells and found to be about 60 hours (Zafarullah, Olsson & Gedamu, 1990). This half-life is longer than that observed in mammals (about 2 hours). The protein half-life in fish has been determined to be about 12 days in untreated rainbow trout, and has been shown to increase to 30 days upon Cu exposure. A half-life of about 30 days has also been shown for the plaice (Overnell, McIntosh & Fletcher, 1987).

Besides heavy metals, several other agents have been shown to induce metallothionein synthesis (Table 1). Metallothionein has been shown to be elevated following stress and inflammation in fish (Baer & Thomas, 1990; Maage et al., 1990). Stress induction of metallothionein

Table 1. *Agents tested for inducibility of teleost metallothionein*

Inducer	Tissue	Cell line	Primary culture
Cadmium	+	+	+
Zinc	+	+	+
Copper	+	+	+
Mercury	+		+
Cortisol	−	+	+
Corticosterone	−		+
Progesterone		+	
Oestradiol	−	−	−
Noradrenaline			+
Lipopolysaccharide B	−		
NIP_{11}-LPH	+		
TPA		−	
H_2O_2		−	
Turpentine	−		

The different agents have been tested for induction (+) or lack of induction (−) of metallothionein and/or metallothionein mRNA synthesis.
References: Burgess *et al.*, 1993; Gagne *et al.*, 1990a,b; George *et al.*, 1992; Hyllner *et al.*, 1989; Olsson *et al.* 1989; Overnell *et al.*, 1987; Price-Haughey *et al.*, 1986; Zafarullah *et al.*, 1989a; Olsson *et al.* unpublished observations.

has been shown for striped mullet (*Mugil cephalus* L.) after capture in the field (Baer & Thomas, 1990). It was found that the metallothionein levels in striped mullet liver increased ten-fold within one week after catch. Induction in response to inflammation has been shown on Atlantic salmon (*Salmo salar*) injected with NIP_{11}-LPH (Maage *et al.*, 1990). It has been suggested that induction of metallothionein in response to inflammation may be a protective response aimed at minimizing cell damage due to hydroxy radicals released during the acute phase response (Karin, 1985). Contradictory results have been obtained on fish in response to inducers of inflammatory response. In plaice injected with turpentine, no induction of metallothionein was observed (Overnell *et al.*, 1987).

Endogenous regulation of metallothionein

Induction of metallothionein can be brought about by administration of a variety of substances to the animal (George *et al.*, 1992; Hyllner *et al.*, 1989; Olsson, Zafarullah & Gedamu, 1989). The metallothionein levels in different organs from fish have been shown to be regulated in response to both exogenous and endogenous stimuli (Table 1). Thus, the basal levels of metallothionein have been shown to vary with time of year, water temperature, reproductive state and developmental stage in rainbow trout, winter flounder and plaice (Fletcher & King, 1978; Olsson, Haux & Förlin, 1987; Olsson et al., 1989; Olsson et al., 1990b; Overnell, Fletcher & McIntosh, 1988).

Some studies have been aimed at investigating the function and regulation of metallothionein in fish. Although this has led to some insight in the possible functions of metallothionein, there is still a lot of uncertainty as to the reason for the seasonal and developmental variations in metallothionein levels. So far, the results point towards a role for metallothionein in buffering against changes in the free metal levels in the cell. Little or no evidence exists that metallothionein in fish is actively involved in the regulation of metabolism by altering the zinc or copper levels of the cell.

The observed seasonal changes in metallothionein levels in fish have been correlated to either the state of sexual maturation or to the water temperature. Studies on winter flounder have shown that the hepatic zinc and copper levels vary with time of the year, and that there is an increased incorporation of zinc into low molecular weight proteins, likely to be metallothionein, during the summer feeding period (Shears & Fletcher, 1983, 1985). These variations further correlate with the period of sexual maturation in winter flounder (Fletcher & King, 1978). At the onset of sexual maturation in female fish, there is a need for increased metabolic activity due to the onset of exogenous vitellogenesis, during which time enormous quantities of vitellogenin are produced in the liver, transported to the gonads and incorporated in the developing oocytes. When the metallothionein levels were determined during an annual reproductive cycle in rainbow trout, it was observed that the metallothionein levels began to increase at the onset of vitellogenesis, and that the levels peaked when spawning occurred (Olsson et al., 1987, 1990b). The peak in metallothionein correlated to a decrease in liver size, and indicated that metallothionein was induced when the period of exogenous vitellogenesis was over. This was later confirmed in an experiment where juvenile rainbow trout were injected with oestradiol to induce vitellogenin synthesis (Olsson et al., 1989). It is believed that the induction of metallothionein is brought about by an increase in the free pool of zinc in the hepatocytes. Elevated metallothionein levels have also been observed in male fish at the time of spawning (Olsson et al., 1987; 1990b; Baer & Thomas, 1990). The actual increase in metallothionein mRNA levels during spawning is about two-fold in male fish and about seven-fold in female fish. While oestradiol does not induce metallothionein synthesis in fish, there are other hormones that have been indicated to regulate the metallothionein levels *in vitro*. Thus, glucocorticoids and progesterone as well as noradrenaline have been shown to influence the metallothionein levels in primary cultured hepatocytes and in cell lines (Hyllner *et al.*, 1989; Olsson *et al.*, 1990*a*; George *et al.*, 1992; Burgess, Frerichs & George, 1993).

Endogenous regulation of metallothionein has also been observed in the developing rainbow trout embryo (Olsson *et al.*, 1990*b*). In this case, an increase in basal levels of metallothionein is seen as gastrulation, and this is followed by a peak in metallothionein at hatching. At gastrulation, both isoforms of metallothionein increase and are correlated to increased levels of both zinc and copper in metallothionein. At hatch, it is mostly metallothionein isoform A that increases, and this increase is correlated to elevated levels of zinc in metallothionein. The mechanism of regulation of metallothionein during early development in fish has not been shown, and awaits further research.

Regulation of metallothionein has also been observed in response to altered water temperature and day length (Olsson *et al.*, unpublished results). It was observed that the metallothionein levels increased in

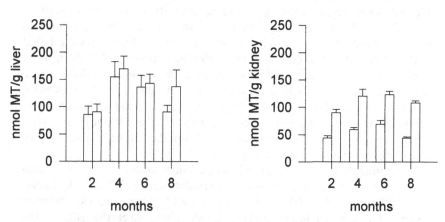

Fig. 2. Metallothionein levels in liver and kidney of rainbow trout during acclimation to lowered water temperature and exposure to cadmium. The water temperature was 12 °C at 2 months, 6 °C at 4 months and 2 °C at 6 and 8 months. The control fish are shown in the left bar and cadmium exposed fish are shown in the right bar. Each group consisted of 12 fish and the results are presented as mean ± S.E. (Olsson, unpublished observations.)

rainbow trout liver when the water temperature dropped and began to return to previous levels once the water temperature had stabilized at a new lower level (Fig. 2).

Most of the regulation of metallothionein has been shown to occur in the liver. When measured, the kidney, gill and intestinal levels of metallothionein have been shown to be far less affected by time of year, water temperature and reproductive state. Further, only a few species have so far been studied with respect to endogenous variability in metallothionein levels.

Metal detoxification and environmental monitoring

When subjected to cadmium exposure, fish show several toxic symptoms. The toxic effects of cadmium may be reduced by pretreatment of animals with sublethal doses of heavy metal (Pascoe & Beattie, 1979; Dixon & Sprague, 1981; Duncan & Klaverkamp, 1983). The pretreatment with heavy metals leads to induction of *de novo* synthesis of metallothionein, which results in increased binding of heavy metals to metallothionein (McCarter & Roch, 1983; Wicklund-Glynn & Olsson, 1991). It has thus been shown that the toxicity of heavy metals is decreased when they are bound to metallothionein.

It has been argued that the toxicity of heavy metals such as Cd and Hg will only occur when the binding capacity of metallothionein is exceeded and an increasing portion of the metal is being bound to other proteins in the cell. This view is based on the observation that pathological changes coincide with the appearance of Cd in non-metallothionein proteins (Winge, Krasno & Colucci, 1974). There would thus be a threshold level for toxicity of heavy metals. However, others have argued that the kinetics of metal uptake and metallothionein synthesis has to be taken into account whereby the pathological effects would appear when the rate of metal uptake exceeds the rate of metallothionein synthesis (McCarter et al., 1982).

Environmental exposure of fish to cadmium results in accumulation of the metal in renal and hepatic metallothionein (Olsson & Haux, 1986; Olsson et al., 1989; Norey et al., 1990). There are two primary routes of uptake of heavy metals in fish, either over the gills or the intestine (Wicklund-Glynn, Haux & Hogstrand, 1992; Sorensen, 1991). Independent of route of uptake, the metals will be in contact with different tissues and compartments before reaching the major sites of detoxification. The metals may thus interact with other cellular molecules before being sequestered by metallothionein.

It has been indicated that cadmium injections reduce the levels of vitellogenin in oestradiol-treated flounder (*Platichthys flesus* L.) (Faaborg

Povlsen, Korsgaard & Bjerregaard, 1990). As discussed above, the period of exogenous vitellogenesis is characterized by a drastic increase in liver size due to increased production of vitellogenin. There is probably an increased need for zinc as cofactor in this process, and the presence of cadmium could thus interfere with the vitellogenin production by replacing zinc in enzymes. In a study on rainbow trout it has been observed that cadmium that is injected in conjunction with oestradiol is initially distributed to high molecular weight proteins (Olsson *et al.*, 1995). In these fish there was no increased *de novo* synthesis of metallothionein during the first five days following cadmium injections. In fish treated with cadmium alone there is a substantial increase in metallothionein mRNA levels within five days of treatment. However, in the oestradiol-treated fish, cadmium was eventually redistributed from the high molecular weight proteins to metallothionein, and increased metallothionein mRNA levels could be observed 15 days post-injection. Components involved in the production of vitellogenin thus appear to compete with metallothionein for the pool of free metal ions, leading to incorporation of cadmium into non-metallothionein proteins in the cell. This indicates that, besides the rate of metal uptake and the rate of metallothionein synthesis, the metal-binding affinity of other proteins as well as the rate of synthesis of these proteins may be involved in the occurrence of toxic effects on the cell. These studies demonstrate the need for a better understanding of the regulation of metallothionein and its interaction with other systems when developing metallothionein assays for biomonitoring purposes.

The use of metallothionein as an early warning system for heavy metal pollution has received a great deal of attention. Considering the extensive knowledge of the induction and regulation of metallothionein in fish and the availability of sensitive and reliable assay methods for metallothionein, these proteins have a good potential as tools in monitoring exposure to toxic heavy metals.

There are certain criteria that have to be met if a system is to be used in biomonitoring. General criteria that have to be met are: (i) the assay system has to be reliable and sensitive; (ii) the species being used should preferably have low and stable basal levels of the measured parameter; (iii) the species should be sufficiently territorial so that past pollution exposure can be assigned to the geographic area being investigated.

As has been discussed earlier in this chapter, there are several assay systems that can be used to determine metallothionein levels, including immunoassays, differential pulse polarography and metal saturation assays. These have all been shown to be sensitive enough to measure tissue levels of metallothionein. The basal levels of hepatic metallothionein are known for a number of teleost species and are generally low except in certain fish with atypical metal metabolism, such as the white perch (Bunion *et al.*, 1987) and certain squirrelfish (Hogstrand & Haux, 1990).

When using metallothionein as a bioindicator some special criteria have to be met. (i) The investigation period should be well defined and periods of rapid change in temperature should be avoided; (ii) periods of sexual maturation should also be avoided and juvenile fish should be used whenever possible.

The levels of metallothionein have been shown to vary with the time of year and have been correlated to both environmental water temperature and the state of sexual maturation in fish. In both fish and mammals, metallothionein induction has been shown to occur in response to a multitude of agents (Hamer, 1986; Zafarullah *et al.*, 1989*b*). Further studies have shown that, besides MREs, there are other putative *cis*-acting elements in fish metallothionein genes (Olsson, 1993; Gedamu & Zafarullah, 1993). These include possible API sites, adenovirus E1A sites, CREB sites and CAAT boxes (Olsson, 1993; Gedamu & Zafarullah, 1993). The presence of these *cis*-acting elements in the two rainbow trout genes indicate that these two genes may be regulated via non-metal pathways. It is therefore important to avoid using fish that are developing, sexually maturing or in a changing environment for biomonitoring purposes. In a study on rainbow trout it was found that the hepatic levels of metallothionein increased as the water temperature dropped and that cadmium exposure did not increase the metallothionein levels any further (Olsson, unpublished data). However, the kidney levels remained fairly stable during this period, and exposure of these fish to cadmium resulted in elevated levels in the kidney. This indicates that, in some cases, it may be of interest to use organs other than liver in biomonitoring experiments.

Conclusions

The increasing amount of information on the induction and regulation of metallothionein in different fish species has shown that the induction of metallothionein by metals in the environment provides a suitable system for monitoring purposes. Studies have shown that the metallothionein regulation in fish is fairly complex, and that many endogenous and exogenous factors will induce metallothionein synthesis. Due to the complex regulation, certain criteria have to be fulfilled before metallothionein can be used in biomonitoring to avoid high variations in the basal levels of metallothionein. It is of particular importance to avoid the period of sexual maturation as well as times with high variation in water temperature. If the necessary criteria are being met, metallothionein has a strong potential for use in monitoring effects of metal exposure in the environment.

Acknowledgements

This study was supported by a grant from the Swedish National Research Council. I would like to thank Dr Lars Förlin for critical reading of the manuscript and valuable comments. I am also grateful to Dr Peter Kille for making unpublished material available for this book chapter.

References

Andrews, G.K. (1990). Regulation of metallothionein gene expression. *Progress in Food and Nutritional Science*, **14**, 193–258.

Baer, K.N. & Thomas, P. (1990). Influence of capture stress, salinity and reproductive status on zinc associated with metallothionein-like proteins in the livers of three marine teleost species. *Marine Environmental Research*, **28**, 157–61.

Bonham, K., Zafarullah, M. & Gedamu, L. (1987). The rainbow trout metallothioneins: molecular cloning and characterization of two distinct cDNA sequences. *DNA*, **6**, 519–28.

Bunion, T., Baksi, S.M., George, S. & Frazier, J.M. (1987). Abnormal copper storage in a teleost fish (*Morone americana*). *Veterinary Pathology*, **24**, 515–24.

Burgess, D., Frerichs, N. & George, S. (1993). Control of metallothionein expression by hormones and stressors in cultured fish cells. *Marine Environmental Research*, **35**, 25–8.

Chan, K.M., Davidson, W.S. & Fletcher, G.L. (1989). Molecular cloning of metallothionein cDNA and analysis of metallothionein gene expression in winter flounder (*Pseudopleuronectes americanus*). *Canadian Journal of Zoology*, **67**, 2520–9.

Dixon, D.G. & Sprague, J.B. (1981). Copper bioaccumulation and hepatoprotein synthesis during acclimation to copper by juvenile rainbow trout. *Aquatic Toxicology*, **1**, 69–81.

Duncan, D.A. & Klaverkamp, J.F. (1983). Tolerance and resistance to cadmium in white suckers (*Catastomus commersoni*) previously exposed to cadmium, mercury, zinc or selenium. *Canadian Journal of Fisheries and Aquatic Sciences*, **40**, 128–38.

Faaborg Povlsen, A., Korsgaard,B. & Bjerregaard, P. (1990). The effect of cadmium on vitellogenin metabolism in estradiol-induced flounder (*Platichthys flesus* L.) males and females. *Aquatic Toxicology*, **17**, 253–62.

Fletcher, G.L. & King, M.J. (1978). Seasonal dynamics of Cu^{2+}, Zn^{2+}, Ca^{2+} and Mg^{2+} in gonads and liver of winter flounder (*Pseudopleuronectes americanus*): Evidence for summer storage of Zn^{2+} for winter gonad development in females. *Canadian Journal of Zoology*, **56**, 284–90.

Gagne, F., Marion, M. & Denizeau, F. (1990*a*). Metal homeostasis and metallothionein induction in rainbow trout hepatocytes exposed to cadmium. *Fundamental and Applied Toxicology*, **14**, 429–37.

Gagne, F., Marion, M. & Denizeau, F. (1990*b*). Metallothionein induction and metal homeostasis in rainbow trout hepatocytes exposed to mercury. *Toxicology Letters*, **51**, 99–107.

Gedamu, L. & Zafarullah, M. (1993). Molecular analysis of rainbow trout metallothionein and stress protein genes: structure, expression and regulation. In *Biochemistry and Molecular Biology of Fishes vol 2. Molecular Biology Frontiers*. Hochachka, P.W. & Mommsen, T.P. (eds.), pp. 241–58. Elsevier Science Publisher, Amsterdam.

George, S., Burgess, D., Leaver, M. & Frerichs, N. (1992). Metallothionein induction in cultured fibroblasts and liver of a marine flatfish, the turbot, *Scopothalmus maximus*. *Fish Physiology and Biochemistry*, **10**, 43–54.

Hamer, D.H. (1986). Metallothionein. *Annual Review of Biochemistry*, **55** 913–51.

Hogstrand, C. & Haux, C. (1990). Metallothionein as an indicator of heavy metal exposure in two subtropical fish species. *Journal of Experimental Marine Biology and Ecology*, **138**, 69–84.

Hylland, K. (1992). Cellular markers for metal exposure in marine invertebrates and fish. PhD Thesis, University of Oslo, Norway.

Hyllner, S.J., Andersson, T., Haux, C. & Olsson, P.-E. (1989). Cortisol induction of metallothionein in primary culture of rainbow trout hepatocytes. *Journal of Cellular Physiology*, **139**, 24–8.

Karin, M. & Richards, R.I. (1984). The human metallothionein gene family: structure and expression. *Environmental Health Perspectives*, **54**, 111–15.

Karin, M. (1985). Metallothionein: protein in search of function. *Cell*, **41**, 9–10.

Kay, J., Cryer, A., Brown, M.W., Norey, C.G., Bremner, I., Overnell, J., Parten, B. & Dunn, B.M. (1987). N-terminal sequence of metallothionein from rainbow trout. *Biochemical Society Transactions*, **15**, 453–4.

Kille, P., Stephens, P.E. & Kay, J. (1991). Elucidation of cDNA sequences for metallothioneins from rainbow trout, stone loach and pike liver using the polymerase chain reaction. *Biochimica Biophysica Acta*, **1089**, 407–10.

Kille, P., Kay, J. & Sweeney, G.E. (1993). Analysis of regulatory elements flanking the metallothionein genes in Cd-tolerant fish (pike and stone loach). *Biochimica Biophysica Acta*, **1216**, 55–69.

Kling, P. & Olsson, P.-E. (1993). Metal regulation of the rainbow trout metallothionein-A gene. In *Proceedings of the Seventh International Symposium on Responses of Marine Organisms to Pollutants.*

Leaver, M. & George, S. (1989). Nucleotide and deduced amino acid sequence of a metallothionein cDNA clone from the marine fish, *Pleuronectes platessa.* EMBL accession, #56743.

Maage, A., Waagbo, R., Olsson, P.-E., Julshamn, K. & Sandnes, K. (1990). Ascorbate-2-sulfate as a dietary vitamin C source for Atlantic salmon (*Salmo salar*): 2. Effects of dietary levels and immunization on the metabolism of trace elements. *Fish Physiology and Biochemistry,* **8**, 429–36.

McCarter, J.A., Matheson, A.T., Roch, M., Olafson, R.W. & Buckley, J.T. (1982). Chronic exposure of coho salmon to sublethal concentrations of copper II: Distribution of copper between high- and low-molecular weight proteins in the liver cytosol and the possible role of metallothionein in detoxification. *Comparative Biochemistry and Physiology,* **72C**, 21–6.

McCarter, J.A. & Roch, M. (1983). Hepatic metallothionein and resistance to copper in juvenile coho salmon. *Comparative Biochemistry and Physiology,* **74C**, 133–7.

Marafante, E. (1976). Binding of mercury and zinc to cadmium-binding protein in liver and kidney of goldfish (*Carassius auratus* L.). *Experientia,* **32**, 149–50.

Murphy, M.F., Collier, J., Koutz, P. & Howard, B. (1990). Nucleotide sequence of the trout metallothionein A gene 5′ regulatory region. *Nucleic Acid Research,* **18**, 4622.

Norey, C.G., Brown, M.W., Cryer, A. & Kay, J. (1990). A comparison of the accumulation, tissue distribution and secretion of cadmium in different freshwater fish. *Comparative Biochemistry and Physiology,* **96C**, 181–4.

Olafson, R.W. & Olsson, P.-E. (1991). Electrochemical detection of metallothionein. In *Methods in Enzymology.* Riordan, J.F. & Vallee, B.L. (eds.), pp. 205–13. Academic Press Inc., San Diego.

Olsson, P.-E. & Haux, C. (1986). Increased hepatic metallothionein content correlates to cadmium accumulation in environmentally exposed perch (*Perca fluviatilis*). *Aquatic Toxicology,* **9**, 231–42.

Olsson, P.-E. (1993). Metallothionein gene expression and regulation in fish. In *Biochemistry and Molecular Biology of Fishes vol 2. Molecular Biology Frontiers.* Hochachka, P.W. & Mommsen, T.P. (eds.), pp. 259–78. Elsevier Science Publisher, Amsterdam.

Olsson, P.-E., Haux, C. & Förlin, L. (1987). Variations in hepatic metallothionein, zinc and copper levels during an annual reproductive cycle in rainbow trout, *Salmo gairdneri. Fish Physiology and Biochemistry,* **3**, 39–47.

Olsson, P.-E. & Hogstrand, C. (1987). Subcellular distribution and binding of cadmium to metallothionein in tissues of rainbow trout after exposure to [109]Cd in water. *Environmental Toxicology and Chemistry*, **6**, 867–74.

Olsson, P.-E., Hyllner, S.J., Zafarullah, M., Andersson, T. & Gedamu, L. (1990a). Differences in metallothionein gene expression in primary cultures of rainbow trout hepatocytes and the RTH-149 cell line. *Biochimica Biophysica Acta*, **1049**, 78–82.

Olsson, P.-E., Zafarullah, M., Foster, R., Hamor, T. & Gedamu, L. (1990b). Developmental regulation of metallothionein mRNA, zinc and copper levels in rainbow trout, *Salmo gairdneri*. *European Journal of Biochemistry*, **193**, 229–35.

Olsson, P.-E., Zafarullah, M. & Gedamu, L. (1989). A role of metallothionein in zinc regulation after oestradiol induction of vitellogenin synthesis in rainbow trout, *Salmo gairdneri*. *Biochemical Journal*, **257**, 555–9.

Olsson, P.-E., Kling, P., Petterson, C. & Silversand, C. (1995). Interaction of cadmium and oestradiol-17β on metallothionein and vitellogenin synthesis in rainbow trout (*Oncorhyncus mykiss*). *Biochemical Journal* **307**, 197–203.

Onosaka, S. & Cherian, M.G. (1982). Comparison of metallothionein determination by polarographic and cadmium saturation methods. *Toxicology and Applied Pharmacology*, **63**, 270–4.

Overnell, J., Berger, C. & Wilson, K.J. (1981). Partial amino acid sequence of metallothionein from the plaice (*Pleuronectes platessa*). *Biochemical Society Transactions*, **9**, 217–18.

Overnell, J., McIntosh, R. & Fletcher, T.C. (1987). The enhanced induction of metallothionein by zinc, its half-life in marine fish, *Pleuronectes platessa*, and the influence of stress factors on metallothionein levels. *Experientia*, **43**, 178–81.

Overnell, J., Fletcher, T.C. & McIntosh, R. (1988). Factors affecting hepatic metallothionein levels in marine flatfish. *Marine Environmental Research*, **24**, 155–8.

Pascoe, D. & Beattie, J.H. (1979). Resistance to cadmium by pretreated rainbow trout alevins. *Journal of Fish Biology*, **30**, 539–46.

Piotrowski, J.K., Bolanowska, W. & Sapota, A. (1973). Evaluation of metallothionein content in animal tissues. *Acta Biochimica Polonica*, **20**, 207–15.

Price-Haughey, J., Bonham, K. & Gedamu, L. (1986). Heavy metal-induced gene expression in fish and fish cell lines. *Environmental Health Perspectives*, **65**, 141–7.

Radtke, F., Heuchel, R., Georgiev, O., Hergersberg, M., Gariglio, M., Dembic, Z. & Schaffner, W. (1993). Cloned transcription factor MTF-1 activates the mouse metallothionein I promoter. *The EMBO Journal*, **12**, 1355–62.

Shears, M.A. & Fletcher, G.L. (1983). Regulation of Zn^{2+} uptake from the gastrointestinal tract of a marine teleost, the winter flounder (*Pseudopleuronectes americanus*). *Canadian Journal of Fisheries and Aquatic Sciences*, **40**, 197–205.

Shears, M.A. & Fletcher, G.L. (1985). Hepatic metallothionein in the winter flounder (*Pseudopleuronectes americanus*). *Canadian Journal of Fisheries and Aquatic Sciences*, **63**, 1602–9.

Sorensen, E.M. (1991). *Metal Poisoning in Fish*. 374 pp. CRC Press, Inc., Boca Raton.

Thiele, D.J. (1988). ACE1 regulates expression of the *Saccharomyces cervisiae* metallothionein gene. *Molecular and Cellular Biology*, **9**, 421–9.

Wei, D. & Andrews, G. (1988). Molecular cloning of chicken metallothionein: deduction of the complete amino acid sequence and analysis of expression using cloned cDNA. *Nucleic Acid Research*, **16**, 537–53.

Wicklund-Glynn, A. & Olsson, P.-E. (1991). Cadmium turnover in minnows (*Phoxinus phoxinus*) exposed to waterborne cadmium. *Environmental Toxicology and Chemistry*, **10**, 383–94.

Wicklund-Glynn, A., Haux, C. & Hogstrand, C. (1992). Chronic toxicity and metabolism of Cd and Zn in juvenile minnows (*Phoxinus phoxinus*) exposed to a Cd and Zn mixture. *Canadian Journal of Fisheries and Aquatic Sciences*, **49**, 2070–9.

Winge, D.R., Krasno, J. & Colucci, A.V. (1974). Cadmium accumulation in rat liver: correlation between bound metal and pathology. In *Trace Element Metabolism in Animals*. Hokstra, W.G., Suttie, J.W., Ganther, H.E. & Mertz, W. (eds.), pp. 500–2. University Park Press, Baltimore.

Xu, C. (1993). cDNA cloning of a mouse factor that activates transcription from a metal responsive element of the mouse metallothionein-I gene in yeast. *DNA and Cell Biology*, **12**, 517–25.

Zafarullah, M., Bonham, K. & Gedamu, L. (1988). Structure of the rainbow trout metallothionein B gene and characterization of its metal-responsive region. *Molecular and Cellular Biology*, **8**, 4469–76.

Zafarullah, M., Olsson, P.-E. & Gedamu, L. (1989a). Endogenous and heavy metal ion metallothionein gene expression in Salmonid tissues and cell lines. *Gene*, **83**, 85–93.

Zafarullah, M., Olsson, P.-E. & Gedamu, L. (1989b). Rainbow trout metallothionein gene structure and regulation. In *Oxford Survey of Eukaryotic Genes*. Maclean, N. (ed.), pp. 111–43. Oxford University Press, New York.

Zafarullah, M., Olsson, P.-E. & Gedamu, L. (1990). Differential regulation of metallothionein gene in rainbow trout fibroblasts, RTG-2. *Biochimica et Biophysica Acta*, **1049**, 318–23.

J.P. SUMPTER, S. JOBLING, and C.R. TYLER

Oestrogenic substances in the aquatic environment and their potential impact on animals, particularly fish

Introduction

In this review we summarize what is known presently about which oestrogenic substances are present in the aquatic environment, at what concentrations they are present, and then discuss what effects these oestrogenic substances might have on aquatic organisms. By 'oestrogenic substances' we mean chemicals that have been shown to mimic the physiological effects of true oestrogens such as 17β-oestradiol (the major physiologically active oestrogen in all classes of vertebrates). Oestradiol is quite a small molecule; its molecular weight is 272 and it is composed of four aromatic rings (Fig. 1). Likewise, chemicals which mimic the effects of oestradiol are often small, and their structures are often based around one or more aromatic rings (Fig. 1 for examples). However, not all oestrogenic substances have structures which superficially resemble that of oestradiol; kepone, for example, does not have a structure which would in any way suggest that it might act as an oestrogen (Fig. 1). Because chemicals which do not resemble oestradiol can function as oestrogens, it has not been possible to predict, based on knowledge of their structures, which chemicals might be oestrogenic; hence, the oestrogenic activity of chemicals has usually been discovered only when effects are noted, by which time the chemical might be in widespread use (DDT provides a good example here). Further, it is suggested from recent findings (see later section on oestrogenic effects on aquatic organisms) that there are chemicals already in widespread use, and therefore present in most (if not all) environments, which are not recognized presently to be oestrogenic, but which might yet turn out to be so. If this seems unlikely, it is as well to remember that alkylphenol polyethoxylate surfactants (APEs) were introduced in the 1940s and have been used in large quantities ever since, yet it was only an unexpected observation in a laboratory 50 years later that led to the realization that nonylphenol, a major and persistent degradation product of APEs, is oestrogenic (Soto *et al.*, 1991).

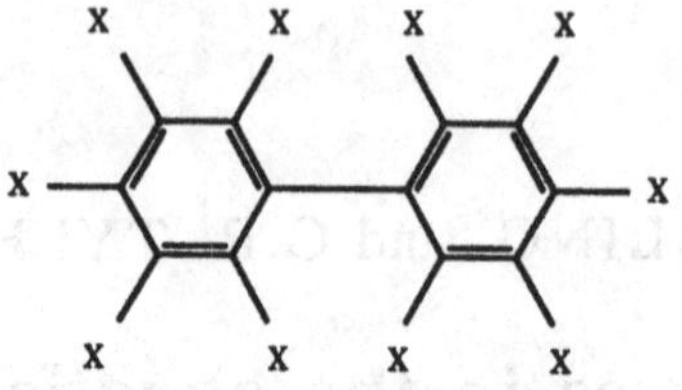

Arochlor 1221 (21% chlorine) (X=H or Cl)

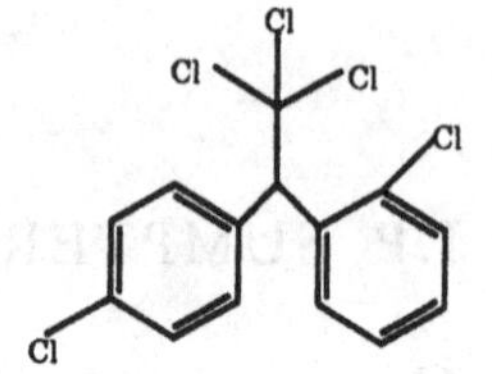

o,p'-DDT

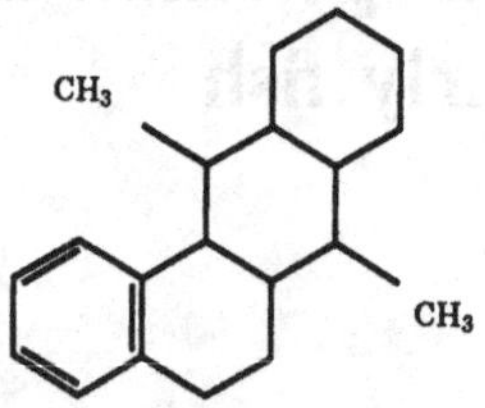

7,12, dimethylbenzanthracene

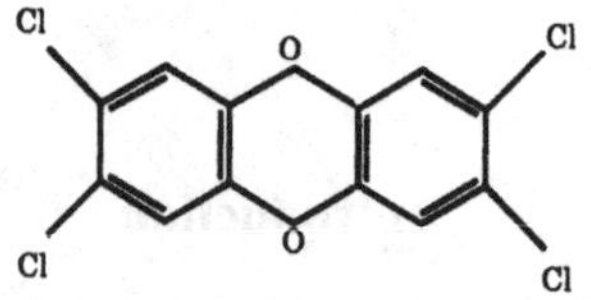

2,3,7,8,TCDD

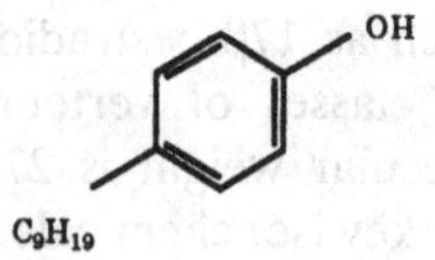

4-nonylphenol

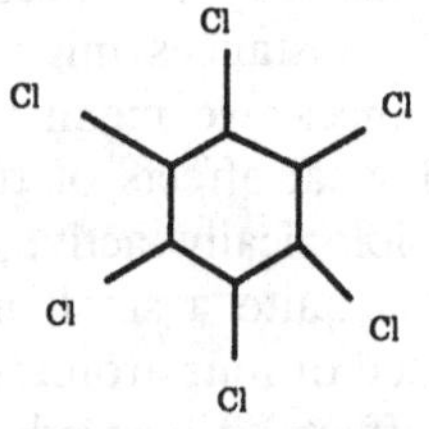

β-Hexachlorocyclohexane

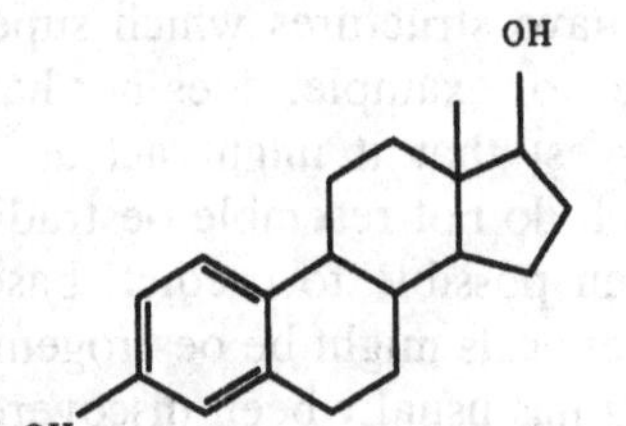

17β-Oestradiol

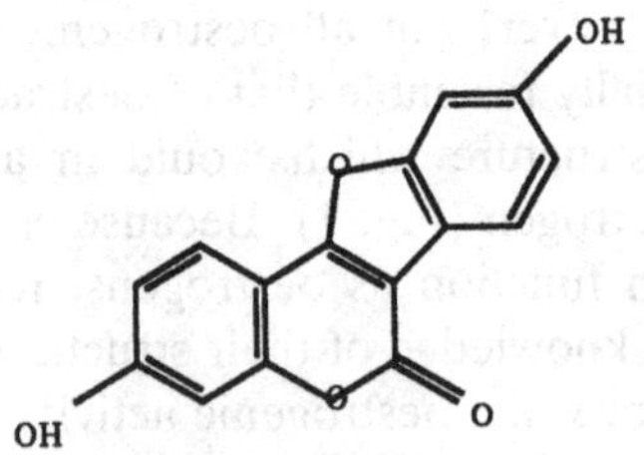

Coumestrol

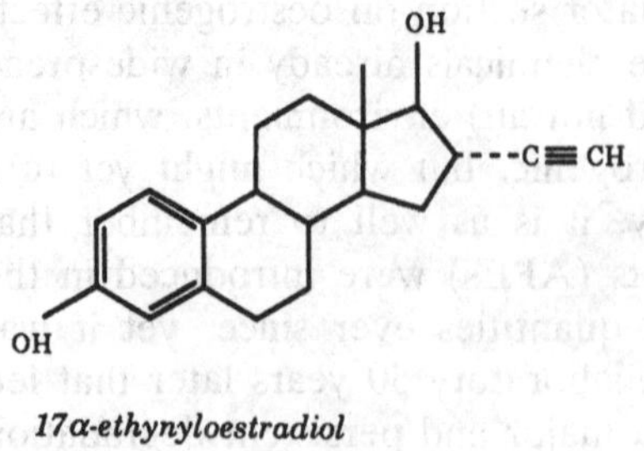

17α-ethynyloestradiol

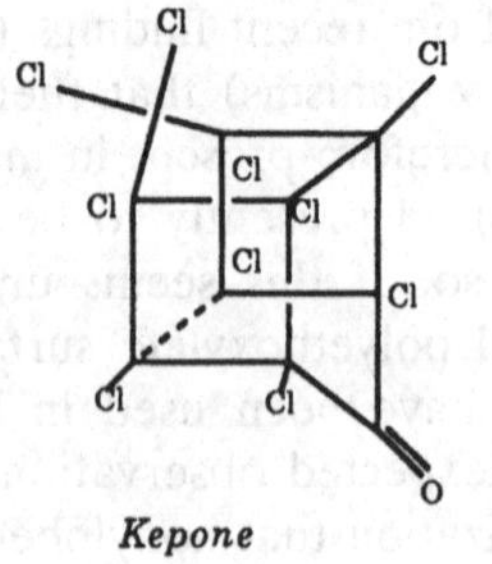

Kepone

Most of the oestrogenic chemicals with which we are concerned here are man-made. There are, however, some natural chemicals which mimic the effects of oestrogens; fungi can produce oestrogenic substances (which are collectively called mycoestrogens) as can some plants, particularly some leguminous plants (the phytoestrogens). It is not known whether representatives of either of these groups of oestrogenic substances are present in the aquatic environment; however, it seems unlikely that, even if present, the concentrations would be high enough to affect aquatic organisms. It is also possible that natural oestrogens (including 17β-oestradiol itself) from animals may enter the aquatic environment. Fish produce and excrete oestrogens, although probably not in amounts likely to affect other organisms. Intensive agriculture, however, can lead to significant amounts of steroids, including oestrogens, being excreted in the urine and faeces, which are most likely to find their way into the aquatic environment. It has been hypothesized that this route is responsible for the appreciable concentrations of oestrogens and androgens present in river and lake water in Israel (Shore, 1993). Unfortunately, extremely little, if anything, is known about the concentrations of these natural oestrogenic chemicals in the aquatic environment.

Finally, by way of introduction, we want to emphasize that the words 'aquatic environment' hide the fact that this is a highly variable environment. Not only does it include freshwater and seawater, but both of these (especially the former) are themselves variable. For instance, freshwater includes lakes, river, reservoirs and groundwater. Even within a single category of freshwater, there will be considerable variability: an upland reservoir situated in mountains is a very different environment from a lowland reservoir supplied primarily from a river.

Fig. 1. The chemical structures of representatives of each of the major groups of oestrogenic xenobiotics: *o,p*-DDT, β-Hexachlorocyclohexane and Kepone are organochlorine pesticides; Arochlor 1221 is a polychlorinated biphenyl (PCB); 2, 3, 7, 8-TCDD is a polychlorinated dibenzodioxin; 7, 12-dimethylbenzanthracene is a polycyclic aromatic hydrocarbon (PAH); 4-nonylphenol is an alkylphenol; coumestrol is a phytoestrogen and 17β-oestradiol and 17α-ethynyloestradiol are natural and synthetic oestrogens, respectively. Note that some of these chemicals which can mimic the effects of oestrogens have structures that resemble that of the major natural oestrogen 17β-oestradiol, but that others do not.

Because most of the oestrogenic chemicals are man-made, waters in urban areas are more likely to contain significant concentrations of the chemicals than are waters in rural areas (unless the oestrogenic chemical is used in agriculture). What follows serves as no more than an introduction to the chemicals known to be oestrogenic; there has been no systematic study of the distribution of any of these chemicals in the aquatic environment. Even less is known about whether these chemicals are present at high enough concentrations to have oestrogenic effects on aquatic organisms, or on organisms that drink the water.

Oestrogenic chemicals

The chemicals discussed below are all recognized to be oestrogenic. Often this oestrogenic activity has been confirmed in mammals (or in tissue culture using mammalian cell lines) only, but it is very likely that, if a chemical can mimic the effects of natural oestrogens, it can do so in all classes of vertebrates (Pelissero et al., 1993; White et al., 1994). The chemicals may also act as oestrogens in plants, though the roles of 'oestrogens' in plants are unclear, making it difficult to know what functions should be monitored in plants to determine whether environmental 'oestrogens' are affecting plants as well as animals. For information to support our statement that the following chemicals are oestrogenic, the reader should consult McLachlan, 1980 and 1985. More recent references on specific groups of oestrogenic chemicals include polychlorinated biphenyls (Jansen et al., 1993); dibenzodioxins and dibenzofurans (Krishnan & Safe, 1993); alkylphenolic compounds (White et al., 1994) and phytoestrogens and mycoestrogens (Mayr, Butsch & Schneider, 1992).

Organochlorine pesticides

A large number of pesticides are approved for use and, in theory, any of these could enter the aquatic environment. Many of the environmentally persistent pesticides are organochlorine compounds, which were introduced in the 1950s. DDT is the best known of these pesticides and is perhaps the most studied; Lindane (and other isomers of hexachlorocyclohexane: HCH), Methoxychlor, Kepone (chlordecone), HCB (BHC) and Mirex are others.

Although their usage is now restricted in the Western world, owing to their toxic effects on both aquatic and terrestrial animals, they are still used widely in developing countries such as India and Africa, where it is not uncommon to find concentrations as high as 1 to 10 µg/litre in well-water, rainwater and groundwater (Begum, Begum &

Alam, 1992). Concentrations in the Western world are much lower, though they are still present in the environment due to their resistance to biodegradation and hence long half-lives.

Viewed on a global scale, the concentrations of organochlorine pesticides in the surface waters of oceans in the northern hemisphere are higher (1 to 2.3 ng/litre) than those in the South (0.25 ng/litre) (Allchin, 1991; Tatsukawa, 1992). In more localized areas, concentrations can be much higher; for example, 63 and 400 ng/litre of α- and β-HCH, respectively, were reported in wet deposition in Northern Italy (Galassi *et al.*, 1987). In the Great Lakes of North America, the concentrations of organochlorine pesticides have been better studied than in any other region. After falling from a peak in 1972, the concentrations of DDT and other pesticides in fish which inhabit the lakes are increasing again, probably due to long-range atmospheric transport of the pesticides to the Lakes (Kelly *et al.*, 1991).

Polychlorinated biphenyls (PCBs)

PCBs, of which Aroclor serves as a good example, are a group of compounds very closely related to DDT. They are used not only as pesticides, but also as sealants, paint additives and in plastics. They are causing a great deal of concern as global aquatic pollutants, particularly as 70% of the world's production of PCBs is still in use or in stock, and could therefore reach the aquatic environment in the future. Like the organochlorine pesticides mentioned above, PCBs are present in higher concentrations in the northern hemisphere than in the southern; for example, it has been estimated that between 11 and 17 tonnes of total PCB enters the North Sea every year (Klamer, Laane & Marquenie, 1991). As contaminants, they are present in similar concentrations to DDT, although their continued heavy usage in developing countries is leading to increasing concentrations in the global environment as a whole.

Polycyclic aromatic hydrocarbons (PAHs) and polychlorinated dibenzodioxins (PCDDs)

There are about 80 known PAH structures with between two and six aromatic rings: they are pollutants from petroleum-related sources, such as ship traffic, oil seepage or spillage, and the combustion of natural fuels. In a typical modern sewage treatment works, PAHs would be dominated by the naphthalenes, such as benzo(a)pyrene, concentrations of which are up to 0.1 µg/litre in the influent and less than 50 ng/litre in the effluent. These compounds are usually, if not exclusively, associated with particles of organic matter, and hence information on the occurrence in the dissolved phase is rare.

Atmospheric transport is one of the major methods by which these compounds are transported around the globe; for example, Kelly *et al.* (1991) estimated that 8740 kg PAHs enter Lake Erie in North America each year via this route. By contrast, the Detroit River, which runs into Lake Erie, contributes 41 000 kg/annum. The concentration of total PAHs in the water of the Detroit river was 200 ng/litre; benzo(a)pyrene alone was only 1% of this total. The Great Lakes, however, appear to be a worst-case scenario, and concentrations in waters of other parts of the world are usually reported in tens of ng/litre, rather than in hundreds (e.g. Boubassi & Sailot, 1991). Another exceptional situation occurs in the region of pulp mills, where concentrations in the effluent of mills can be as high as 30 µg/litre (Merrimen *et al.*, 1991).

Almost any combustion process occurring in the presence of chlorine will cause the emission of PCDDs (and their relatives the polychlorinated dibenzofurans; PCDFs) into the atmosphere. There are approximately 210 PCDD/Fs with chlorine atoms in the 2, 3, 7 and 8 positions; they are the most toxic of the PCDD/Fs and hence most studied, particularly 2, 3, 7, 8- TCDD. They are emitted from motor exhausts, are unwanted products in the production of PCBs, pesticides and herbicides, and will also be produced during the burning of municipal, chemical, hospital and domestic wastes. PCDDs and PCDFs are chemically very stable and have low water solubilities. Because of their hydrophobicity and lipophilicity, PDCCs will adsorb strongly to organic material in the particulate and dissolved phases. For this reason, very few studies have focused on the presence of PCDDs in solution (see Fletcher & McKay, 1993, for a review). Concentrations in the aquatic environments are generally between 0.3 and 1.4 µg/litre, although in heavily polluted areas they may reach 10 to 40 µg/litre.

Alkylphenolic compounds

The two major classes of surfactants are the non-ionic (including both alkylphenol polyethoxylates and alcohol ethoxylates) and anionic (linear alkylbenzene sulphonates). These compounds are present in the environment owing to their very high usage in industry and households. They are used primarily in detergents, but are also widely used in many other products such as paints, formulated pesticides and herbicides and other agricultural products, personal care products and in food packaging.

Only products derived from one group of surfactants, the alkylphenol polyethoxylates (APEs), have been shown to be oestrogenic (Jobling & Sumpter, 1993). Approximately 80% of the APEs used are nonyl-phenol-polyethoxylates (NPEOs), whilst most of the remaining 20% are octylphenol-polyethoxylates. These compounds enter the aquatic environment after use, often going via sewage treatment works, where they may undergo biological degradation before entering rivers and other waters. APEs are biodegraded to variable degrees, leading to persistent metabolites (Ahel, Conrad & Giger, 1987; Giger, Brunner & Schaffner, 1984). Some of these metabolites are hydrophobic and lipophilic (e.g. the alkylphenols nonylphenol and octylphenol) and tend to accumulate in sediment and sludge, whereas others, such as the longer-chain APEs and the carboxylic acid derivatives (APECs) are more soluble (see Ahel and Giger, 1993, for further details on solubilities of these chemicals).

Domestic effluents contain up to hundreds of µgs APEs/litre (Ahel & Giger, 1985; Naylor *et al.*, 1992), whereas some industrial effluents, such as those originating from pulp mills and textile companies can contain significantly higher concentrations. Concentrations of these chemicals in rivers seem generally to be in the low µg/litre range (Ahel *et al.*, 1987; Naylor *et al.*, 1992; Paxeus, Robinson & Balmér, 1992; Thoumelin, 1991), but may be higher in countries such as Israel where water availability is limited (Zoller *et al.*, 1990; Zoller, 1993). So widespread are these chemicals that, like many other persistent pollutants, they have even found their way into drinking water; for example, Clark *et al.* (1992) reported the presence of many alkylphenolic compounds in drinking water in New Jersey, USA, the total concentration being approximately 1 µg/litre. Schröder (1992) also describes non-ionic surfactant contaminants in drinking water in Germany, and Ventura *et al.* (1992) identified NPECs in the drinking water of Barcelona.

Natural and synthetic oestrogens

No information is available as to whether 'oestrogens' of plant and fungal origin (phytoestrogens and mycoestrogens respectively) are present in the aquatic environment. Natural oestrogens and their metabolites will be present, because they are synthesized and subsequently secreted by aquatic animals, particularly fish. In general, it seems likely that their concentrations in water will again be very low, although it is possible that in specific circumstances they could be present in significant concentrations. One such circumstance occurs in Israel, where oestrogens present in urine and faeces of farmed animals appear to be responsible for concentrations of oestradiol between 50 and 150 ng/litre in raw sewage, and 5 to 20 ng/litre in Lake Kiri adjacent to a large city, Tel Aviv (Shore, 1993).

A variety of synthetic oestrogens are used in agriculture (to fatten animals) and as pharmaceuticals. Of particular concern is ethinyl oestradiol, because it is a very potent oestrogen and is very widely used, being the active ingredient of the contraceptive pill, as well as being used in post-menopausal hormone-replacement therapy. Although a little may be excreted unaltered, most is metabolized to inactive conjugates prior to excretion. Nevertheless, it has been reported to be present in sewage treatment works at concentrations up to 15 ng/litre (Aherne & Briggs, 1989), although to us this concentration seems much higher than the theoretical maximum derived from calculations based on the number of people taking 'the pill' and water flow rates through sewage treatment works.

Bioconcentration and bioaccumulation

Bioconcentration refers to the concentration of a chemical in an organism such that the concentration in the tissues of the organism is greater than that in the water they live in. Bioaccumulation (or biomagnification as it is sometimes called), on the other hand, refers to the progressive accumulation of a chemical up the food chain.

Most of the oestrogenic substances discussed above are lipophilic and hydrophobic. They have strong tendencies to bioaccumulate in aquatic organisms, and are often transferred through the food chain. For example, bioconcentration factors (BCFs) for many PCBs and other organochlorine pesticides are well documented in aquatic organisms, especially fish; for most of these compounds the BCF in fish is between 1000 and 100 000 (Saito, Tanoue & Matsuo, 1992). This has led to present levels of PCBs and other organochlorine pesticides in Great Lakes fish of the order of 2 to 5 µg/g of fat. Bioaccumulation factors for these compounds appear to be of the order of 10 to 30 from plankton to fish (Evans, Noguchi & Rice, 1991). Marine mammals may bioaccumulate organochlorine pesticides to a greater extent, probably due to the high percentage of body weight in the form of subcutaneous fat, leading to concentrations of various organochlorine pesticides in these animals between 1 and 500 µg/g of blubber (Muir et al., 1992; Tatsukawa, 1992).

PAHs and PCDDs behave in a similar manner. Frakes, Zeeman & Mower (1993) reported that the concentrations of TCDD in Maine river fish in the USA were high enough to prompt health departments in 15 States to issue fish consumption advisory notices. This particular study reported bioaccumulation factors (BAFs) in fish of between 3000 and 28 300. The magnitude of the BAF depended greatly on the diet and microhabitat of the fish species in question; bottom feeding fish being those with the highest BAFs.

It is also likely that surfactants, and their degradation products, bioaccumulate and/or bioconcentrate in aquatic organisms, although published studies are not as numerous as those on the pesticides, PAHs and PCDDs. Alkylphenolic compounds originating from the biodegradation of nonylphenol polyethoxylates, such as nonylphenol (NP), NPEOs and nonylphenol carboxylic acids (NPECs), have an unusual tendency to bioaccumulate to a very high degree in certain species of aquatic macroalgae (Marcomini *et al.*, 1990; Ahel, McEvoy & Giger, 1993). In these plants, concentrations up to 38, 80 and 28 mg of NP, NPIEO and NP2EO, respectively, per kg have been reported (BCF = 10 000 for NP). BCFs for NP in fish have been reported to be between 13 and 410 (Ahel *et al.*, 1993) and 1300 in sticklebacks (Ekelund *et al.*, 1990). Adsorption of hydrophobic compounds to organic particles will increase their bioavailability to filter-feeding organisms such as mussels. Thus, BCFs for some alkylphenolic compounds in these animals are high (3400 in one report).

The important points are that, because most of the oestrogenic xenobiotics discussed here are hydrophobic, they bioconcentrate and bioaccumulate in aquatic organisms, both plants and animals. Different organisms will bioconcentrate to different extents. Even with a single organism, the bioconcentrated compound is unlikely to be equally spread through all tissues; it is much more likely to be concentrated preferentially in a few tissues, such as fat. What happens to these compounds once bioconcentrated within an organism is essentially unknown: they may be physiologically inactive whilst stored in adipose tissue, but when this fat is mobilized (such as often occurs during reproduction) the compounds may be 'freed' to act elsewhere, or they may be metabolized into other compounds which may or may not be active as oestrogens.

Oestrogenic effects on aquatic organisms

From the outset it should be emphasized that there is very little evidence that the oestrogenic compounds mentioned above have had any effects on aquatic organisms that can with certainty be ascribed to their oestrogenic activity. However, it is equally true to say that very few studies have explored this possibility. The vast majority of the studies which have assessed the effects of these compounds on aquatic organisms have been concerned with toxicity, with death of the test organism often being the end-point. Few studies have investigated sub-lethal effects, and fewer still have focused on a particular parameter or group of parameters that, if affected, could be ascribed to a particular activity of the compound under investigation.

To address these questions, two types of study are possible: one is a field study, where aquatic organisms are potentially exposed to a wide range of oestrogenic compounds, the nature and concentration of each often being unknown, and the other is a laboratory study, where an aquatic organism is usually exposed to a single oestrogenic compound at a specific concentration. The first type of study reflects the 'real world', and allows us to assess whether the cumulative simultaneous exposure of organisms to a battery of oestrogenic compounds is likely to produce any deleterious effects. The second type of study is much more controlled, and enables oestrogenic compounds to be detected, and their potencies determined.

Before summarizing the (to date) few studies which have demonstrated that some persistent xenobiotics are oestrogenic to fish (and probably other aquatic organisms), we will discuss the roles that endogenous oestrogens play in normal physiology.

Role of endogenous oestrogens

This section will concentrate exclusively on the role of oestrogens in fish, which we are using as an example. These hormones have similar, if not identical, effects in all classes of vertebrates, but their functions, if any, in invertebrates and plants are unclear at present.

Oestrogens are hormones concerned primarily with reproduction in females. Oestradiol-17β is the most active natural oestrogen, and it plays many roles in reproduction. It is synthesized by, and secreted from, the follicle cells surrounding the oocytes of the ovary. As the ovary grows, the amount of oestradiol produced, and hence the blood concentration, rises steadily, although it usually falls before the ovary is fully grown, and concentrations can be quite low at ovulation (Scott & Sumpter, 1983). One of the main roles of oestradiol in oviparous animals (such as most species of fish) is to stimulate the liver to produce vitellogenin, the precursor of the major constituents of yolk (Tyler, Sumpter & Bromage, 1988). It also stimulates the liver to synthesize and secrete the vitelline envelope proteins (Hyllner et al., 1991), which will form the eggshell. Thus, oestradiol is vital for oocyte growth, egg formation, and provision of yolk for the developing embryo. Oestradiol also regulates its own synthesis by a negative feedback loop; gonadotrophins from the pituitary gland stimulate the synthesis of oestradiol, but in turn oestradiol acts back on the hypothalamus and pituitary gland to suppress gonadotrophin secretion.

Oestradiol also affects metabolism (as do many steroid hormones); these effects are not well investigated, although it appears that oestradiol stimulates both protein and lipid deposition, and hence growth (Haux & Norberg, 1985; Washburn *et al.*, 1993). Oestradiol is also one of a number of hormones involved in the control of various behaviours. It is difficult to place a specific function on oestradiol, but there seems little doubt that the hormone has a role to play in the various behaviours associated with successful reproduction (for review see Liley, Cardwell & Rouger, 1987).

One final effect of oestradiol, which may or may not be a natural one, concerns its role in sex determination. In lower vertebrates, including fish, exposure of eggs and/or embryos to sex steroid at the critical period can affect the sex ratio of the offspring; androgens masculinize and oestrogens feminize. This feature has been exploited by the aquaculture industry, who expose eggs to oestrogens to produce all-female offspring (Bye & Lincoln, 1986). The critical period, when sex is labile and can be altered by exposure to hormones is a week or two either side of hatching (Piferrer & Donaldson, 1992). Although there is no doubt that exposure of eggs to high concentrations of potent oestrogens can feminize the offspring, it is unclear whether this is a natural function of endogenous oestrogen such as 17β-oestradiol. The important point is that oestradiol plays many important roles in reproduction. Without appropriate concentrations of oestrogens at the appropriate time, reproduction would fail for many reasons.

How oestrogens work

Oestrogens are small, lipophilic molecules; they diffuse through cell membranes and bind to a protein in the nucleus called the oestradiol receptor. The sequence of the oestradiol receptor in rainbow trout, and the organization of the gene encoding it, are known (Pakdel *et al.*, 1990; Le Roux *et al.*, 1993). Once the oestrogen and receptor are tightly coupled, they interact, as a dimer, with an oestrogen-responsive element (ERE) on the DNA. As we have discussed, many genes, such as those coding for vitellogenin and the vitelline envelope proteins, are regulated to varying degrees by the activation of EREs.

The question of concern here is how substances which mimic the effects of natural oestrogens, such as those described in this review, cause their effects. Work in higher vertebrates has shown that, in general, they do so by binding to the oestrogen receptor. This implies that the oestrogen receptor does not show the high degree of specificity shown by most hormone receptors. Studies on fish are few, but nevertheless it has been shown that some xenobiotics produce their oestrogenic effects by interacting with the oestrogen receptor in a way indistinguishable from that of oestradiol itself (White *et al.*, 1994).

Relative potencies of oestrogenic substances

Many factors will influence the potential effects of an oestrogenic substance on an aquatic organism, including environmental concentration, bioconcentration, bioaccumulation and potency. Although a few of the oestrogenic compounds which are known to be, or might be, present in the aquatic environment are very potent oestrogens, most are not. Ethinyl oestradiol is one such very potent oestrogen that, if present in water even in very low concentrations, would produce pronounced oestrogenic effects (Purdom *et al.*, 1994; Sheahan *et al.*, 1994).

The available data, although limited, suggest that the known oestrogenic xenobiotics, e.g. organochlorine pesticides, PAHs, and alkylphenolic compounds, are only weakly oestrogenic, with potencies three or more orders of magnitude less than that of oestradiol-17β. Further, it appears that the potency of a specific oestrogenic compound relative to oestradiol-17β is similar in whichever organism it is tested. For example, the alkylphenols octylphenol and nonylphenol have about 0.1% and 0.005% the potency of oestradiol whether they are tested in mammalian, avian, or piscine-based systems (White *et al.*, 1994).

Presently no dose–response studies have been reported, and hence it is not possible to assess whether the concentrations of oestrogenic compounds known to be present in the aquatic environment are high enough to produce oestrogenic effects in aquatic organisms.

Evidence that some xenobiotics are oestrogenic to aquatic organisms

As stated earlier, very few studies have investigated whether substances which can mimic the effects of oestradiol in higher vertebrates also do so in fish. However, one very recent study (Purdom *et al.*, 1994) does demonstrate clearly that there are oestrogenic substances in river water, and in concentrations high enough to produce pronounced effects.

In this study, the concentration of vitellogenin in the plasma was used as a measure of the degree of exposure to 'oestrogens' (see Bromage & Cumaranatunga, 1988, for data demonstrating that the synthesis of vitellogenin is controlled to a large extent by oestrogens). Male rainbow trout, which do not usually express the vitellogenin gene and hence have no vitellogenin in their plasma, were held in cages placed in effluent channels of sewage treatment works close to where effluent entered the river, and at various distances downstream, including sites where water was abstracted for domestic use. After only one week in effluent, plasma vitellogenin concentrations had risen one thousand-fold or more, and by three weeks the concentrations were in tens of milligrams per ml, at which time vitellogenin was the major blood protein, comprising more than 50% of the total plasma protein. Some of the results from this study of ours are presented in Fig. 2. A nationwide survey in the UK demonstrated that effluent from all sewage treatment works tested produced this very strong oestrogenic effect (Purdom *et al.*, 1994), suggesting that at least a significant proportion of the effect was due to domestic, rather than industrial, waste (because not all the sewage treatment works received industrial waste). Fish placed in the rivers, rather than the effluent channels, also often had raised plasma vitellogenin concentrations, but the response was not as great as observed in fish placed in effluent. Positive effects were sometimes observed at water abstraction sites.

This study demonstrated very clearly that effluent from sewage treatment works, which can make a very significant contribution to the flow of rivers in the UK (and other densely populated countries), contains a substance, or substances, oestrogenic to fish. Although the nature of the substance(s) is unknown, it appears to be of domestic origin. As far as we are aware, this single study is the only one which demonstrates that oestrogenic xenobiotics are present in the aquatic environment at concentrations high enough to produce pronounced effects in aquatic organisms. It seems likely that similar effects will be observed in other countries: we are aware of studies conducted by colleagues in France, using eels instead of trout, that have shown a similar problem in the River Seine. Presently we do not know whether this 'feminization' of male fish has serious consequences as far as successful reproduction is concerned; studies are under way in an attempt to answer this important question.

There are a few, and only a few, laboratory-based studies which show that some xenobiotics are oestrogenic to fish, as they are to mammals. A series of papers by Wester and colleagues (Wester, Canton & Bisschop, 1985; Wester & Canton, 1986; Wester, 1991) has shown that the pesticide β-hexachlorocyclohexane (β-HCH) is oestrogenic to fish, as it is in higher vertebrates. These authors reported that after three months' exposure to concentrations of β-HCH of 32 μg/litre

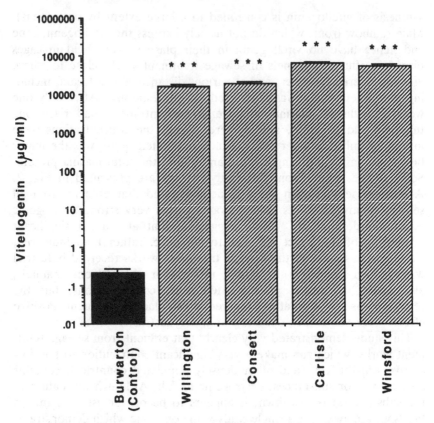

Fig. 2. The oestrogenic response of male rainbow trout placed in the effluent channels of sewage treatment works. Groups of 20 male rainbow trout were maintained either on a fish farm (Burwarton) which was supplied with well-water, or in the effluent channels of four large sewage treatment works (situated near the cities of Willington, Consett, Carlisle and Winsford in northern England) for three weeks. At the end of the experiment a blood sample was collected and the concentration of vitellogenin in the plasma determined. Male trout usually have very low concentrations of vitellogenin in the plasma; high concentrations are indicative of exposure to oestrogens, and normally occur only in females. The results shown here suggest strongly that the effluent contained an oestrogenic chemical or chemicals. Results are means ± SEM ($n = 20$). (***$P<0.001$).

and above, excessive vitellogenesis was stimulated in both male and female guppies – the liver hypertrophied, vitellogenin was present in the plasma, the gonadotrophs of the pituitary gland (which synthesize and secrete the gonadotrophins LH and FSH) underwent hyperplasia, and yolk was present in 'premature' oocytes (Wester *et al.*, 1985; Wester, 1991). Similar effects were noted in medaka after long-term exposure to β-HCH; particularly noticeable was the occurrence of ovo–testis (hermaphroditism). As Wester and colleagues note, all the effects are indicative of the oestrogenic activity of this widely used pesticide. In an unrelated study, some observations on insecticide-resistant mosquito fish (*Gambusia affinis*) have been interpreted, probably correctly, to imply that *o,p*-DDT can induce vitellogenin synthesis in this species, thus acting as an oestrogen (Denison, Chambers & Yarbrough, 1981).

The only other study we are aware of where the aquatic pollutant has been shown to be oestrogenic to fish is that of Jobling and Sumpter (1993). These authors used an *in vitro* technique to demonstrate that a variety of different environmentally persistent degradation products of alkylphenol polyethoxylates stimulated vitellogenin production from cultured hepatocytes, a response known to be oestrogen dependent. They also showed that the oestrogenic activities of these chemicals were due to their interaction with the oestradiol receptor.

Only a single study has been published which investigated the mechanism of action of oestrogenic xenobiotics in fish (Thomas & Smith, 1993). It was reported that the pesticide Kepone (chlordecone) was an effective competitor for the oestradiol receptor (ER), whereas *o,p*-DDE, *o,p*-DDT and Methoxychlor did not inhibit the binding of oestradiol to its receptor. These latter results are surprising, because all of these xenobiotics produce their oestrogenic effects in higher vertebrates by interacting directly with the ER. Our own (unpublished) work suggests that oestrogenic pesticides do interact directly with the hepatic ER of the rainbow trout, as do some of the oestrogenic alkylphenolic compounds (White *et al.*, 1994).

Conclusions

Many xenobiotics present in significant concentrations in the aquatic environment are known to be weakly oestrogenic. Although the effects of exposure to many of these chemicals has received considerable attention, almost all of this has focused on the toxicity of the chemicals to different aquatic organisms. Little attention has been directed at non-lethal effects, and then usually not at effects which might be caused by the oestrogenic activities of the chemicals. Thus, often we do not know whether xenobiotics known to be oestrogenic to mammals are also oestrogenic to common aquatic organisms (particularly fish). However, the limited data available suggest that these xenobiotics will also be oestrogenic to fish – the situation regarding invertebrates (and plants) is unknown. Because oestrogens are fundamental to successful reproduction of all vertebrates, exposure of fish to oestrogenic xenobiotics, if the latter are present in high enough concentrations, could have very serious implications for fish stocks. The presence of many different oestrogenic chemicals in the aquatic environment, their persistence and hence accumulation, together with the regular discovery that yet another group of chemicals is oestrogenic, make it likely that situations already exist where 'oestrogenic effects' are occurring, with potentially deleterious consequences.

References

Ahel, M. & Giger, W. (1985). Determination of alkylphenols and alkylphenol mono- and diethoxylates in environmental samples by high-performance liquid chromatography. *Analytical Chemistry*, **57**, 1577–83.

Ahel, M. & Giger, W. (1993). Aqueous solubility of alkylphenols and alkylphenol polyethoxylates. *Chemosphere*, **26**, 1461–70.

Ahel, M., Conrad, T. & Giger, W. (1987). Persistent organic chemicals in sewage effluents. 3. Determination of alkylphenoxy carboxylic acids by high-resolution gas-chromatography/spectrometry and high-performance liquid chromatography. *Environmental Science and Technology*, **21**, 697–703.

Ahel, M., McEvoy, J. & Giger,W. (1993). Bioaccumulation of the lipophilic metabolites of nonionic surfactants in freshwater organisms. *Environmental Pollution*, **79**, 243–8.

Aherne, G.W. & Briggs, R. (1989). The relevance of the presence of certain synthetic steroids in the aquatic environment. *Journal of Pharmacy and Pharmacology*, **41**, 735–6.

Allchin, C.R. (1991). Concentrations of alpha- and gamma-hexachlorocyclohexane (Lindane) in the coastal waters of England and Wales. *Water Science and Technology*, **24**, 143–6.

Begum, S., Begum, Z. & Alam, M.S. (1992). Organochlorine pesticide contamination of rain-water, domestic tap water and well-water of Karachi City. *Journal of the Chemical Society of Pakistan*, **14**, 8–11.

Boubassi, I. & Sailot, A. (1991). Composition and sources of dissolved and particulate PAH in surface waters from the Rhone delta (N.W. Mediterranean). *Marine Pollution Bulletin*, **22**, 588–94.

Bromage, N.R. and Cumaranatunga, R. (1988). Egg production in the rainbow trout. In Muir, J.F. & Roberts, R.J. (eds.). *Recent*

Advances in Aquaculture, Vol. 3, pp. 63–138. Groom-Hill, Westview Press, London and Sydney.

Bye, V.J. & Lincoln, R.F. (1986). Commercial methods for the control of sexual maturation in rainbow trout (*Salmo gairdneri* R.). *Aquaculture*, **57**, 299–309.

Clark, L.B., Rosen, R.T., Hartman, T.G., Louis, J.B., Suffet, I.H., Lippincott, R.L. & Rosen, J.D. (1992). Determination of alkylphenol ethoxylates and their acetic acid derivatives in drinking water by particle beam liquid chromatography/mass spectrometry. *International Journal of Environmental and Analytical Chemistry*, **47**, 169–80.

Denison, M.S., Chambers, J.E. & Yarbrough, J.D. (1981). Persistent vitellogenin-like protein and binding of DDT in the serum of insecticide-resistant mosquito fish (*Gambusia affinis*). *Comparative Biochemistry and Physiology*, **69C**, 109–12.

Ekelund, R., Bergman, A., Granmo, A. & Bergren, M. (1990). Bioaccumulation of 4-nonylphenol in marine animals – a reevaluation. *Environmental Pollution*, **64**, 107–20.

Evans, M.S., Noguchi, G.E. & Rice, C.P. (1991). The biomagnification of polychlorinated biphenols, Toxaphene, and DDT compounds in a Lake Michigan offshore food web. *Archives of Environmental Contamination and Toxicology*, **20**, 87–93.

Fletcher, C.L. & McKay, W.A. (1993). Polychlorinated dibenzo-*p*-dioxins (PCDDs) and dibenzofurans (PCDFs) in the aquatic environment – a literature review. *Chemosphere*, **26**, 1041–69.

Frakes, R.A., Zeeman, C.Q.T. & Mower, B. (1993). Bioaccumulation of 2, 3, 7, 8-tetrachlorodibenzo-*p*-dioxin (TCDD) by fish downstream of pulp and paper mills in Maine. *Ecotoxicology and Environmental Safety*, **25**, 244–52.

Galassi, S., Batlaglia, C., Camusso, M., De Paolis, A. & Tartari, G. (1987). Microinquinanti organici nelle deposizioni umide di Brugherio (Milano). *Inquinamento*, **19**, 3–7.

Giger, W., Brunner, P.H. & Schaffner, C. (1984). 4-nonylphenol in sewage sludge: accumulation of toxic metabolites from nonionic surfactants. *Science*, **225**, 623–5.

Haux, C. & Norberg, B. (1985). Effect of sexual maturation and estradiol-17β on liver content of protein, lipids, glycogen and nucleic acids in juvenile rainbow trout, *Salmo gairdneri*. *Comparative Biochemistry and Physiology*, **81B**, 275–9.

Hyllner, S.J., Oppen-Berntsen, D.O., Helvik, J.V., Walther, B.T. & Haux, C. (1991). Oestradiol-17β induces the major vitelline envelope proteins in both sexes in teleosts. *Journal of Endocrinology*, **131**, 229–36.

Jansen, H.T., Cooke, P.S., Porcelli, J., Liu, T.-C. & Hansen, L.G. (1993). Estrogenic and antiestrogenic actions of PCBs in the female rat: *in vitro* and *in vivo* studies. *Reproductive Toxicology*, **7**, 237–48.

Jobling, S. & Sumpter, J.P. (1993). Detergent components in sewage effluent are weakly oestrogenic to fish: an *in vitro* study using rainbow trout (*Oncorhynchus mykiss*) hepatocytes. *Aquatic Toxicology*, **27**, 661–72.

Kelly, T.J., Czuczura, J.M., Sticksel, P.R., Sverdrup, G.M., Koval, P.J. & Hodanbosi, R.F. (1991). Atmospheric and tributary inputs of toxic substances to Lake Erie. *Journal of Great Lakes Research*, **17**, 504–16.

Klamer, J., Laane, R.W.P.M. & Marquenie, J.M. (1991). Sources and fate of PCBs in the North Sea: a review of available data. *Water Science and Technology*, **24**, 77–85.

Krishnan, V. & Safe, S. (1993). Polychlorinated biphenyls (PCBs), dibenzo-*p*-dioxins (PCDDs) and dibenzofurans (PCDFs) as antiestrogens in MCF-7 human breast cancer cells: quantitative structure–activity relationships. *Toxicology and Applied Pharmacology*, **120**, 55–61.

Le Roux, M.-G., Thézé, N., Wolff, J. & Le Pennee, J.-P. (1993). Organisation of a rainbow trout estrogen receptor gene. *Biochimica et Biophysica Acta*, **1172**, 226–30.

Liley, N.R., Cardwell, J.R. & Rouger, Y. (1987). Current status of hormones and sexual behaviour in fish. In Idler, D.R., Crim, L. & Walsh, J.M. (eds.). *Reproductive Physiology of Fish*, pp. 142–9. Memorial University of Newfoundland Publication.

McLachlan, J.A. ed. (1980). *Estrogens in the Environment*. Elsevier, North Holland.

McLachlan, J.A. ed. (1985). *Estrogens in the Environment II*. Elsevier, New York.

Marcomini, A., Pavoni, B., Sfriso, A. & Orio, A.A. (1990). Persistent metabolites of alkylphenol polyethoxylates in the marine environment. *Marine Chemistry*, **29**, 307–23.

Mayr, U., Butsch, A. & Schneider, S. (1992). Validation of two *in vitro* test systems for estrogenic activities with zearalenone, phytoestrogens and cereal extracts. *Toxicology*, **74**, 135–49.

Merriman, J.C., Anthony, D.H.J., Kraft, J.A. & Wilkinson, R.J. (1991). Rainy river water quality in the vicinity of bleached kraft mills. *Chemosphere*, **23**, 1605–15.

Muir, D.C.G., Ford, C.A., Grift, N.P. & Stewart, R.E.A. (1992). Organochlorine contaminants in narwhal (*Monodon monoceros*) from the Canadian Arctic. *Environmental Pollution*, **75**, 307–16.

Naylor, C.G., Mieure, J.P., Weeks, J.A., Castaldi, F.J. & Romano, R.R. (1992). Alkylphenol ethoxylates in the environment. *Journal of the American Oil Chemists Society*, **69**, 695–703.

Pakdel, F., Le Gac, F., Goff, P. Le & Valotaire, Y. (1990). Full-length sequence and *in vitro* expression of rainbow trout estrogen receptor cDNA. *Molecular and Cellular Endocrinology*, **71**, 195–204.

Paxéus, N., Robinson, P. & Balmér, P. (1992). Study of organic pollutants in municipal wastewater in Göteborg, Sweden. *Water Science and Technology*, **25**, 249–56.

Pelissero, C., Flouriot, G., Foucher, J.L., Bennetau, B., Dunogues, J., Le Gac, F. & Sumpter, J.P. (1993). Vitellogenin synthesis in cultured hepatocytes: an *in vitro* test for the estrogenic potency of chemicals. *Journal of Steroid Biochemistry and Molecular Biology*, **44**, 263–72.

Piferrer, F. & Donaldson, E.M. (1992). The comparative effectiveness of the natural and a synthetic estrogen for the direct feminisation of chinook salmon (*Oncorhynchus tshawytscha*). *Aquaculture*, **106**, 183–93.

Purdom, C.E., Hardiman, P.A., Bye, V.J., Eno, N.C., Tyler, C.R. & Sumpter, J.P. (1994). Estrogenic effects of effluent from sewage-treatment works. *Chemistry and Ecology*, **8**, 275–85.

Saito, S., Tanoue, A. & Matsuo, M. (1992). Applicability of the i/o-characters to a quantitative description of bioconcentration of organic chemicals in fish. *Chemosphere*, **24**, 81–7.

Schröder, H.Fr. (1992). Polar organic pollutants on their way from waste water to drinking water. *Water Science and Technology*, **25**, 241–8.

Scott, A.P. & Sumpter, J.P. (1983). A comparison of the female reproductive cycle of autumn-spawning and winter-spawning strains of rainbow trout (*Salmo gairdneri*). *General and Comparative Endocrinology*, **52**, 79–85.

Sheahan, D.A., Bucke, D., Matthiessen, P., Sumpter, J.P., Kirby M.F., Neall, P. & Waldock, M. (1994). Chapter 9. The effects of low levels of 17α-ethynyloestradiol upon plasma vitellogenin levels in male and female rainbow trout, *Oncorhynchus mykiss* held at two acclimation temperatures. In *Sublethal and chronic effects of pollutants on freshwater fish*. Muller, R. & Lloyd, T. (eds.). pp. 99–112. Cambridge: FAO, Fishing News Books, Blackwell Scientific.

Shore, L.S. (1993). Estrogen as an environmental pollutant. *Bulletin of Environmental Contamination and Toxicology*, **51**, 361–6.

Soto, A.M., Justica, H., Wray, J.W. & Sonnenschein, C. (1991). *p*-Nonyl-phenol: an estrogenic xenobiotic released from 'modified' polystyrene. *Environmental Health Perspectives*, **92**, 167–73.

Tatsukawa, R. (1992). Contamination of chlorinated organic substances in the ocean ecosystem. *Water Science and Technology*, **25**, 1–8.

Thomas, P. & Smith, J. (1993). Binding of xenobiotics to the estrogen receptor of spotted seatrout: a screening assay for potential estrogenic effects. *Marine Environmental Research*, **35**, 147–51.

Thoumelin, G. (1991). Presence et comportement dans le milieu aquatique des alkylbenzenesulfonates lineaires (LAS) et des alkylphenols ethoxyles (APE); mise au point. *Environmental Technology*, **12**, 1037–45.

Tyler, C.R., Sumpter, J.P. & Bromage, N.R. (1988). *In vivo* ovarian uptake and processing of vitellogenin in the rainbow trout, *Salmo gairdneri*. *Journal of Experimental Zoology*, **246**, 171–9.

Ventura, F., Caixach, J., Romero, J., Espadaler, I. & Rivera, J. (1992). New methods for the identificaion of surfactants and their acidic metabolites in raw and drinking water: FAB-MS and MS/MS. *Water Science and Technology*, **25**, 257–64.

Washburn, B.S., Krantz, J.S., Avery, E.H. & Freedland, R.A. (1993). Effects of estrogen on gluconeogenesis and related parameters in male rainbow trout. *American Journal of Physiology*, **264**, R720–5.

Wester, P.W. (1991). Histopathological effects of environmental pollutants β-HCH and methyl mercury on reproductive organs in freshwater fish. *Comparative Biochemistry and Physiology*, **100C**, 237–9.

Wester, P.W. & Canton, J.H. (1986). Histopathological study of *Oryzias latipes* (medaka) after long-term β-hexachlorocyclohexane exposure. *Aquatic Toxicology*, **9**, 21–45.

Wester, P.W., Canton, J.H. & Bisschop, A. (1985). Histopathological study of *Poecilia reticulata* (guppy) after long-term β-hexachlorocyclohexane exposure. *Aquatic Toxicology*, **6**, 271–96.

White, R., Jobling, S., Hoare, S.A., Sumpter, J.P. & Parker, M.G. (1994). Environmentally persistent alkylphenolic compounds are estrogenic. *Endocrinology*, **135**, 175–82.

Zoller, U. (1993). Groundwater contamination by detergents and polycyclic aromatic hydrocarbons – a global problem of organic contaminants: is the solution locally specific? *Water Science and Technology*, **27**, 187–94.

Zoller, U., Ashash, E., Ayali, G., Shafir, S. & Azinon, B. (1990). Nonionic detergents as tracers of groundwater pollution caused by municipal sewage. *Environment International*, **16**, 301–6.

J.W. NUNN, D.R. LIVINGSTONE and
J.K. CHIPMAN

Effect of genetic toxicants in aquatic organisms

Introduction

In recent years, attention has been given to investigating the occurrence of genotoxic agents in the aquatic environment. The increasing concern of the general public and of governments for the welfare of the natural environment has warranted an upsurge in such studies which have often been designed to investigate the cause of tumour production in various aquatic organisms in areas of raised pollution status. A causal relationship between pollutants and cancer in aquatic organisms is by no means proven. There are studies, however, that raise considerable concern. For example, results of a series of experiments on the association between sediment pollutants and hepatic tumours in English sole (*Parophrys vetulus*) of the Puget Sound, USA, have provided convincing evidence of a causative relationship (see Stein *et al.*, 1990). Similarly, although much less is known of the situation in invertebrates, experimental studies have demonstrated chemical induction of tumours in the oyster (*Crassostrea virginica*) by a mixture of polycyclic aromatic hydrocarbons (PAHs), polychlorobiphenyls (PCBs), amines and metals (Gardner, Pruell & Malcolm, 1992), and in planarians by a range of mammalian carcinogens (Schaeffer, 1993). The development of new, sensitive, molecular techniques, so-called biomarkers (Livingstone, 1993), for assessing genetic toxicity have also played an important role in driving the new era of environmental monitoring.

Why should we be concerned about exposure of organisms to mutagens? An increased tumour incidence is of concern, not only regarding the organisms directly exposed but also regarding human beings exposed to bioaccumulated chemicals via the diet. Of the various classes of chemicals known to cause cancer, those that operate via a genotoxic (rather than an epigenetic) mechanism are of the greatest concern regarding relatively low dose exposures. This is because a threshold dose may exist for many of the latter agents (see Purchase, 1994). Levelled against this concern must be an appreciation that humans and other organisms are constantly exposed to a 'background' level of genetic toxicity imposed by, for example, reactive oxygen species (ROS) and naturally occurring plant products (Ames, 1989). Highly efficient protective detoxifying enzymes and DNA repair systems have evolved (Caldwell & Jakoby, 1983; Friedburg, 1985) to combat this assault, thereby lessening the potential threat of genotoxic pollutants. It is when these protective systems become overloaded that the threat may become significant. Also, somewhat paradoxically, some of these 'protective enzymes' can themselves activate pollutants to genotoxic products.

Cancer is not the only consequence of mutagen exposure. Mutations in germ line cells causing hereditary defects are also an important consequence. Although cancer is a major concern regarding human exposure to mutagens, it could be argued that this disease is of little impact at the population level of a feral species. If cancer weakens or kills an individual organism at a stage in its life after reproductive success, it may have little impact at the population level. Indeed, genotoxicity might enhance genetic diversity. Detrimental mutations may be selected against and lost within only several generations provided a species has a reproductive surplus (Würgler & Kramers, 1992).

A further aspect to consider, however, is that genotoxic substances rarely damage only DNA. Their interaction is often with cellular nucleophiles such as protein thiols, and the requirement of cells to protect against and to repair these damaging effects may also exert a considerable stress on the energy reserves and normal function of organisms thereby compromising their survival in the environment. Exposures may also affect reproductive capacity. Thus, examples of pollutant effects on animal fitness include reduced growth size in fish exposed to bleached kraftmill effluent (see Livingstone, Förlin & George, 1994), reduced egg masses and egg numbers in freshwater snail (*Lymnaes stagnalis*) exposed to PCBs (Wilbrink *et al.*, 1987), and higher field mortality in clams (*Mya arenaria*) with haematopoietic neoplasia than without neoplasia (Brousseau & Baglio, 1991).

This chapter reviews the interaction of genotoxic compounds with aquatic organisms and considers the progress made in developing sensitive biomarkers for the monitoring of such damage in the environment.

Biotransformation of pollutants and generation of mutagenic species

Organic pollutants in the aquatic environment comprise a vast and ever-increasing range of compounds, including PAHs, PCBs, dioxins, nitroaromatics, aromatic amines, organophosphate and organochlorine pesticides, and phthalate ester plasticisers. Such foreign compounds (xenobiotics) are fat soluble and are therefore readily taken up from the water, sediment and food sources into the tissues of aquatic organisms (Walker & Livingstone, 1992). A schematic representation of the possible molecular fate and effects of xenobiotics once in the tissues is given in Fig. 1. It identifies the major pathways of biotransformation and elimination (phase I and II metabolism), four major possible sources of toxic molecular species (parent compound, metabolites, organic radicals, ROS), other protective systems (antioxidant enzymes, scavengers), and potential deleterious molecular effects, including genotoxicity.

The parent compound of the pollutant can exert toxicity in a number of ways, including membrane disturbance, steroid effects and (in some cases) direct damage to DNA (direct mutagens). It is detoxified by biotransformation to water-soluble, excretable metabolites. Phase 1 metabolism (oxidation, reduction, hydration, hydrolysis) introduces a

Metabolic fate and molecular toxicity of natural and pollutant xenobiotics in marine organisms

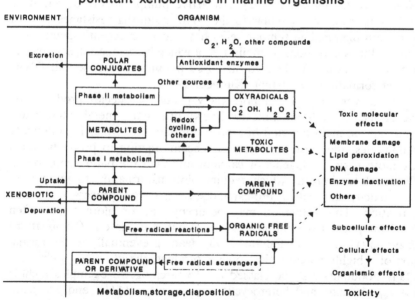

Fig. 1. Schematic representation of the major pathways of biotransformation, detoxication and activation, and the resultant molecular effects of organic xenobiotics in animals. (Livingstone, 1991.)

functional group, e.g. $-OH$, $-NH_2$, $-COOH$, into the xenobiotic, to which phase II metabolism (conjugation, acetylation, others) attaches a large water-soluble polar moiety derived from various endogenous molecules (glutathione, sugars, amino acids, sulphate). Most major phase I enzymes (mixed-function oxygenase (MFO) system, flavoprotein monooxygenase, epoxide hydrolase, various hydrolases) and phase II enzymes (glutathione S-transferases, UDP-glucuronyl transferases, UDP-glucosyl transferases, sulphotransferases) are present to varying degrees in aquatic organisms, but not surprisingly the levels are generally lower in invertebrates than vertebrates (Livingstone, 1991; Walker & Livingstone, 1992; Stegeman, 1993; George, 1994), as for example is seen for the cytochrome P450-dependent MFO system (Table 1). Given that pollutant uptake is largely passive and therefore similar for different organisms, one consequence of this enzyme difference is that readily metabolizable pollutants, such as PAHs, bioaccumulate to highest tissue levels in aquatic invertebrates at the bottom of food chains, whereas poorly metabolizable ones, such as certain PCB congeners, bioaccumulate along food chains and reach highest tissue levels in top predators (Table 2). Such increased tissue levels, or residence times, of pollutants may increase their potential for toxicity (Walker & Livingstone, 1992). An active pump capable of removing pollutants from cells (similar to the P170 multidrug resistance protein of certain mammalian tumour cells) has also been recently detected in fish, molluscs and sponges, inhibition of which by competing xenobiotics has been indicated to lead to increased pollutant presence and DNA-adduct formation (see later) (Kurelec, 1992).

Of the various types of potentially genotoxic species produced by metabolism of pollutants, most is known of reactive metabolites. Thus, although the main function of biotransformation enzymes is to detoxify xenobiotics, they also convert a significant fraction to products which are more toxic, mutagenic or carcinogenic than the parent compound. Chemical carcinogens which require biotransformation to exert their genotoxicity are called pro-carcinogens, and the process is termed activation. This activation may occur by one or more steps, often involving phase I (oxidation, reduction) but also phase II (conjugation) enzymes (Goldstein & Faletto, 1993), leading eventually to the formation of a highly reactive electrophilic product, the ultimate carcinogen. Many pollutants may be classed as pro-carcinogens; examples include PAHs, aromatic and heterocyclic amines, nitro-PAHs, and azo compounds. The genotoxic risk to marine organisms, therefore, depends on their capacity to biotransform such pro-carcinogenic pollutants to ultimate carcinogens. However, although in general this will be lower

Table 1. *Levels of microsomal mixed-function oxygenase system in different aquatic organisms[a]*

Animal group	Microsomal protein (mg/g wet wt)	Cytochrome P450 (pmol/mg)[b]	Benzo[a]pyrene hydroxylase (pmol/min/mg)
Whales/dolphins	22±1	230±41	395±167
Seals	—[c]	177±48	–
Birds[d]	13±3	318±48	150
Fish	14±4	322±35	307±149
Echinoderms	5–10	63±12	8.4±5.5
Crustaceans	5–10	328±70	40±31
Polychaetes	–	82	8.5±6.8
Molluscs	5–10	73±10	21.2±3.4

[a] From Walker & Livingstone (1992): means ±SEM or ± range for different species within each group; data are generally for tissues with highest cytochrome P450 content, namely liver (vertebrates), pyloric caeca (echinoderms), hepatopancreas (crustaceans), intestines (polychaetes) and digestive gland (molluscs). [b] Microsomal protein. [c] no information. [d]includes seabirds.

Table 2. *Patterns of bioaccumulation of poorly metabolizable (polychlorobiphenyls, PCBS) and readily metabolizable (polycyclic aromatic hydrocarbons, PAHs) organic pollutants in different aquatic organisms[a]*

Pollutant	Mollusc	Crustacean	Fish	Bird	Sea mammal
Total PCBs	0.002–14	0.03–7.7	0.44–7.3	0.4–340	1–800
Total PAHs	0.1–75	0.1–60	0.01–0.1	—[b]	—

[a] From Livingstone *et al.* (1994a); pollutant ranges in µg/g wet wt. for various whole tissues, individual tissues, e.g. liver, eggs (birds) and blubber (whales). [b] No information.

for lower vertebrates than mammals, and invertebrates than vertebrates, the eventual genotoxic outcome may depend on many factors, including, for example, the capacity to repair DNA which may also be lower in lower organisms. Biotransformation enzymes have a wide tissue distribution, but are generally highest in tissues concerned with the processing of food and ports of entry or removal of xenobiotics, which will therefore be potential sites of genotoxic effect, for example, the liver, intestines, kidneys and gills of fish, pyloric caeca of echinoderms, hepatopancreas of crustaceans, digestive gland of molluscs, and intestines of polychaetes (Livingstone, 1991, 1993).

Of particular importance to the activation of many pro-carcinogens to reactive metabolites is the cytochrome P450-dependent MFO system, which, for example, catalyses the N-oxidation of 2-acetylaminofluorene (AAF), one of the steps responsible for the transformation of this aromatic amide into a potent carcinogen in mammals (Thoreirsson, McManus & Glowinski, 1984). Cytochrome P450 is the terminal catalytic component of the MFO system, capable of catalysing epoxidation, hydroxylation, dealkylation and other monooxygenase reactions. It exists in a number of isoenzyme and gene forms (termed the P450 supergene family) with different biotransformation and endogenous functions (Livingstone, 1991; Stegeman, 1993). Cytochrome P4501A1 (CYP1A1) in particular, but also other P450s, are of central importance in the metabolism and activation of environmentally important PAHs, planar PCBs and dioxins, and are localized in the endoplasmic reticulum (microsomes) of the cell. For example, the initial step in the activation of the PAH benzo[a]pyrene (BaP) involves CYP1A1-catalysed oxidation to form an arene oxide. Further metabolism by epoxide hydrolase (EC 4.2.1.64) yields a dihydrodiol which undergoes further CYP1A1-catalysed oxidation to form a diol epoxide. Thus, the suspected ultimate carcinogenic metabolite of BaP is (+) anti-BaP-7,8-dihydrodiol 9,10-epoxide, with (−) BaP-7,8-dihydrodiol as its precursor. Examples of the metabolic activation of pro-carcinogens are shown in Fig. 2. Studies *in vitro* with hepatic microsomes of English sole (*Parophrys vetulus*) and starry flounder (*Platichthys stellatus*) show the proximate carcinogen, BaP-7,8-dihydrodiol to be a major metabolite of BaP (Stein *et al.*, 1990). The production of this metabolite by fish liver microsomes indicates the potential for fish to metabolize pro-carcinogens to reactive electrophilic products which can bind covalently to DNA. Cytochrome P450-catalysed DNA-adduct formation from BaP and certain other PAHs can also occur in mammals via another pathway, termed one-electron oxidation giving rise to a reactive cation radical (Cavalieri *et al.*, 1993), and the same may be possible for aquatic organisms. The constitutive levels of CYP1A1 isoenzymes in liver of vertebrates are very low, but it is readily inducible by exposure to PAHs, planar PCBs, dioxins and a wide range of other organic xenobiotics. This induction has been observed in many laboratory and field studies with over 30 species of fish, and results in increased metabolism of xenobiotics such as BaP, increased mutagen production, and increased DNA damage (Livingstone, 1993). Much less is known of the situation in marine invertebrates, but, for example, the ability to metabolize BaP is widespread (Table 1), including production of the BaP-7,8-dihydrodiol (Livingstone, 1991), and binding of BaP to macromolecules, including DNA has been indicated (see below). Also, recent enzymological (Schlenk & Buhler, 1989; Porte *et al.*, 1994), molecular biological (Wootton *et al.*, 1994) and field studies (Livingstone *et al.*, 1994*b*) have indicated the presence of an inducible CYP1A-like enzyme in molluscs, although its exact functional nature is unknown.

The *Salmonella* bacterial mutagenicity assay (Maron & Ames, 1983) has proved a useful tool for studying the activation of xenobiotics by substituting the classical rat liver post-mitochondrial supernatant (PMS) with subcellular fractions of tissues of marine and other aquatic organisms of interest. Using this assay it has been demonstrated that both fish and invertebrates are capable of activating a variety of compounds from different classes of pollutants, including PAHs, aromatic amines/amides and nitro-PAHs (Livingstone, 1991). Recent examples include the activation of BaP by PMS of liver of sea bass (*Dicentrarchus labrax*)

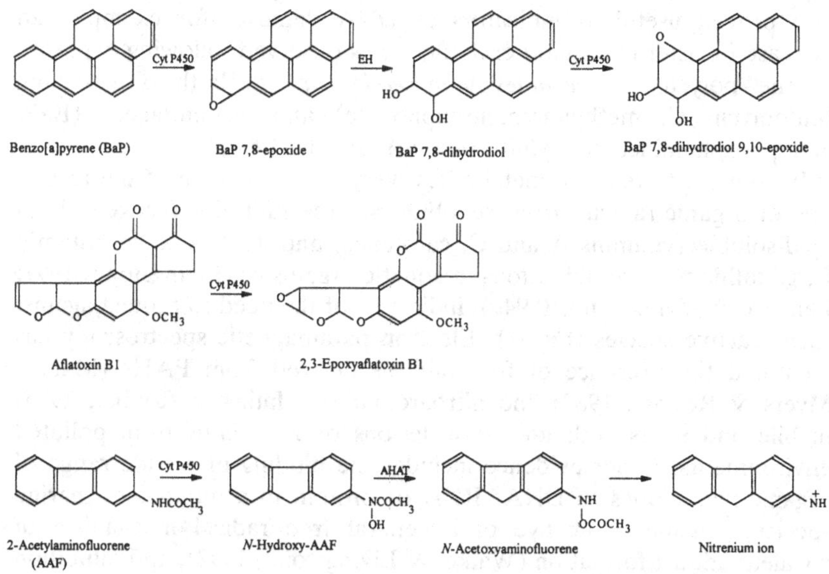

Fig. 2. Examples of metabolic activation of pro-carcinogens. EH= epoxide hydrolase, AHAT=N,O-arylhydroxamic acid acyltransferase.

and digestive gland of mussel (*Mytilus galloprovincialis*) (Michel *et al.*, 1992; Michel, Cassand & Narbonne, 1993), and of 1-nitropyrene and AAF by PMS of digestive gland of mussel (*Mytilus edulis*), hepatopancreas of shore crab (*Carcinus maenas*) and pyloric caeca of starfish (*Asterias rubens*) (Marsh, Chipman & Livingstone, 1992). This approach, in conjunction with the use of specific enzyme inhibitors, has also proved useful in determining mechanisms of activation of some pro-carcinogens. This has included the demonstration that such mechanisms can differ for different animal groups, making the identification of such trends very important in the prediction of pollutant effects in aquatic ecosystems (Chipman & Marsh, 1991; Livingstone, 1991; Walker & Livingstone, 1992). For example, the principal pathway of activation of AAF by rat liver is via *N*-hydroxylation mediated by cytochrome P450, whereas in digestive gland of *M. edulis* it is indicated to be primarily via cytosolic deacetylation to the amine followed by its *N*-hydroxylation by a microsomal flavoprotein monooxygenase (Marsh *et al.*, 1992). More recently, another bacterial mutagenicity assay, the *Salmonella typhimurium umu*-test (Oda *et al.*, 1985) has also been found useful for studying activation of xenobiotics by aquatic species. For example, the nitroaromatic 4-nitroquinoline *N*-oxide was activated by both cytosolic and microsomal fractions of digestive gland of *M. edulis* (Garcia Martinez *et al.*, 1992). Cytogenetic assays have also proved useful as indicators of DNA damage, for example, an increase in sister chromatid exchanges was observed following exposure of the polychaete *Neanthes arenaceodentata*, to both direct-acting (mitomycin C, methylmethanesulphonate) and pro-mutagens (BaP, cyclophosphamide, diethylnitrosamine) (Pesch, 1990).

In contrast to reactive metabolites, very little is known of the formation of organic radicals from xenobiotics. Free radical scavengers, both lipid-soluble (vitamins A and C, carotenes) and water-soluble (vitamin C, glutathione), are ubiquitous in aquatic organisms (Livingstone, 1991; Lemaire & Livingstone, 1994a), indicative of the need to protect against such reactive species (Fig. 1). Electron paramagnetic spectroscopy has identified the presence of free radicals derived from PAHs (Malins, Myers & Roubal, 1983) and nitroaromatics (Malins & Roubal, 1985) in bile and livers with idiopathic lesions of *P. vetulus* from polluted environments. Other evidence includes the binding of a wide range of organic xenobiotics to DNA, RNA or protein in a number of marine species, possibly indicative of a general free radical mechanism of covalent adduct formation (Walker & Livingstone, 1992), and inhibition by NADPH of the binding of BaP to protein in digestive gland microsomes of *M. edulis*, which has been interpreted in terms of the conversion of the BaP cation radical back to the parent compound by the reducing agent (Livingstone, 1991).

A further source of reactive species, ROS, are produced continually in biological systems as toxic bi-products of normal oxidative metabolism (Livingstone, 1991; Lemaire & Livingstone, 1994a). They include the superoxide anion radical ($O_2^-\cdot$), hydrogen peroxide (H_2O_2) and the highly reactive hydroxyl radical ($\cdot OH$). ROS production can be enhanced by both organic xenobiotics, via redox cycling (see below), uncoupling of electron-transfer systems, and induction of cytochrome P450, and by certain transition metals which via redox reactions catalyse the formation of $\cdot OH$ from $O_2^-\cdot$ and H_2O_2. ROS and the products of their oxidative reactions with biological molecules, e.g. aldehydes, are detoxified by free radical scavengers (see above) and specific antioxidant enzymes (Fig. 1), principally superoxide dismutase (SOD; converts $O_2^-\cdot$ to H_2O_2; EC 1.15.1.1), catalase (converts H_2O_2 to H_2O; EC 1.1.1.6), glutathione peroxidase (converts H_2O_2 to H_2O utilizing reduced glutathione; EC 1.11.1.9), aldehyde dehydrogenase (ADH; converts aldehydes to acids; EC 1.2.1.3) and DT-diaphorase (DTD; prevents quinone-mediated $O_2^-\cdot$ production; EC 1.6.99.2). Increased ROS production can lead to increased synthesis of antioxidant enzymes. Failure to remove ROS by antioxidant defences can result in oxidative damage to key biological molecules, including DNA (Sahu, 1991; Kehrer, 1993).

A similar scenario of pro-oxidant (ROS generation, oxidative damage), and antioxidant (scavengers, antioxidant enzymes) processes in relation to exposure to organic xenobiotics and metals is indicated for aquatic invertebrates and vertebrates as described above for mammals (Livingstone, Förlin & George, 1990; Winston & DiGiulio, 1991; Lemaire & Livingstone, 1994a, Livingstone *et al.*, 1994a). Antioxidant enzymes are widespread, and some increases in hepatic activities are seen with exposure to organic xenobiotics, indicative of enhanced ROS generation and/or oxidative damage, e.g. SOD and catalase activities in channel catfish (*Ictalurus punctatus*) (DiGiulio, Habig & Gallagher, 1993) and dab (*Limanda limanda*) (Livingstone *et al.*, 1993) exposed to sediment containing PAHs and PCBs, and ADH and DTD activities in rainbow trout (*Oncorhynchus mykiss*) exposed to the carcinogen aflatoxin B_1 (Parker *et al.*, 1993). The redox cycling of particular organic xenobiotics (X: quinones, nitroaromatics, aromatic amines) involves univalent reduction to the cation radical ($\cdot X$), catalysed by NAD(P)H-dependent flavoprotein reductases, followed by autoxidation

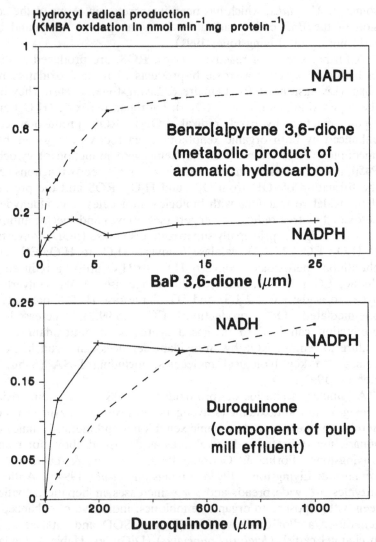

Fig. 3. Stimulation of NAD(P)H-dependent reactive oxygen species (ROS) production (i.e. iron/EDTA-mediated hydroxyl radical formation) in hepatic microsomes of flounder (*Platischthys flesus*) by benzo[a]pyrene 3,6-dione and duroquinone. Data from Lemaire *et al.* (1994); hydroxyl radical formation measured by the oxidation of the scavenging agent 2-keto-4-methiolbutyric acid (KMBA) to ethylene.

to yield $O_2^-\cdot$ and regenerate the parent compound, i.e.

$$1\ e^- + X = \cdot X$$
$$\cdot X + O_2 = X + O_2^-\cdot$$

The net result is the consumption of NAD(P)H and the production of $O_2^-\cdot$ which can then give rise to H_2O_2 via dismutation ($2H^+ + 2O_2^-\cdot = H_2O_2 + O_2$) and $\cdot OH$ via a metal-catalysed Haber-Weiss reaction ($H_2O_2 + O_2^-\cdot = \cdot OH + OH^- + O_2$) and other free-radical interactions. The ability of sub-cellular fractions, particularly hepatic (or equivalent tissue) microsomes, to catalyse the redox cycling of quinones, nitroaromatics and other compounds has been studied extensively in aquatic organisms, and indicated to be a widespread phenomenon (Table 3). Redox cycling and ROS generation has been shown for both pollutant parent compounds, e.g. duroquinone (tetramethyl-1,4-benzoquinone) in pulp mill effluent, and pollutant metabolites, e.g. BaP 3,6-dione (Fig. 3). Recent studies have also demonstrated the involvement of cytochrome P450 reductase in NADPH-dependent redox cycling of nitrofurantoin in hepatic microsomes of founder (*Platichthys flesus*) (Lemaire & Livingstone, 1994*b*), and stimulation of ROS production by metals and non-redox cycling compounds, namely the pesticide lindane (Table 3). The latter type of free-radical interaction, although minimal compared to that of redox cycling xenobiotics, could, if repeated for a wide range of pollutants, have a considerable cumulative effect on ROS production. Also, recently identified is the possibility of a toxic cycle in fish liver of exposure to PAHs leading to ROS production and lipid peroxidation, leading to enhanced dione production from PAHs (via the hydroperoxide-dependent peroxidase activity of cytochrome P450), leading to redox cycling and enhanced ROS production (Livingstone *et al.*, 1993, 1994*b*; Lemaire & Livingstone, 1994*a*).

Mechanisms of DNA damage by mutagenic species (laboratory studies)

Interaction of the parent compound, electrophilic metabolite, organic radical or ROS with critical sites in the genome can lead to the mutational activation of cellular oncogenes (cancer-causing genes), and the inactivation of tumour suppressor genes, leading in combination to the production of a cancer cell (Purchase, 1994). The fact that many marine organisms can convert procarcinogens to mutagenic species, including enhanced ROS production, indicates that there is a potential risk of genotoxicity following *in vivo* exposure. That there is indeed

Table 3. *Stimulation of NAD(P)H-dependent reactive oxygen species (ROS) production of sub-cellular fractions of aquatic organisms by model and pollutant xenobiotics*[a]

Animal	Species	Compound[b]	Reference
Fish	*Ictalurus punctatus Micropterus salmoides Salmo gairdneri*	nitrofurantoin (M) *p*-nitrobenzoic and *m*-dinitrobenzoic acids (P)	Washburn & Di Giulio (1988, 1989)
Fish	*Platichthys flesus*	nitrofurantoin and menadione (M) duroquinone, BaP 1,6-, 3,6- and 6,12- diones, lindane, Cr^{VI}, Ni^{II}(P)	Lemaire *et al.* (1994), Birmelin & Livingstone (1994)
Fish	*Perca fluviatilis*	duroquinone (P)	Lemaire *et al.* (1994)
Bivalve molluscs	*Geukensia demissa Rangia cuneata*	paraquat (P)	Wenning & Di Giulio (1988)
Bivalve molluscs	*Mytilus edulis*	nitrofurantoin, 4NQO and menadione (M) BaP 1,6-, 3,6- and 6,12-diones (P)	Garcia Martinez *et al.* (1989, 1992), Livingstone *et al.* (1989), Garcia Martinez & Livingstone (1994)

[a] All organic compounds are redox cyclers except for lindane; for details of tissues, coenzymes, etc. see references. [b] M. model compound; P. pollutant; 4NQO, 4-nitroquinoline *N*-oxide; BaP, benzo[a]pyrene.

such a risk is demonstrated by laboratory studies on the extent of covalent interaction of ultimate carcinogenic products to nucleophilic nitrogen and oxygen atoms in DNA, as well as with other macromolecules. A higher level of DNA strand breakage was detected in fathead minnows (*Pimephales promelas*) exposed to BaP (1 µg/l) for 6 days compared to unexposed fish (Shugart, 1988). DNA strand breaks may be formed via quinone metabolites (see below) or indirectly via excision repair of DNA-adducts (Waters, 1984), therefore, this study provides indirect evidence for the formation of BaP-derived DNA adducts in fish *in vivo*. A more direct measure of the formation of macromolecular adducts following *in vivo* exposure has been demonstrated, in both fish and invertebrates. Varanasi *et al.* (1986) showed that DNA isolated from the liver of *P. vetulus* dosed with [^{3}H]-BaP (2 mg/kg body weight) contained covalently bound radiolabel. Low levels of macromolecular adducts have also been detected in several marine invertebrates following *in vivo* exposure to [^{3}H]-BaP (James *et al.*, 1992; Marsh, Chipman & Livinstone, 1992). The characterization of BaP–DNA adducts and AAF–DNA adducts has also been further studied using the technique of ^{32}P-postlabelling. This assay has enabled the extremely sensitive detection of a wide range of chemically altered nucleotides following exposure to non-radioactive aromatic compounds in mammals (Randerath *et al.*, 1985). Briefly, the procedure involves enzymic digestion of DNA to deoxyribonucleoside 3'-monophosphates, which are then labelled in the 5' position by polynucleotide kinase catalyzed transfer of ^{32}P from ^{32}P-ATP. The adducted nucleoside bisphosphates are then separated from the non-adducted nucleotides on the basis of different chromatographic characteristics on PEI-cellulose thin layer plates (see Fig. 4). Various modifications have been introduced to enhance the sensitivity of the assay to detect 1 adduct in 10^9–10^{10} nucleotides (Reddy & Randerath, 1986; Gupta, 1985). A clear dose response in the level of DNA adducts was detected, using the postlabelling assay, following treatment of the larvae of the fresh water amphibian *Xenopus laevis* with BaP (Jones & Parry, 1992). Using this assay, one major adduct, which co-chromatographed with the standard anti-BPDE bound to deoxyguanosine-3'5'-biphosphate, was detectable following analysis of hepatic DNA isolated from BaP-exposed *P. vetulus* (Varanasi *et al.*, 1989) and brown bullheads (Sikka *et al.*, 1990). Using improved chromatographic techniques, Varanasi *et al.* (1989) could also detect the syn-BPDE DNA adduct following exposure of English sole to high doses of BaP (100 mg/kg). A major band of radioactivity close to the chromatogram origin was located following post-labelling analysis of digestive gland DNA from mussels (*M. edulis*) treated with AAF which was not present in controls, indicating that *M. edulis* can metabolize AAF to genotoxic products *in vivo* (Marsh, Chipman & Livingstone, 1993).

Generally, there is a good correlation between the ability of PAHs (and a wide range of other chemicals) to form covalent DNA adducts in mammals and their carcinogenic potency (Lutz, 1986). Studies have, therefore, been conducted in the hope of ascertaining whether the susceptibility of some fish to chemically induced tumours is related to DNA binding. For example, BaP bound to hepatic DNA was 2.4-fold

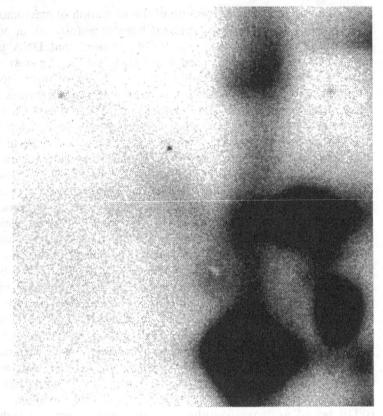

Fig. 4. Typical chromatogram showing carcingen–DNA adducts detected by ^{32}P-postlabelling. (C. Mitchelmore, unpublished data.)

higher in English sole than Starry flounder, which seemed to correlate with the relatively high incidence of cancer found in the former species (Varanasi *et al.*, 1986). Further studies suggested that the detoxication process is more effective in flounder than the English sole as the activity of glutathione-S-transferase (which converts reactive BaP metabolites to water-soluble conjugates) was 3-fold higher in the former (Collier *et al.*, 1986). Such studies may eventually lead to a better understanding of the relative risks (if any) of some species of marine organisms to pollutant-related genotoxicity.

That fish are susceptible to chemical carcinogenesis has been well established with the use of aflatoxin B_1. Nanogram quantities of this procarcinogen induce liver neoplasia in, for example, Rainbow trout following administration to the yolk sac of embryos (Black *et al.*, 1988).

The covalent binding of aflatoxin B_1 to trout liver DNA following intraperitoneal injection correlated well with data on tumour incidence and was modelled by studies on DNA binding of the aflatoxin epoxide metabolite in trout hepatocytes (Loveland *et al.*, 1987; Bailey, Taylor & Selivonchick, 1982). Ten out of 14 liver tumours from rainbow trout (*O. mykiss*) treated with aflatoxin B_1 contained DNA with point mutations in the c-ki-*ras* cellular oncogene. Of these, nine were G to T transversions in either codon 12 or codon 13: changes characteristic of those induced by aflatoxin B_1 in rodent liver carcinogenesis (Chang *et al.*, 1991).

ROS formation has been associated with various forms of DNA damage, including strand breaks, interstrand and DNA-protein cross-links, base modifications, and the formation of apurinic/apyrimidic sites. Increases in DNA strand breakage and sister chromatid exchanges, caused by oxidative damage, have been reported in fathead minnows (*P. promelas*) (Shugart *et al.*, 1989) and *N. arenacoedenta* (Pesch, 1990), respectively, following exposure to gamma-rays. Oxidative DNA damage has been observed following intramuscular injection of juvenile *P. vetulus* with nitrofurantoin (0.1–10 mg/kg) in the form of a significant increase in 8-hydroxydeoxyguanosine (8-OHdG) levels in hepatic DNA (Nishimoto *et al.*, 1991) but no such increase was detected in digestive gland DNA of mussels (*M. edulis*) exposed to menadione (100 ppb) and nitrofurantoin (100 ppb) in their water for 2 or 4 days (Marsh, Chipman & Livinstone, 1993).

Biomarkers of genetic toxicity (field studies)

Knowledge of the ability of pollutants to cause genetic toxicity in aquatic organisms leads us to consider ways in which the detection of DNA damage can help as a biomarker of environmental contamination. It can also lead to identification of causative mechanisms and implicate possible groups of chemicals responsible. Biomarkers of environmental pollution necessarily need to be multiple. They can be indicative of either exposure or damage and they need to be functionally and hierarchically linked (Peakal & Shugart, 1993).

A gradient of biomarkers from exposure to severe effects is indicated in Fig. 5. DNA damage, such as adduct formation, represents a biomarker of intermediate severity, i.e. a change clearly indicative of an adverse effect but of uncertain eventual consequence. In this respect, the analysis of DNA damage should be useful as one of several early and sensitive indicators of environmental pollution relating possibly both to exposure and effect (Livingstone, 1993).

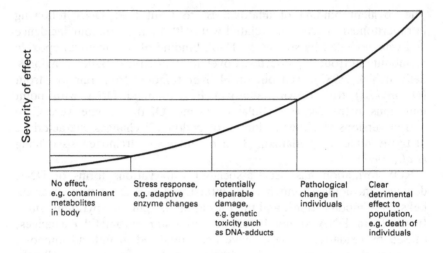

| No effect, e.g. contaminant metabolites in body | Stress response, e.g. adaptive enzyme changes | Potentially repairable damage, e.g. genetic toxicity such as DNA-adducts | Pathological change in individuals | Clear detrimental effect to population, e.g. death of individuals |

Biomarkers of exposure

Fig. 5. Genetic toxicity within a gradient of biomarkers.

In the assessment of genotoxicity, the ^{32}P-postlabelling technique for detection of DNA adducts (see above) has the advantage of high sensitivity and broad specificity capable of detecting a wide range of chemical–DNA interactions. Discrete DNA adducts are rarely discernible in relation to pollution exposure. This probably reflects the wide range of contaminant components to which organisms may be co-exposed. Thus, diffuse areas of post-labelled DNA adducts appear often in chromatographic analyses of contaminated organisms. This response, indicative of multiple, bulky, hydrophobic adducts has been reported in the livers of brown bullheads (*Ictalurus nebulosus*) from the Buffalo and Detroit rivers contaminated with PAHs in comparison to aquarium-raised fish (Dunn, Black & Maccubbin, 1987). Varanasi *et al.* (1989) have also reported similar findings on analysis of DNA adducts in the liver of *P. vetulus* sampled from the Duwamish waterway and Eagle Harbour, Puget Sound, USA with sediment contamination including PAH. Up to three 'zones' of postlabelled adducts were detected on chromatograms which were not present in liver DNA from fish taken from a reference site (Useless Bay). The level of adducts was in the region of 20 nmol per mol of nucleotides. *P. vetulus* treated with extracts of contaminated sediments from the same sites had similar adduct profiles. Bottom-feeding freshwater fish (*Tilapia mossambicca*) exposed to high concentrations of PAHs in the Damsui river in China have also been shown to contain an apparent wide variety of adducts compared to samples from the relatively non-polluted Fe-Tsui reservoir (Liu *et al.*, 1991).

Marine mammals are also susceptible to DNA adduct formation. This has been demonstrated in the Beluga whale (*Delphinapterus leucas*) from two sites in the Canadian Arctic and the Saint Lawrence estuary in the north-west Atlantic Ocean (Ray *et al.*, 1991). The level of DNA adduction ranged from 16–158 nmol per mol of nucleotide. Although the cause of the adducts could not be ascertained, there did not appear to be a correlation with the level of PAH exposure.

Mussels are of particular interest in environmental monitoring since they are sessile organisms. Evidence of pollution-related DNA adducts has been presented for the digestive gland of juvenile mussels collected from a site close to an oil refinery (Kurelec *et al.*, 1990). Four distinct adducts were detected in these mussels that were not found in similar mussels from an unpolluted site. ^{32}P-post-labelling of DNA from foot, mantle and gill of mussels (which may be metabolically less active than the digestive gland) did not, however, reveal DNA adducts related to contamination by PAH in an area of Vancouver harbour in Canada (Dunn, 1991).

The level of metabolic activation and the extent of DNA repair as well as the extent of pollution are critical determinants. In studies in which no contaminant-related adducts have been detected in marine (e.g. Chipman *et al.*, 1992) or freshwater (Kurelec *et al.*, 1989) fish, the combined effect of these parameters may have resulted in insufficient damage to have been present for its detection. Despite this, the latter study and several others have indicated adducts of apparent 'endogenous' origin not related to pollution status. Garg *et al.* (1992) found distinct DNA adducts to be present in different vertebrate and invertebrate aquatic species and these were influenced by season and reproductive stage. These adducts were apparently not pollution related but appear to be controlled by endogenous factors.

Assays measuring cytogenetic endpoints, such as chromosomal aberrations, sister chromatid exchange, micronucleation and aneuploidy, have been modified for *in situ* monitoring using a variety of aquatic species and cell types. Flow cytometry has recenty been applied to pollution monitoring, showing aneuploidy in cottonmouth snakes (*Agkistron piscivorus leucostoma*) inhabiting contaminated aquatic environments (Tiersch *et al.*, 1990). This technique is useful as an initial screen for genotoxicity, as it is both rapid and accurate, detecting changes in DNA content as low as 2–3% (McBee & Bickham, 1988). Increases in the number of chromosomal aberrations in gill cells

(Al-Sabti & Kurelec, 1985) and sister chromatid exchanges in developing eggs (Brunetti, Gola & Majone, 1986) have been seen in mussels from polluted sites compared to those from unpolluted sites. In addition the transfer of control mussels to a polluted site for 2 days also led to an increase in chromosomal aberrations (Al-Sabti & Kurelec, 1985). The application of chromosomal aberrations and sister chromatid exchange assays to pollution monitoring has, however, been hindered by the time length and technical difficulties of the assays (Kadhim, 1990).

The micronucleus assay is proving to be a particularly useful biomonitoring test as it is both fast and sensitive and is able to detect genomic damage due to both clastogenic effects and alterations in the mitotic spindle. Pollution-associated increases in frequency of micronuclei have been observed in erythrocytes of perch (*Perca flaviatitis*) (Al-Sabti & Hardig, 1990), haemocytes of mussel (*M. edulis*) (Wrisberg & Rhemrev, 1990) and gill cells of mussel (*M. galloprovincialis*) (Scarpato *et al.*, 1990). Exposure of *M. edulis* to sediments contaminated with PAHs induced a significant induction in micronuclei, although the effect was not conclusively linked to PAH concentration (Wrisberg & Rhemrev, 1990). The micronucleus test has also been adapted, using *Xenopus laevis* larvae, to be used for monitoring fresh water pollution (Van Hummelen *et al.*, 1989).

The effect of pollution on oxidative DNA damage in marine organisms has only recently been considered. Relatively high levels of the oxidative damage products, 8-OHdG and 8-hydroxydeoxyadenosine, and the ring opened products, 2,6-diamino-4-hydroxy-5-formamido-pyrimidine (Fapy-G), 4,6-diamino-5-formamidopyrimidine (Fapy-A) were detected in hepatic tumours of *P. vetulus* from a habitat highly contaminated with carcinogenic creosote components (Malins & Haimainot, 1990). Such oxidative DNA lesions were also present at lower levels in histologically normal livers from the same tumour-bearing population, but at higher levels, than for fish from non-contaminated sites (Malins & Haimainot, 1991). Thus, the DNA lesions may occur prior to tumour formation, which is consistent with the concept that they are causally linked to tumourogenesis. High levels of oxidative DNA damage (1-8oHdG per 400 dG) were also observed in normal tissue from healthy fish from Eagle Harbour, a moderately polluted environment compared to levels seen in fish from the 'pristine' site, Elger Bay (1-8oHdG per 660 dG) (Malins, 1993). No increase in 8OHdG was, however, seen in hepatic DNA from dab (*L. limanda*) sampled along a pollutant gradient in the North Sea (Chipman *et al.*, 1992) despite an apparent increase in antioxidant enzyme activity (Livingstone *et al.*, 1992) perhaps indicating that, in these fish, the inducible antioxidant defences are coping with an increase in oxyradical generation. Assays measuring DNA-strand breakage could also prove useful for monitoring oxidative damage in the field, although strand breakage is not specific to oxidative damage and could result from the excision repair of DNA lesions. Redbreast sunfish (*Lepomis auritus*), sampled from freshwater creeks contaminated with polychlorinated biphenyls and PAHs, showed a significant increase in strand breaks in DNA compared to the levels in fish from the reference sites (Everaats *et al.*, 1993).

Different classes of genotoxic agents give rise to a characteristic pattern of transitions or transversions in the cellular oncogenes that they activate (Bos, 1989). It may be possible, therefore, to start to incriminate specific classes of genotoxicant in the mutational changes seen in an organism's tumour. Hepatic tumours from feral Winter flounder (McMahon *et al.*, 1988) in Boston Harbour have been found to contain mutational activated *ras* oncogene. Certain activated c-ki-*ras* genes were analysed and found to contain GC to AT transitions and GC to TA transversions suggesting that the agents responsible for the tumour mutations are in a chemical class able to give this characteristic change.

Conclusions

A wide range of potentially carcinogenic organic and metal pollutants are present in the aquatic environment which are readily taken up into the tissues of resident organisms. Activation to mutagenic species can occur via a number of biotransformation-mediated mechanisms giving rise to reactive electrophilic metabolites, free organic radicals and ROS. Interaction of such species with DNA leads to various detectable end-points of damage, including base modifications, DNA strand breakage, mutagenesis and chromosomal aberrations. Changes in critical genes involved in regulation of cell growth and differentiation then contribute to the process of carcinogenesis. The study of a battery of end-points for genotoxicity in environmentally exposed aquatic species will be useful biomarkers as part of the monitoring programmes, particularly since certain parameters such as DNA adducts can persist in organisms relative to the residence time of contaminants and their metabolites, thus allowing for cumulative exposure at target sites to be detected. It is important, however, that account is taken of the variation seen for example in the level of background DNA adducts (Garg *et al.*, 1992) and DNA strand breaks (Shugart, 1990) before effects can be definitively related to effects of pollutants.

References

Al-Sabti, K. & Hardig, J. (1990). Micronucleus test in fish for monitoring the genotoxic effects of industrial waste products in the Baltic sea, Sweden. *Comparative Biochemistry and Physiology*, **97C**, 179–82.

Al-Sabti, K. & Kurelec, B. (1985). Induction of chromosomal aberrations in the mussel *Mytilus galloprovincialis* watch. *Bulletin of Environmental Contamination and Toxicology*, **35**, 660–5.

Ames, B.N. (1989). Mutagenesis and carcinogenesis: endogenous and exogenous factors. *Environmental Molecular Mutagenicity*, **14**, 66–77.

Bailey, G.S., Taylor, M.J. & Selivonchick, D.P. (1982). Aflatoxin B1 metabolism and DNA binding in isolated hepatocytes from rainbow trout (*Salmo gairdneri*). *Carcinogenesis*, **3**, 511–18.

Birmelin, C. & Livingstone, D.R. (1994). Effects of pollutant transition metals (V, Cr, Ni, Co, Cu) on NAD(P)H-dependent hydroxyl radical production by hepatic microsomes of flounder *Platichthys flesus*. Abstract, *Proceedings of the 15th International Symposium of the European Society for Comparative Physiology and Biochemistry*, Univ. Genoa, Italy, 20–23rd Sept., 1994.

Black, J.J., Maccubbin, A.E., Myers, H.K. & Zeigel, R.F. (1988). Aflatoxin B1 induced hepatic neoplasia in Great Lakes Coho salmon. *Bulletin of Environmental Contamination and Toxicology*, **41**, 742–5.

Bos, J.L. (1989). *ras* Oncogenes in human cancer: a review. *Cancer Research*, **49**, 4682–9.

Brousseau, D.J. & Baglio, J.A. (1991). Disease progression and mortality in neoplastic *Mya arenaria* in the field. *Marine Biology*, **110**, 249–52.

Brunetti, R., Gola, I. & Majone, F. (1986). Sister-chromatid exchange in developing eggs of *Mytilus galloprovincialis* Lmk. (Bivalvia). *Mutation Research*, **174**, 207–11.

Caldwell, J. & Jakoby, W.B. (eds.) (1983). *Biological Basis of Detoxication*. Academic Press, NY, USA.

Cavalieri, E.L., Rogan, E.G., Murray, W.J. & Ramakrishna, N.V.S. (1993). Mechanistic aspects of benzo[a]pyrene metabolism. *Poly. Aromatic Compounds*, 3(supplement), 1047–54.

Chang, Y.-J., Mathews, C., Mangold, K., Marien, K., Hendricks, J. & Bailey, G. (1991). Analysis of *ras* gene mutations in rainbow trout liver tumours initiated by aflatoxin B1. *Molecular Carcinogenesis*, **4**, 112–19.

Chipman, J.K. & Marsh, J.W. (1991). Bio-techniques for the detection of genetic toxicity in the aquatic environment. *Journal of Biotechnology*, **17**, 199–208.

Chipman, J.K., Marsh, J.W., Livingstone, D.R. & Evans, B. (1992). Genetic toxicity in the dab (*Limanda limanda*) from the North Sea. *Marine Ecology Programme Series*, **91**, 121–6.

Collier, T.K., Stein, J.E., Wallace, R.J. & Varanasi, U. (1986). Xenobiotic metabolizing enzymes in spawning English sole (*Parophrys vetulus*) exposed to organic solvent extracts of marine sediments from contaminated and reference areas. *Comparative Biochemistry and Physiology*, **84C**, 291–8.

DiGiulio, R.T., Habig, C. & Gallagher, E.P. (1993). Effects of Black Rock Harbor sediments on indices of biotransformation, oxidative stress, and DNA integrity in channel catfish. *Aquatic Toxicology*, **26**, 1–22.

Dunn, B.P. (1991). Carcinogen adducts as an indicator for public health risks of consuming carcinogen-exposed fish and shellfish. *Environmental Health Perspectives*, **90** 111–16.

Dunn, B.P., Black, J.J & Maccubbin, A. (1987). ^{32}P-Postlabeling analysis of aromatic DNA adducts in fish from polluted areas. *Cancer Research*, **47**, 6543–8.

Everaats, J.M., Shugart, L.R., Gustin, M.K., Hawkins, W.E. & Walker, W.E. (1993). Biological markers in fish: DNA integrity, hematological parameters and liver somatic index. *Marine Environmental Research*, **35**, 101–7.

Friedberg, E.C. (ed.) (1985). *DNA Repair*. W.H. Freeman and Company, USA.

Garcia Martinez, P. & Livingstone, D.R. (1994). Benzo[a]pyrene-dione-stimulated oxyradical production by microsomes of digestive gland of the common mussel, *Mytilus edulis* L. *Marine Environmental Research*. In press.

Garcia Martinez, P., O'Hara, S., Winston, G.W. & Livingstone, D.R. (1989). Oxyradical generation and redox cycling mechanisms in digestive gland microsomes of the common mussel, *Mytilus edulis*. *Marine Environmental Research*, **28**, 271–4.

Garcia Martinez, P., Hajos, A.K.D., Livingstone, D.R. & Winston, G.W. (1992). Metabolism and mutagenicity of 4-nitroquinoline *N*-oxide by microsomes and cytosol of digestive gland of the mussel *Mytilus edulis* L. *Marine Environmental Research*, **34**, 303–7.

Gardner, G.R., Pruell, R.J. & Malcolm, A.R. (1992). Chemical induction of tumors in oysters by a mixture of aromatic and chlorinated hydrocarbons, amines and metals. *Marine Environmental Research*, **34**, 59–63.

Garg, A., Krca, S., Kurelec, B. & Gupta, R.C. (1992). Endogenous DNA modifications in aquatic organisms and their probable biological significance. *Comparative Biochemistry and Physiology*, **102B**, 825–32.

George, S.G. (1994). Enzymology and molecular biology of phase II detoxication systems in fish. In *Aquatic Toxicology: Molecular, Biochemical and Cellular Perspectives*. Malins, D. & Ostrander, G. (eds.), pp. 37–85. Lewis Publishers, Boca Raton, Florida.

Goldstein, J.A. & Faletto, M.B. (1993). Advances in mechanisms of activation and deactivation of environmental chemicals. *Environmental Health Perspectives*, **100**, 169–79.

Gupta, R.C. (1985). Enhanced sensitivity of ^{32}P-postlabeling analysis of aromatic carcinogen:DNA adducts. *Cancer Research*, **45**, 5656–62.

James, M.O., Altman, A., Li, C.-L.J. & Boyle, S.M. (1992). Dose and time dependent formation of benzo[a]pyrene metabolite DNA adducts in the spiny lobster, *Panulirus argus*. *Marine Environmental Research*, **34**, 299–302.

Jones, N.J. & Parry, J.M. (1992). The detection of DNA adducts, DNA base changes and chromosomal damage for the assessment of exposure to genotoxic pollutants. *Aquatic Toxicology*, **22**, 323–44.

Kadhim, M.A. (1990). Methodologies for monitoring the genetic effects of mutagens and carcinogens accumulated in the body tissues of marine mussels. *Reviews in Aquatic Science*, **2**, 83–107.

Kehrer, J.P. (1993). Free-radicals as mediators of tissue injury and disease. *Critical Reviews in Toxicology*, **23**, 21–48.

Kurelec, B. (1992). The multixenobiotic resistance mechanism in aquatic organsms. *Critical Reviews in Toxicology*, **22**, 23–43.

Kurelec, B., Garg, A., Krca, S., Chacko, M. & Gupta, R.A. (1989). Natural environment surpasses polluted environment in inducing DNA damage in fish. *Carcinogenesis*, **10**, 1337–9.

Kurelec, B., Garg, A., Krca, S. & Gupta, R.A. (1990). DNA adducts in marine mussel *Mytilus galloprovincialis* living in polluted and unpolluted environments. In *Biomarkers of Environmental Contamination*. McCarthy, J.F. & Shugart, L.R. (eds.), pp. 217–27. Lewis Publishers, Chelsea.

Lemaire, P. & Livingstone, D.R. (1994*a*). Pro-oxidant/antioxidant processes and organic xenobiotic interactions in marine organisms, in particular the flounder *Platichthys flesus* and the mussel *Mytilus edulis*. *Trends in Comparative Biochemistry and Physiology*. In press.

Lemaire, P. & Livingstone, D.R. (1994*b*). Inhibition studies on the involvement of flavoprotein reductases in menadione- and nitrofurantoin-stimulated oxyradical production by hepatic microsomes of flounder (*Platichthys flesus*). *Journal of Biochemistry and Toxicology*. In press.

Lemaire, P., Matthews, A., Förlin, L. & Livingstone, D.R. (1994). Stimulation of oxyradical production of hepatic microsomes of flounder (*Platichthys flesus*) and perch (*Perca fluviatilus*) by model and pollutant xenobiotics. *Archives of Environmental Contamination and Toxicology*, **26**, 191–200.

Liu, T.-Y., Cheng, S.-L., Ueng, T.-H., Ueng, Y.-F. & Chi, C.-W. (1991). Comparative analysis of aromatic DNA adducts in fish from polluted and unpolluted areas by the ^{32}P-postlabeling analysis. *Bulletin of Environmental Contamination and Toxicology*, **47**, 783–9.

Livingstone, D.R. (1991). Organic xenobiotic metabolism in marine invertebrates. In *Advances in Comparative and Environmental Physiology*. Gilles, R. (ed.), Vol. 7, pp. 45–185. Springer-Verlag, Berlin.

Livingstone, D.R. (1993). Biotechnology and pollution monitoring: use of molecular biomarkers in the aquatic environment. *Journal of Chemical Technology and Biotechnology*, **57**, 195–211.

Livingstone, D.R., Garcia Martinez, P. & Winston, G.W. (1989). Menadione-stimulated oxyradical formation in digestive gland microsomes of the common mussel, *Mytilus edulis* L. *Aquatic Toxicology*, **15**, 213–36.

Livingstone, D.R., Garcia Martinez, P., Michel, X., Narbonne, J.F., O'Hara, S., Ribera, D. & Winston, G.W. (1990). Oxyradical production as a pollution-mediated mechanism of toxicity in the common mussel, *Mytilus edulis* L., and other molluscs. *Functional Ecology*, **4**, 415–24.

Livingstone, D.R., Archibald, S., Chipman, J.K. & Marsh, J.W. (1992). Antioxidant enzymes in liver of dab *Limanda limanda* from the North Sea. *Marine Ecology Programme Series*, **91**, 97–104.

Livingstone, D.R., Lemaire, P., Matthews, A., Peters, L.D., Bucke, D. & Law, R.J. (1993). Pro-oxidant, antioxidant and 7-ethoxyresorufin *O*-deethylase (EROD) activity responses in liver of dab (*Limanda limanda*) exposed to sediment contaminated with hydrocarbons and other chemicals. *Marine Pollution Bulletin*, **26**, 602–6.

Livingstone, D.R., Förlin, L. & George, S. (1994*a*). Molecular biomarkers and toxic consequences of impact by organic pollution in aquatic. In Sutcliffe, D. (ed.). *Water Quality and Stress Indicators: Linking Levels of Organisation*. Freshwater Biological Association Special Publication.

Livingstone, D.R., Lemaire, P., Matthews, A., Peters, L.D., Porte, C., Fitzpatrick, P.J., Förlin, L., Nasci, C., Fossato, V., Wootton, A.N. & Goldberg, P.S. (1994*b*). Assessment of the impact of organic pollutants on goby (*Zosterisessor ophiocephalus*) and mussel (*Mytilus galloprovincialis*) from Venice Lagoon, Italy: Biochemical studies. *Marine Environmental Research*. In press.

Loveland, P.M., Wilcox, J.S., Pawlowski, N.E. & Bailey, G.S. (1987). Metabolism and DNA-binding of aflatoxicol and aflatoxin B1 in vivo and in isolated hepatocytes from rainbow trout (*Salmo gairdneri*). **8**, 1065–70.

Lutz, W.K. (1986). Quantitative indicator of DNA binding data for risk estimation and for classification of direct and indirect carcinogens. *Journal of Cancer Research and Clinical Oncology*, 112, 85–91.

Malins, D.C. (1993). Identification of hydroxyl radical-induced lesions in DNA base structure: biomarkers with a putative link to cancer development. *Journal of Toxicology and Environmental Health*, 40, 247–61.

Malins, D.C. & Haimainot, R. (1990). 4,6-Diamino-5-form-amido-pyrimidine, 8-hydroxyguanine, and 8-hydroxyadenine in DNA from neoplastic liver of English sole exposed to carcinogens. *Biochemical and Biophysical Research Communications*, 173, 614–19.

Malins, D.C. & Haimainot, R. (1991). The etiology of cancer: hydroxy radical-induced DNA lesions in histologically normal lesions of English sole (*Parophrys vetulus*) from polluted marine environments. *Environmental Science and Technology*, 17, 679–85.

Malins, D.C., Myers, M.S. & Roubal, W.T. (1983). Organic free radicals associated with idiopathic lesions of English sole (*Parophrys vetulus*) from polluted marine environments. *Environmental Science and Technology*, 17, 679–85.

Malins, D.C. & Roubal, W.T. (1985). Free radicals derived from nitrogen-containing xenobiotics in sediments and liver and bile of English sole from Puget Sound, Washington. *Marine Environmental Research*, 17, 205–10.

Maron, D.M. & Ames, B.N. (1983). Revised methods for the *Salmonella* mutagenicity test. *Mutation Research*, 113, 173–215.

Marsh, J.W., Chipman, J.K. & Livingstone, D.R. (1992). Activation of xenobiotics to reactive and mutagenic products by the marine invertebrates *Mytilus edulis*, *Carcinus maenus* and *Asterias rubens*. *Aquatic Toxicology*, 22, 115–28.

Marsh, J.W., Chipman, J.K. & Livingstone, D.R. (1993). Formation of DNA-adducts following laboratory exposure of the mussel, *Mytilus edulis*, to xenobiotics. *Science Total Environment, Suppl.*, 567–72.

McBee, K. & Bickman, J.W. (1988). Petrochemical-related DNA damage in wild rodents detected by flow cytometry. *Bulletin of Environmental Contamination and Toxicology*, 40, 343–9.

McMahon, G., Huber, L.J., Stegeman, J.J. & Wogan, G.N. (1988). Identification of a c-K-*ras* oncogene in a neoplasm isolated from winter flounder. *Marine Environmental Research*, 24, 345–50.

Michel, X.R., Cassand, P.M., Ribera, D.G. & Narbonne, J.F. (1992). Metabolism and mutagenic activation of benzo[a]pyrene by subcellular fractions from mussel (*Mytilus galloprovincialis*) digestive gland and sea bass (*Dicentrarchus labrax*) liver. *Comparative Biochemistry and Physiology*, 103C, 43–51.

Michel, X.R., Cassand, P.M. & Narbonne, J.F. (1993). Activation of benzo[a]pyrene and 2-aminoanthracene to bacteria mutagens by mussel digestive gland postmitochondrial fraction. *Mutation Research*, **301**, 113–19.

Nishimoto, M., Roubal, W.T., Stein, J.E. & Varanasi, U. (1991). Oxidative DNA damage in tissues of English sole (*Parophrys vetulus*) exposed to nitrofurantoin. *Chemical–Biological Interactions*, **22**, 392–6.

Oda, Y., Nakamura, S., Oki, I., Kato, T. & Shinagawa, H. (1985). Evaluation of the new system (*umu*-test) for the detection of environmental mutagens and carcinogens. *Mutation Research*, **147**, 219–29.

Parker, L.M., Laurén, D.J., Hammock, B.D., Winder, B. & Hinton, D.E. (1993). Biochemical and histochemical properties of hepatic tumors of rainbow trout, *Oncorhynchus mykiss. Carcinogenesis*, **14**, 211–17.

Peakal, D.B. & Shugart, L.R. (1993). Biomarkers: research and application in the assessment of environmental health. *NATO Advanced Science Institutes Series*, Springer-Verlag, Berlin.

Pesch, G.G. (1990). Sister chromatid exchange and genotoxicity measurements using polychaete worms. *Reviews in Aquatic Sciences*, **2**, 19–25.

Porte, C., Lemaire, P., Peters, L.D. & Livingstone, D.R. (1994). Partial purification and properties of cytochrome P450 from digestive gland microsomes of the common mussel, *Mytilus edulis* L. *Marine Environmental Research.* In press.

Purchase, I.F.H. (1994). Current knowledge of mechanisms of carcinogenicity: genotoxins versus non-genotoxins. *Human Experimental Toxicology*, **13**, 17–28.

Randerath, K., Randerath, E., Agrawal, H.P., Gupta, R.C., Schurdak, M.E. & Reddy, M.V. (1985). Postlabeling methods for carcinogen-DNA adduct analysis. *Environmental Health Perspectives*, **62**, 57–65.

Ray, S., Dunn, B.P., Payne, J.F., Fancey, L., Helbig, R. & Beland, P. (1991). Aromatic DNA-carcinogen adducts in Beluga whales (*Delphinapterus leucas*) from the Canadian Arctic and the Gulf of St. Lawrence. *Marine Pollution Bulletin*, **22**, 392–6.

Reddy, M.V. & Randerath, K. (1986). Nuclease P1-mediated enhancement of sensitivity of ^{32}P-postlabeling test for structurally diverse DNA adducts. *Carcinogenesis*, **7**, 1543–51.

Sahu, S.C. (1991). Role of oxygen free radicals in the molecular mechanisms of carcinogenesis: a review. *Environmental Carcinogenesis Ecotoxicology Review*, **C9**, 83–112.

Scarpato, R., Migliore, L., Alfinito-Cognetti, G. & Barale, R. (1990). Induction of micronuclei in gill tissue of *Mytilus galloprovincialis* exposed to polluted marine waters. *Marine Pollution Bulletin*, **21**, 74–80.

250   J.W. NUNN, D.R. LIVINGSTONE AND J.K. CHIPMAN

Schaeffer, D.J. (1993). Planarians as a model system for *in vivo* tumorigenesis studies. *Ecotoxicological Environmental Safety*, **25**, 1–18.

Schlenk, D. & Buhler, D.R. (1989). Determination of multiple forms of cytochrome P-450 in microsomes from the digestive gland of *Cryptochiton stelleri*. *Biochemical Biophysical Research Communications*, **163**, 476–80.

Shugart, L. (1988). An alkaline unwinding assay for the detection of DNA damage in aquatic organisms. *Marine Environmental Research*, **24**, 321–5.

Shugart, L.R. (1990). Biological monitoring: testing for genotoxicity. In *Biomarkers of Environmental Contamination*. McCarthy, J.F. & Shugart, L.R. (eds.), pp. 217–27. Lewis Publishers, Chelsea.

Shugart, L.R., Gustin, M.K., Laird, D.M. & Dean, D.A. (1989). Susceptibility of DNA in aquatic organisms to strand breakage: effects of X-rays and gamma radiation. *5th International Symposium on 'Responses of Marine Organisms to Pollutants'*, Plymouth, UK.

Sikka, H.C., Rutowski, J.P., Kandaswami, C., Kumar, S., Earley, K. & Gupta, R.C. (1990). Formation and persistence of DNA adducts in the liver of brown bullheads exposed to benzo[a]pyrene. *Cancer Letters*, **49**, 81–7.

Stegeman, J.J. (1993). The cytochromes in fish. In *Biochemistry and Molecular Biology of Fishes*. Hochachka, P.V. & Mommsen, T.P. (eds.), Vol. 2, pp. 137–58. Elsevier Science Publisher BV, Amsterdam.

Stein, J.E., Reichert, W.L., Nishimoto, M. & Varanasi, U. (1990). Overview of studies on liver carcinogenesis in English sole from Puget Sound; evidence for a xenobiotic chemical etiology II: biochemical studies. *Science Total Environment*, **94**, 51–69.

Thoreirsson, S.S., McManus, M.E. & Glowinski, I.B. (1984). Metabolic processing of aromatic amines. In *Drug Metabolism and Drug Toxicity*. Mitchell, J.R. & Horning, M.G. (eds.), pp. 183–97. Raven Press, NY, USA.

Tiersch, T.R., Figiel, C.R., Lee, R.M., Chandler, R.W. & Houston, A.E. (1990). Use of flow cytometry to screen for the effects of environmental mutagens: baseline DNA values in cottonmouth snakes. *Bulletin Environmental Contamination Toxicology*, **45**, 833–9.

Van Hummelen, P., Zoll, C., Paulussen, J., Kirsch-Volders, M. & Jaylet, A. (1989). The micronucleus test in *Xenopus*: a new and simple '*in vivo*' technique for detection of mutagens in fresh water. *Mutagenesis*, **4**, 12–16.

Varanasi, U., Nishimoto, M., Reichart, W.L. & Le Ebarhar, B.-T. (1986). Comparative metabolism of benzo[a]pyrene and covalent binding to hepatic DNA in English sole, starry flounder, and rat. *Cancer Research*, **46**, 3817–24.

Varanasi, U., Reichart, W.L., Le Eberhart, B.-T. & Stein, J.E. (1989). Formation and persistence of benzo[a]pyrene-diolepoxide–DNA adducts in liver of English sole (*Parophrys vetulus*). *Chemical–Biological Interactions*, **69**, 203–16.

Walker, C.H. & Livingstone, D.R. (eds.) (1992). *Persistent Pollutants in Marine Ecosystems*. pp. 272, Pergamon Press, Oxford.

Washburn, P.C. & DiGiulio, R.T. (1988). Nitrofurantoin-stimulated superoxide production by channel catfish (*Ictalurus punctatus*) hepatic microsomes and soluble fractions. *Toxicol. Applied Pharmacology*, **95**, 363–77.

Washburn, P.C. & DiGiulio, R.T. (1989). Stimulation of superoxide production by nitrofurantoin, *p*-nitrobenzoic acid and *m*-dinitrobenzoic acid in hepatic microsomes of three species of freshwater fish. *Environmental Toxicological Chemistry*, **8**, 171–80.

Waters, R. (1984). DNA repair tests in cultured mammalian cells. In *Mutagenicity Testing: A Practical Approach*. Venitt, S. & Parry, J.M. (eds.), pp. 99–117. IRL Press, Oxford.

Wilbrink, M., De Vries, J., Vermeulen, N.P.E., Janse, C. & De Vleiger, T.A. (1987). Effects of dihalogenated biphenyls on various functional parameters in the pond snail *Lymnaea stagnalis*. *Comparative Biochemistry and Physiology*, **87A**, 1025–31.

Winston, G.W. & DiGiulio, R.T. (1991). Prooxidant and antioxidant mechanisms in aquatic organisms. *Aquatic Toxicology*, **19**, 137–61.

Wootton, A.N., Herring, C., Spry, J.A., Wiseman, A., Livingstone, D.R., Goldfarb, P.S. (1994). Evidence for the existence of cytochrome P450 gene families (CYP1A1, CYP3A, CYP4A1, CYP11A1) and modulation of gene expression (CYP1A1) in the mussel *Mytilus* sp. *Marine Environmental Research*. In press.

Wrisberg, M.N. & Rhemrev, R. (1990). Detection of genotoxins in the aquatic environment with the mussel *Mytilus edulis*. *Water Science Technology*, **25**, 317–24.

Würgler, F.E. & Kramers, P.G.N. (1992). Environmental effects of genotoxins (eco-genotoxicology). *Mutagenesis*, **7** 321–7.

S.G. GEORGE

In vitro toxicology of aquatic pollutants: use of cultured fish cells

Introduction

The human health disasters such as the severe abnormalities and fatalities due to consumption of Cd-contaminated rice (Itai-itai disease) and methyl-mercury contaminated shellfish (Minimata disease) in Japan, infantile liver cirrhosis from Cu-contaminated rice in India, the environmental contamination caused by the industrial accidents at Bhopal in India and Seveso in Italy (2,4,5-T and dioxin, respectively), as well as those in petroleum production and transport (platform blowouts in the Gulf of Mexico and North Sea, the wrecks of Exxon Valdez and Braer and the Gulf War), have raised awareness in the problems of environmental contamination by man's activities. Hundreds of new man-made chemicals are introduced into the environment each year and assessment of the potential impact of these xenobiotic ('foreign to life') contaminants on aquatic organisms is therefore being afforded a much higher priority than in past decades.

The effects of man-made contaminants are alarmingly widespread, one such example being the effects of polychlorinated biphenyls (PCBs). First produced in the 1920s, their chemical inertness, thermal conductivity and dielectric properties led to the use of around a million tonnes in electrical transformers and capacitors before it was realized that they could 'switch on' genes involved in generating carcinogenic compounds from polyaromatic hydrocarbons (PAHs) and that a combustion product, dioxin (also present as a contaminant of production) was one of the most toxic compounds known. They are no longer produced in significant amounts. However, they have been so pervasive that they have recently been identified, and their impact has been detected, in livers of fish from one of the oceans deepest abyssal plains. Another example is the effect of organo-tin based antifoulants whose use was not severely restricted until they were found to be responsible for causing widespread defects in oyster shells and decimation of dog-whelk populations due to sex reversal and sterilization. The latter effect, which occurred at very low concentrations (<100 ng l^{-1}), was entirely unpredictable and is particularly alarming since it was not noticed until population level effects were observed. More recently, it has been found that sewage effluents contain both oestrogenic and anti-oestrogenic substances; one of the former group has been identified as nonyl-phenol which is present in plastics. Predictably, testing for these reproductive, or hormonal, effects will assume high priority in the near future.

To carry out field- or even laboratory-based population or whole animal toxicity testing for the myriad of xenobiotics produced would obviously be impossible; therefore the development of cellular methods would be useful for screening individual compounds or even effluents for 'effect specific' or 'compound class specific' effects. It is here that *in vitro* methods using isolated cultured cells may have great potential in environmental monitoring.

Cytotoxicity testing

Initially, it was hoped that *in vitro* methods would replace standard whole animal LC50 tests, i.e. were 'alternatives to whole animal testing' and several test systems using established cell lines have been investigated. A variety of end-points of cytotoxicity have been used, including colony formation, cell replication, mitotic index, uridine uptake, cell detachment as well as two easily measured end points, cell death and cell viability. Cell death can be estimated either by cell counting in a haemocytometer (or flow cytometer) or by spectrophotometric determination of total cellular protein with a dye binding assay. Cell viability can be determined by the ability to take up vital dyes such as trypan blue (measured in a cytometer) or neutral red (determined spectrophotometrically). Spectrophotometric assays have a major advantage in that they can be performed in 96-well cell culture plates, and thus the entire experiment can be carried out on a small scale with many replicates (typically 8). Another strength of the *in vitro* test system is the avoidance of the severe outbreeding which is found in natural fish populations, thus increasing reproducibility and the ease of performing many replicates. The availability of a simple, reproducible system makes it ideal for screening a number of chemically related compounds and ranking their relative acute toxicities to determine quantitative structure activity relationships (QSARs). Such cytotoxicity determinations have been carried out successfully with fish cells. Using a cell detachment assay, Bols *et al.* (1985) determined QSARs for a number of substituted phenols in RTG-2 cells and found a significant correlation with their water-borne toxicity, although they concluded that this method was tedious and would be biased towards substances which affect cell membrane integrity. Borenfreund and co-workers developed the neutral red cytotoxicity assay, initially for mammalian use, which is based upon the ability of cells to take up this vital dye and they have investigated its applicability to determination of the toxicity of many compounds with a number of fish cell lines. Many of these applications have been reviewed by Babich & Borenfreund (1991). They favoured the BF-2 cell line, largely due to its high growth rate (optimal at 34 °C) which enabled a high throughput of tests. However, they did note differences in toxicity with temperature and thus, in my opinion, selection of the test cell line should be based on target genus and its environmental temperature rather than convenience. Examples of two QSAR studies carried out in my laboratory are shown in Fig. 1 where the relative toxicities of metal ions to BF-2 and TF cells have been determined by the neutral red cytotoxicity test. Fig. 2 shows that the relative toxicities (NR EC 50s) are similar to the LC50s observed *in vivo*.

Extrapolation of *in vitro* data to toxicity *in vivo* can be hazardous since the physicochemical properties of the compound can also contribute to its ultimate toxicity and uptake rates (i.e. bioavailability of compounds) are often highly dependent upon these properties. In the case of divalent metals, it has been shown that it is the free metal ion that is transported across the cell membrane, thus bioavailability will be a function of complexation. For a given group of metals, this can be related to the Irving Williams-series, i.e. the chemical hardness. Thus it would be predicted that acute cytotoxicity would follow the same series. The data shown in Fig. 1 can be replotted as NR 50 cytotoxicity v. chemical softness, σ_p (Fig. 3) to confirm this. Similar relationships to chemical properties have also be observed for inorganic anions (Babich & Borenfreund, 1991).

In the case of organic compounds, which are generally lipophilic, uptake from water is often governed by their partitioning into the cell membrane lipid, and this can be modelled by the logarithm of the octanol/water partition coefficient (log P) of the chemical. Thus it may be predicted that acute cytotoxicity of a series of organic chemicals might show a correlation with their log P values. Indeed, a number of investigators, including Bols *et al.* (1985) and Babich & Borenfreund (1991) have shown that this holds true for a variety of compounds, including series of substituted phenolics, substituted toluenes, chlorinated anilines and chlorinated benzenes. This same relationship is generally followed in acute LC50 tests *in vivo* where substances are accommodated in the water. For many substances there is a definite

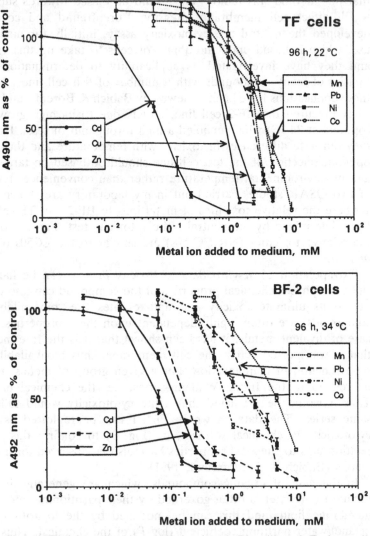

Fig. 1. Toxicity of various metal salts to Bluegill and Turbot cell lines. *In vitro* cytotoxicity as determined by 96 h neutral red assay in BF-2 and TF cells. Data expressed as percentage of control ±SEM.

lack of correlation between LC50 values obtained *in vivo* and their cytotoxicity to cultured cells *in vitro*, a notable exception being the chlorinated pesticides (see Bols *et al.*, 1985; Babich & Borenfreund, 1991) where there are probably marked differences in exposure con-

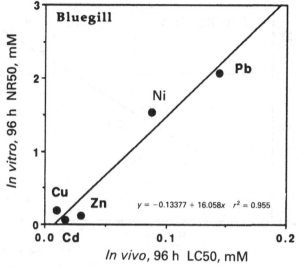

Fig. 2. Comparison of neutral red cytotoxicity (96 h NR_{50} values) in BF-2 cells with *in vivo* 96 h LC_{50} values for bluegill.

ditions, and whose action may involve interaction with cell specific protein factors.

The route of uptake of a toxicant is often important. Toxic effects of water-borne substances may be due to membrane damage to gill epithelial cells, whilst ingested (or injected substances) may be targeted at specific organs such as the liver or kidney and may also interfere with physiological processes. Thus the mechanism of cytotoxicity must be considered in interpretation of *in vitro* data. If the substance is present in dietary lipid, it will partition into chylomicrons and be carried in the blood, as is the case for DDT. In many fish, lipids are stored in the liver, and the liver is therefore likely to be the site of cytotoxicity. Nutritional status will also be a compounding factor in toxicity of dietary pollutants. In all culture media nutritional status is controlled; however, lipids are often taken up by receptor-mediated endocytosis, and the presence of the appropriate cell surface LDL receptor and associated pinocytic process are prerequisites for cellular uptake and toxicity of a compound in an established cell line. When established in culture, cells tend to de-differentiate and often this not only involves loss of morphological characteristics but there may also be a loss of transport and metabolic abilities. For compounds which are directly cytotoxic, often acting by direct cell membrane effects or as enzyme inhibitors, this may not be a problem and acute LC50s will

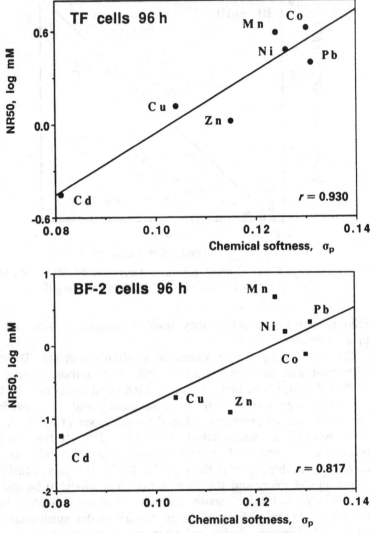

Fig. 3. NR$_{50}$ cytotoxicity values for a series of metal ions to BF-2 and TF cells as a function of their chemical softness (σ_p).

reflect their *in vivo* toxicities. In other cases, the functionality of the receptor-mediated uptake mechanisms are critical for cytotoxicity. The toxicity of many compounds is due to their interference with physiological or biochemical processes which are tissue-specific functions and metabolism (activation) may be required to generate the cytotoxic compound. Commonly, this activation (often in liver or kidney) involves the mixed function oxygenase (MFO or CYP450) system. Notable examples of compounds requiring metabolism are benzo(a)pyrene (BaP) which is metabolized to its carcinogenic derivative by the CYP450 system and epoxide hydrolase, and 2-acetylaminofluorene which is activated by CYP450 and then conjugated with sulphate, glucuronic acid or acetate to form reactive toxicants. When studying effects of this type of compound, a metabolically competent cell line is required. A possible lack of these differentiated functions of a particular established cell line is probably the most serious disadvantage encountered in *in vitro* cytotoxicity testing.

Metabolism of pollutant compounds by cultured fish cells

Primary cell cultures

One way of avoiding the effects of this dedifferentiation and loss of metabolic functionality in established cell lines is to utilize primary cultures of cells freshly isolated from animals. Another advantage of this approach is that pharmacokinetic considerations due to the effects of differing routes of exposure (ip or po) and excretion (bile or urine) encountered *in vivo* are avoided. Moreover, it also enables metabolite excretion to be studied easily and enables studies to be carried out under strictly controlled conditions where differences due to nutrient status, temperature and sexual differences can be avoided. During isolation and primary culture of mammalian hepatocytes, a spontaneous decline in many hepatic functions has been observed and expression of the xenobiotic metabolizing enzymes, especially the mixed function oxygenases (CYP450s), is altered drastically. In contrast, expression of xenobiotic metabolizing enzymes appears to be stable for several days in primary culture of hepatocytes from several fish species and metabolism of the procarcinogen BaP has been studied in a freshly isolated hepatocytes derived from a number of fish species (e.g. Zaleski, Steward & Sikka, 1991). The carcinogenic action of BaP is by formation of a (+)-*anti*-BaP7,8-diol 9,10 epoxide–dGuo adduct of DNA. The metabolite profile obtained from isolated hepatocytes of many species, including the plaice (shown in Table 1) shows that isolated hepatocytes metabolize BaP by a CYP450-dependent mechanism to a variety of phenols, dihydrodiols and quinones (BaP-7,8- and 9,10-diols are the major metabolites) thus demonstrating retention of metabolic capability in short-term culture.

Table 1. *Metabolism of benzo(a)pyrene by isolated plaice hepatocytes*

| Metabolite class | % | % of metabolites | | |
		Dihydrodiols	Quinones	Phenols
Unconjugated	6±2	60	6	34
Water soluble				
Glucuronide	70±9	34	9	57
Thioether	19±5	n.d.	n.d.	n.d.
Sulphate	5±3	n.d.	n.d.	n.d.

Metabolites were extracted and determined after culture of isolated hepatocytes for 16 h in the presence of benzo(a)pyrene. n.d., not determined.

Metabolic studies with isolated hepatocytes have enabled the relative importance of alternative detoxication pathways (phase II) in metabolism of PAHs. 1-naphthol is a simple model PAH that is derived from naphthalene by CYP450-dependent metabolism and after incubation with freshly isolated primary hepatocytes of plaice we were able to show that glucuronidation was the major pathway and exceeded sulphation by over 20-fold. When metabolism of BaP was studied in this system, water-soluble metabolites were produced (Leaver, Clarke & George, 1992). Incubation in the presence of inhibitors showed that it was dependent upon both CYP1A1 activity and conjugative (phase II) metabolism. Quantitative analysis of metabolite classes by derivatization or enzymic hydrolysis showed that the major metabolites (95%) were conjugates of glucuronic acid (predominantly the glucuronide of 7,8-dihydrodiol BaP) and glutathione, whilst the competing pathway, sulphation, only accounted for 3% of the metabolites (Table 1).

CYP1A1-dependent metabolism of PAHs can be induced transcriptionally by exposure of animals to PAHs and PCBs. Inducibility of CYP1A1 enzyme activity (measured with synthetic substrates) has also been demonstrated in isolated hepatocytes of trout (Devaux et al., 1992). In a recent study (Hitchman, Leaver & George, 1995) we utilized primary cultures of plaice hepatocytes to investigate the ability of various PAHs and PCBs to induce CYP1A1 mRNA levels using a plaice cDNA probe. The results (Fig. 4) demonstrate that only polyaromatics are functionally active and that this experimental system can be very useful both in defining modes of action of compounds and also

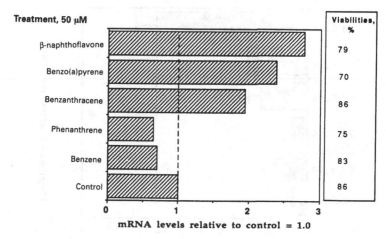

Fig. 4. Induction of CYP1A mRNA in isolated hepatocytes of plaice by PAHs and PCBs. Cells were exposed to agents for 24 h.

for definition of SARs. Not all the enzymes involved in metabolism of compounds such as BaP are induced either by the same compounds or to the same degree. Thus exposure to inducers could lead to alterations in the metabolite profile of a given chemical. *In vitro* studies of this nature will prove extremely useful for this type of investigation.

Thus freshly isolated primary cell cultures are very useful for studies of metabolism and effects of enzyme inducers, definition of mechanisms of action and also in determination of SARs.

Established cell lines

Similar metabolic studies of organic compound metabolism, particularly BaP, have been carried out with several established cell lines (see Varanasi *et al.*, 1989). RTG (rainbow trout gonad), BF-2 (bluegill fry), FHM (fathead minnow) all appear to metabolize BaP and it is cytotoxic to these cells, whilst BaP does not appear to be metabolized by or cytotoxic to the STE (steelhead trout embryo) cells. Thus cell lines which have the ability to carry out metabolism are required for toxicity studies of protoxicants. Generally, the rate of metabolism in established cell lines is lower than that in primary cultures.

Mechanistic studies of toxic action

The use of fish cell lines for studies of the mode of action of toxicants and factors controlling expression of metabolism and detoxication sys-

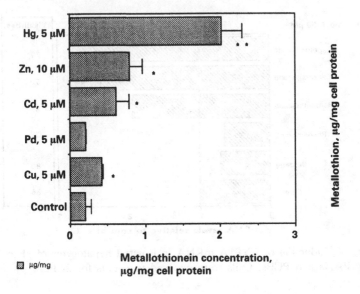

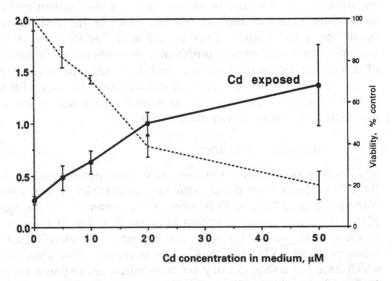

Fig. 5. Induction of metallothionein in TF cells by divalent metal ions. Specificity of induction by various metals and dose dependence of Cd. Viability was measured by NR cytotoxicity assay.

tems is a relatively recent application of *in vitro* techniques which will expand markedly. The formation of DNA adducts of BaP and associated cell repair processes are being investigated (Varanasi *et al.*, 1989). Two other important research areas where they have been applied are in the regulation of PAH and PCB metabolism, which has been studied by Stegeman and co-workers (Hahn *et al.*, 1993) and regulation of the expression of metallothionein, a protein involved in Cu/Zn homeostasis and Cd detoxication, which has been studied by Gedamu and co-workers and by my own group (see George & Olsson, 1994).

Current studies have shown that there are taxon specific differences in expression of CYP450s, including CYP1A1 which is important in PAH metabolism. In a recent study, Hahn *et al.* (1993) investigated CYP1A induction in a cell line (PLHC-1) derived from a hepatocellular carcinoma of *Poeciliopsis lucida*. The presence of the cytosolic Ah receptor, known to control CYP1A induction in mammals, was demonstrated by photoaffinity labelling. Dose-dependent induction of CYP1A by 3,3',4,4'-tetrachlorobiphenyl exposure was demonstrated immunochemically and by enzyme activity measurements. This study demonstrated that this established cell line appears to retain all the characteristic features of the CYP1A system, and establishes its suitability for further mechanistic studies of regulation of this system. It is also being used for QSAR studies of polychlorinated biphenyls (PCBs) as well as for studies of compounds which require metabolic activation.

Intracellular Cu and Zn levels are regulated by a low molecular weight cysteine-rich binding protein, metallothionein (MT). Mechanistic studies of the regulation of MT synthesis in mammals have been carried out almost exclusively with cell lines. MT was purified and characterized from fish, sensitive assay methods for the protein (including an immunoassay) developed, and its cDNA cloned from a number of species, including trout and plaice. Initial *in vivo* studies established that fish MT levels were induced by metal exposure and appeared to be sexually dependent. By use of established cell lines from trout (RTG-2 and RTH-149) and turbot (TF), we were able to show that MT expression is controlled transcriptionally, that there is a strong dose-dependent induction by Cd, Cu, Hg and Zn exposure (Fig. 4), and that there is a weak induction by corticosteroids, whilst hepatic MT levels in sexually mature female fish are highly elevated, indicating the possibility that oestrogen may be an inducer of MT. *In vitro* studies with isolated trout hepatocytes and the TF cell line showed that this was not due to a direct hormonal effect but to an oestrogen-induced mobilization of Zn from other tissues to the liver.

Mechanistic studies of regulation-binding proteins and enzymes involved in pollutant metabolism will of necessity involve both gene cloning and *in vitro* experimentation with cultured cells. Whilst cDNA probes are often sufficiently homologous to be used across species (e.g. plaice MT cDNA and CYP1A cDNA probes will cross-hybridize at high stringency with most fish MT mRNA and CYP1A mRNAs, respectively) only fish cell lines are likely to be of use for gene promoter analysis owing to both the temperature requirements and possible lack of homologous protein co-factors in mammalian cell lines. Presently, we are investigating the functional elements of the plaice glutathione S-transferase gene promoter region by transfection assays with the TF cell line.

Futures

Toxicological studies using *in vitro* techniques with either primary cell cultures or established cell lines of aquatic animals are in their infancy. Most cell lines have been established for viral diagnosis work and thus they have not been characterized biochemically. Generally, much more characterization is required for them to be useful for toxicological studies. Existing QSAR studies show promise as an acute toxicity model; however, until longer-term studies are carried out with cells retaining metabolic functionality then they will always have limited use. At present, most cell lines are fibroblastic in character, and there are not enough available fish cell lines which retain differentiated characteristics to enable organ-specific toxicities to be investigated.

In my opinion, the great role that *in vitro* toxicology will play is in the provision of information on the toxicological profile of a given chemical, particularly in elucidating mechanisms of action. They are not a panacea which can entirely replace environmental or whole animal studies but may form an important component of safety evaluation which will aid legislative measures.

References

Babich, H. & Borenfreund, E. (1991). Cytotoxicity and genotoxicity assays with cultured fish cells: a review. *Toxicology* in Vitro, **5**, 91–100.

Bols, N.C., Bolinska, S.A., Dixon, D.G., Hodson, P.V. & Kaiser, K.L.E. (1985). The use of fish cell cultures as an indication of contaminant toxicity to fish. *Aquatic Toxicology*, **6**, 147–55.

Devaux, A., Pesonen, M., Monod, G. & Andersson, T. (1992). Glucocorticoid mediated potentiation of P450 induction in primary culture of rainbow trout hepatocytes. *Biochemical Pharmacology*, **43**, 898–901.

George, S.G. & Olssen, P.-E. (1994). Metallothioneins as indicators of trace metal pollution. In *Biological Monitoring of Coastal Waters and Estuaries*. Kramer, K. (ed.), Chapter 5, pp. 151–78. CRC Press, Boca Raton.

George, S., Burgess, D., Leaver, M. & Frerichs, N. (1992). Metallo-thionein induction in cultured fibroblasts and liver of a marine flatfish, the turbot, *Scopthalmus maximus*. *Fish Physiol. Biochemistry*, **10**, 43–54.

Hahn, M.E., Lamb, T.M., Schultz, M.E., Smolowitz, R.M. & Stege-man, J.J. (1993). Cytochrome P4501A induction and inhibition by 3,3′,4,4′-tetrachlorobiphenyl in an Ah receptor containing fish hepa-toma cell line (PLHC-1). *Aquatic Toxicology*, **26**, 185–208.

Hitchman, N., Leaver, M. & George, S. (1995). Alternatives to whole animal testing: use of cDNA probes for studies of Phase I and II enzyme induction in isolated plaice hepatocytes. *Marine Environ-mental Research*, **39**, 289–92.

Leaver, M.J., Clarke, D.J. & George, S.G. (1992). Molecular studies of the phase II xenobiotic conjugating enzymes in marine Pleuronec-tid flatfish. *Aquatic Toxicology*, **21**, 265–76.

Morrison, H., Young, P. & George, S.G. (1985). Conjugation of organic compounds in isolated hepatocytes from a marine fish, the plaice, *Pleuronectes platessa*. *Biochemical Pharmacology*, **34**, 3973–8.

Varanasi, U., Nishimoto, M., Baird, W. & Smolarek, T.A. (1989). Metabolic activation of PAH in subcellular fractions and cell cultures from aquatic and terrestrial species. In *Metabolism of Polycyclic Aromatic Hydrocarbons in the Aquatic Environment*. Varanasi, U. (ed.), Chapter 6, pp. 203–252. CRC Press, Boca Raton.

Zaleski, J., Steward, A.R. & Sikka, H.C. (1991). Metabolism of benzo(a)pyrene and (-)-*trans*-benzo(a)pyrene-7,8-dihydrodiol by freshly isolated hepatocytes from mirror carp. *Carcinogenesis*, **12**, 167–74.

M.J. LEAVER

Principles governing the use of cytochrome P4501A1 measurement as a pollution monitoring tool in the aquatic environment

Introduction

Pollution of the aquatic environment is a source of mounting concern not only in areas adjacent to industrial centres but in the oceans as a whole which are increasingly acting as a sink for the products of industrialized societies. Of particular concern are a group of structurally related compounds, the halogenated polyaromatic hydrocarbons (HPAHs) which include chlorinated biphenyls, dibenzo-*p*-dioxins and dibenzofurans (Fig. 1). These compounds readily become adsorbed to sediments and are only slowly metabolized in biological systems. Their lipophilic and persistent natures result in their bioaccumulation in organisms and magnification of their concentrations during passage up the food chain. These processes have been studied extensively in the Great Lakes of North America and in the Baltic, and there is strong evidence that they result in reproductive failure in fish-eating birds (Walker, 1990). In addition, HPAHs have been found in a variety of marine mammals, and much has been talked of the possibility that these compounds have contributed to reproductive failure and immune suppression in these animals, although direct evidence to support these contentions is scanty. Another group of common organic pollutants are the polyaromatic hydrocarbons (PAHs) (Fig. 1), which may be derived from the exploitation of petroleum or from the deposition of combustion products to rivers and oceans. These compounds are less persistent than HPAHs but are known to give rise to highly mutagenic and hence carcinogenic entities.

As with higher animals, there is very little evidence to connect pollutant body burdens of any of these chemicals directly with either reproductive failure or cancerous conditions in wild fish populations (Mix, 1986). Furthermore, the monitoring of body burdens is difficult and expensive, while the assessment of reproductive capacity is fraught with difficulty owing to many natural compounding variables. Whilst the estimation of population numbers may provide answers, there is

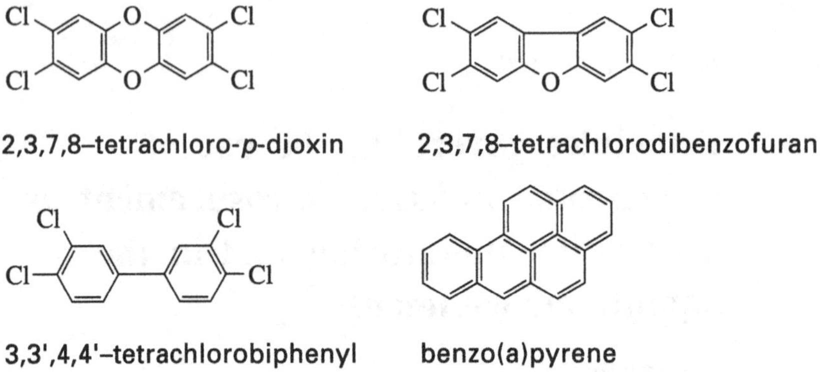

2,3,7,8–tetrachloro-*p*-dioxin

2,3,7,8–tetrachlorodibenzofuran

3,3',4,4'–tetrachlorobiphenyl

benzo(a)pyrene

Fig. 1. Chemical structures of representative HPAHs and PAHs.

an argument that, when pollution induced changes in numbers of animals can be determined, the damage has already been done. In addition, for a number of species, the determination of such changes is extremely difficult to lay at the door of pollution due to intense hunting and fishing pressures. For these reasons there have been increasing attempts to develop some early warning measures of sub-lethal pollutant impact in aquatic animals. Efforts in this direction have pursued a wide range of behavioural, physiological and cellular responses in exposed individuals (Bayne, Clarke & Gray, 1988). It is well established that specific chemicals act in specific ways to produce toxicity, and the primary interfaces between an organism and its chemical environment are the biochemical constituents of the cell, particularly those enzyme systems which are capable of sensing, responding to, and metabolizing xenobiotics. Changes in these xenobiotic enzyme metabolizing systems have become the most widely mooted biochemical measure of an environmental pollutant-induced response and the cytochrome P4501A1 system has been particularly closely studied in this regard (Stegeman & Hahn, 1994; Payne *et al.*, 1987; Goksøyr & Förlin, 1992). This review is an attempt to explain and justify the use of such a measure as an early indicator of deleterious pollutant impact, especially that due to HPAHs and PAHs, in fish species.

HPAHs, PAHs, TCDD and the Ah receptor

The term dioxin is often used for the most notorious HPAH, more correctly named 2,3,7,8-tetrachlorodibenzo-*p*-dioxin (TCDD), and this compound has been the subject of intense toxicological research since its discovery as the teratogenic agent of the defoliant herbicide 'Agent Orange'. The toxic effects and potency of TCDD vary enormously depending on the species, for example, the guinea pig, which is exquisitely sensitive, is 2500 times more susceptible than the relatively insensitive hamster. A debate has been raging for many years over the toxicological consequences of exposure in humans (Roberts, 1991). Certain other structurally related HPAHs produce similar patterns of toxicity as TCDD, albeit at higher doses, and these classes of compound include chlorinated biphenyls and chlorinated dibenzofurans, both of which are widespread environmental contaminants (Safe, 1990). As TCDD is the most potent representative of this class of compound known, it has been used as a prototype and has led to the proposal that a toxic equivalency factor (TEF) can be deduced for any HPAH based upon the relative dose required to give the equivalent response effect to that seen for TCDD in a particular species. These TEFs have been proposed as tools to assess risk associated with known mixtures of HPAHs (Safe, 1990).

The biological properties and toxic potencies of HPAHs are related strongly to the degree of lateral substitution of the halogen groups, and the large number of possible congeners in these families of compounds greatly complicates environmental analysis. Although TCDD is extremely toxic to susceptible animals, death is not immediate but follows after a number of other symptoms, which may vary widely depending on the affected species, but often include a wasting syndrome, thymic involution, immune suppression and tumour promotion (Poland & Knutson, 1982). There is no evidence to implicate TCDD in any direct insult to cellular macromolecules. However, most possibly all of the effects of TCDD have been shown to proceed only after binding to, and activating, a cytosolic protein complex known as the Ah (*Aryl h*ydrocarbon) receptor (also known as the dioxin receptor), and it is probable that a very similar situation exists for the actions of other laterally substituted HPAHs (Whitlock, 1990). Another group of chemicals known to activate the Ah receptor are the polyaromatic hydrocarbons (PAHs), which have been known as potent tumour inducers for many years, and again these compounds are most active after activation of the Ah receptor (Gelboin, 1980).

Evidence for the existence of an Ah receptor protein was first demonstrated by cross-breeding two strains of mice, namely DBA/2 mice, which show a relatively low biochemical and toxicological response to PAHs, and PAH responsive, susceptible C57BL/6J mice. This led to the conclusion that PAH responsiveness was governed by a single dominant gene which was regulatory in nature (Nebert &

Gonzalez, 1987). Subsequently, the protein product of this gene was isolated on the basis of its ability to bind radiolabelled TCDD and its cDNA was more recently cloned and its amino acid sequence deduced. C57BL/6J mice, which contain high levels of functional Ah receptor show greater susceptibility to TCDD and related compounds than the DBA/2 unresponsive strain, which only contain low levels of functional Ah receptor, thus implicating activation of the Ah receptor as a necessary early event in TCDD induced toxicity (Poland & Knutson, 1982). The unactivated Ah receptor is now known to be a multimeric complex consisting of at least two proteins, hsp90 and the Ah receptor protein. Compounds such as HPAHs or PAHs are believed to enter the cell by passive diffusion and are then specifically bound by the Ah receptor protein. This binding allows the dissociation of hsp90, a 90 kD protein induced by heat shock but also present in normal cells, leaving the liganded Ah receptor free to dimerize with the closely related protein, Arnt (*Ah receptor nuclear translocator*). Both the Ah receptor and Arnt are members of a family of DNA binding proteins, the basic helix–loop–helix family. The dimerized activated complex is able to enter the nucleus and interact directly with a variety of target genes containing a specific regulatory sequence motif to which the complex binds (Fig. 2). Binding of the activated complex to the regulatory regions of these genes then results in a modulation of their transcription

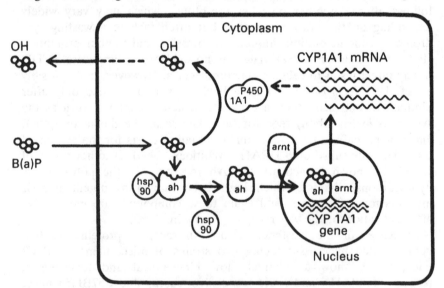

Fig. 2. Schematic representation of Ah receptor activation and P450 1A1 gene expression by benzo(a)pyrene.

and thus an alteration in the levels of the protein products of these genes (Swanson & Bradfield, 1993). The toxic effects of HPAHs are believed to be due to the actions of the products of these genes, most of which have yet to be identified. However, a number of xenobiotic metabolizing enzymes are known to be upregulated by PAHs or HPAHs and these include glutathione S-transferase, UDP-glucuronosyltransferase and cytochrome P4501A1 (Gonzalez & Nebert, 1990).

P4501A1 and the Ah receptor

The gene for which the interaction of the Ah receptor complex has been most studied, and indeed the gene which led to the discovery of the receptor in mice, is that for cytochrome P4501A1 (Nebert & Gonzalez, 1987). Cytochrome P4501A1 is a membrane-bound hemoprotein belonging to a vast family of enzymes, the cytochromes P450. These enzymes generally act to insert molecular oxygen onto a bewildering array of both exogenous and endogenous lipophilic chemicals. P4501A1 is involved primarily in the metabolism of lipophilic planar aromatic type compounds, such as PAHs, and acts to insert oxygen on to these molecules to produce hydroxylated and epoxidated products, which are thus rendered more polar, and can enter the animals excretion pathways (Parke, Ioannides & Lewis, 1990). Some of these products are potent mutagens; however, for example, the product of P4501A1-dependent metabolism, benzo(a)pyrene-7,8-dihydrodiol-9,10-epoxide reacts specifically with guanine residues to produce covalent DNA adducts, and thus can lead to cancer (Gelboin, 1980). Unlike PAHs, HPAHs are metabolized relatively poorly and tend to persist for longer periods than compounds metabolized more readily such as PAHs. The persistent nature of TCDD and related compounds is thought to cause a sustained activation of the Ah receptor, resulting in a prolonged alteration of the levels of a wide variety of cellular constituents, which manifests as a range of toxic symptoms. Cytochrome P450 induction is not believed to be a major cause of TCDD induced toxicity, but the levels of P4501A1 rise dramatically, up to 100× from very low basal levels, shortly after exposure to PAHs and HPAHs, and this rise is due to an increase in the transcription of the gene for this enzyme, mediated by the Ah-receptor complex (Whitlock, 1990). No known endogenous compound has been found yet to activate the Ah receptor and be metabolized subsequently by P4501A1, which raises the possibility that this system has evolved specifically for the metabolism of naturally occurring toxins which may be part of the diet, for example, plant secondary metabolites (Gonzalez & Nebert, 1990). The advent of industrial chemical synthesis appears to have produced a range of manmade xenobiotics for which the system is unprepared.

Fish species have long been known to respond to PAH and HPAH exposure by increasing P4501A1 dependent activities, and various P450 proteins including P4501A1 have been purified and characterized from fish. Recently, the Ah receptor has been shown to be present in a variety of fish, and the cloning of P4501A1 cDNAs from rainbow trout and plaice has confirmed the essential similarity between the piscine and mammalian systems (Stegeman & Hahn, 1994). In addition, a number of the toxic effects of TCDD seen in mammals are also observed in fish and, furthermore, fish are at least as sensitive as mammals such as the mouse or rat; indeed the carp appears to be as sensitive as the most susceptible animal known, the guinea-pig (van der Weiden et al., 1994a). The relevance of the Ah-receptor systems to toxicity in fish is further emphasized by the observations that a variety of other halogenated aromatics and PCBs, when administered to fish species, produce some of the typical symptoms of TCDD toxicity elicited in birds and mammals and the relative toxic potencies of these compounds highly correlate with their ability to induce piscine P4501A1 (Safe, 1990; van der Weiden et al., 1994b). This strongly suggests that an observed increase in P4501A1 represents a toxicologically relevant threshold of exposure to such chemicals, and this concept has recently been discussed in relation to setting limits for human exposure to TCDD (Roberts, 1991).

Applications

Given that P4501A1 levels have been shown to rise in fish experimentally exposed to HPAHs and PAHs, and that such an induction is also an indication of a toxic condition that is deleterious to the animal, then an observed increase in the levels of P4501A1 messenger RNA, protein or enzymic activity in fish exposed to environmental pollution would be a strong indicator of a health risk to that animal.

Enzymic activity is determined most effectively by measuring the de-ethylation of an artificial substrate, ethoxyresorufin (EROD), which is highly specific for P4501A1 in both mammals and fish. The availability of antibodies and cDNAs allows the determination of P4501A1 protein and mRNA in fish. The kinetics of induction and clearance of mRNA, protein and activity will clearly be slightly different (Fig. 3), and the choice of method(s) used to measure P4501A1 must be chosen with this in mind.

P450 induction kinetics

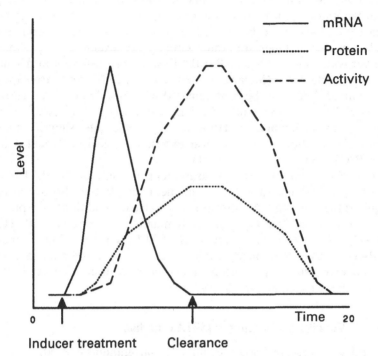

Fig. 3. Induction kinetics of P450 1A1 mRNA, protein and enzymic activity. (Adapted from studies by Haasch *et al.*, and Celander *et al.*)

As biomonitoring exercises most workers have obtained fish caught from a defined area and compared P4501A1 levels with fish caught from another defined area known to be, or believed to be, clean. As fish move about freely from one location to another, their P4501A1 levels would vary considerably. Thus, such a programme would give a measure of the mean P4501A1 level of a sub-population of fish, and an indication of the potential risk to that population. The results of a number of such studies have been reviewed by Payne *et al.* (1987). A 1987 workshop (Bayne *et al.*, 1988), which investigated the efficacy of a variety of potential pollution responsive parameters in fish living over a defined pollution gradient, showed P4501A1 activity to be the most sensitive method tested. Similarly, increases in P4501A1 have been shown to be the most sensitive indicators of pollution exposure in fish caught from areas affected by bleached pulp mill effluents and subjected to a variety of assays for biological effects (Andersson *et al.*, 1988).

As a bioassay, a number of studies have exposed 'clean' fish to potentially contaminated sediments or water samples, *in situ* in cages or in laboratory controlled conditions, and the levels of P4501A1 after such experiments have been used as a measure of the environmental quality of that sediment or water sample. For example, carp exposed to river water showed higher P4501A1 levels than fish kept in dechlorinated drinking water (Melancon, Yeo & Lech, 1987), and carp exposed to sediments known to be contaminated with HPAH and PAH showed elevated P4501A1 (Van der Weiden *et al.*, 1993). Similarly, tomcod exposed to contaminated Hudson Bay sediments showed similar P4501A1 induction profiles to fish exposed to clean sediment spiked with PAHs (Kreamer *et al.*, 1991).

It is also possible that the measurement not only of P4501A1 induction by a contaminated water or sediment sample but also of the time required to clear P4501A1 after transfer to clean conditions may provide an indication of the type of contamination present (i.e. PAH or HPAH). Because of the lower level of HPAH metabolism compared to PAH, HPAH compounds are likely to remain in the animals tissues for longer causing a prolonged induction even after transfer to clean surroundings (Kreamer *et al.*, 1991).

Variables affecting P4501A1 in fish

Several variables are known to affect basal uninduced P4501A1 levels in fish, and these need to be taken into account during monitoring or bioassay programs.

First, large interspecific variation in the levels of P4501A1 exist in fish, and particular species are usually experimentally tested for HPAH and PAH responsiveness to determine basal and maximally induced P4501A1 levels before any bioassay or monitoring programme is undertaken (Goksøyr & Förlin, 1992).

Secondly, a great deal of intraspecific variation has been reported, both due to natural individual variation and due to environmental and life history influences. Juvenile salmonids and pleuronectids and probably all species, at least in temperate regions, show a marked seasonal cycle, possibly as a result of temperature adaptation, which is similar in both males and females. Thus, when assayed at constant temperature, fish in winter exhibit low levels of P4501A1, with rising levels during spring and summer. However, whereas sexually mature males exhibit similar cycles to immature fish, sexually mature females during egg production and spawning have very low, uninducible P4501A1 levels which are associated with high oestrogen levels (Fig. 4). The use of such females in P4501A1 monitoring studies would be uninformative.

Most investigators have used liver as the organ of choice for P4501A1 measurement; however, a number of other tissues are also responsive including intestine, kidney and heart. With the exception of kidney, the use of these tissues has not been tested in monitoring programs but it is possible that they may avoid some of the seasonal and spawning affects seen in liver. Indeed, some investigators have advocated the use of kidney in preference to liver for environmental monitoring studies (Payne *et al.*, 1987). The route of exposure to a particular contaminant, for example, the intestine from food or the skin from sediment may also influence the site of greatest P4501A1 response.

In some situations where inducers act as inhibitors of P4501A1 activities, either experimentally at very high exposure levels or possibly in fish from grossly and multiply polluted sites, induction may not be apparent from enzymic activity measurements. In these cases, the use of antibody-based detection methods for P4501A1 have proved to be

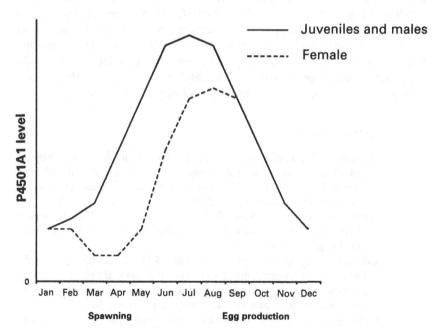

Fig. 4. Plaice cytochrome P4501A1 (EROD) annual cycle. (Adapted from George *et al.*, 1990.)

more reliable techniques. Thus measurement of levels by activity and immunological methods is frequently adopted when undertaking field monitoring studies. Recently, the availability of cDNAs for piscine P4501A1 has provided a means for the measurement of P4501A1 mRNA, a method which avoids the phenomenon of enzymic inhibition but is at least as sensitive.

Concluding remarks

The use of sub-lethal indicators of pollutant-induced stress has been criticized because of the lack of evidence to link such measures with perturbations in higher levels of organization such as populations and communities. However, whilst hard evidence is not forthcoming, increasing aquatic environmental problems are occurring, for example, reproductive failure in fish eating birds and mammals, particularly in those adjacent to heavily industrialized areas. The weight of evidence available, albeit circumstantial, points strongly to chemical contamination as the primary cause, and it may therefore be sensible to adopt a more precautionary approach to industrial pollution in future. This review has attempted to justify the use of biochemical indicators of sub-lethal pollution impact such as P4501A1 in fish species, by arguing that our current knowledge of the mechanisms by which a wide range of chemical contaminants can affect individuals is sufficient to predict that alterations in biochemical parameters such as P4501A1 are indicators of a serious health risk to individuals with potentially deleterious consquences to reproductive capacity and population numbers.

References

Andersson, T., Förlin, L., Härdig, J. & Larsson, Å. (1988). Physiological disturbances in fish living in coastal water polluted with bleached kraft pulp mill effluents. *Canadian Journal of Fisheries and Aquatic Sciences*, 45, 1525–36.

Bayne, B.L., Clarke, K.R. & Gray, J.S. (eds.) (1988). Biological effects of pollutants. Results of a practical workshop. *Marine Ecology Progress Series*, 46, 1–278.

Celander, M., Leaver, M.J., George, S.G. & Förlin, L. (1993). Induction of cytochrome P450 1A1 and conjugating enzymes in rainbow trout (*Oncorhyncus mykiss*) liver: a time course study. *Comparative Biochemistry and Physiology*, 106C, 343–9.

Gelboin, H.V. (1980). Benzo(a)pyrene metabolism, activation, and carcinogenesis: role and regulation of mixed-function oxidases and related enzymes. *Physiological Reviews*, 60, 1107–65.

George, S.G., Young, P., Leaver, M. & Clarke, D. (1990). Activities of pollutant metabolising and detoxication systems in the liver of plaice, *Pleuronectes platessa*: sex and seasonal variations in non-induced fish. *Comparative Biochemistry and Physiology*, **96C**, 185–92.

Goksøyr, A. & Förlin, L. (1992). The cytochrome P-450 system in fish, aquatic toxicology and environmental monitoring. *Aquatic Toxicology*, **22**, 287–312.

Gonzalez, F.J. & Nebert, D.W. (1990). Evolution of the P450 gene superfamily. *Trends in Genetics*, **6**, 182–6.

Haasch, M.L., Kleinow, K.M. & Lech, J.J. (1988). Induction of cytochrome P-450 mRNA in rainbow trout: *in vitro* translation and immunodetection. *Toxicology and Applied Pharmacology*, **94**, 246–53.

Kreamer, G.-L., Squibb, K., Gioeli, D., Garte, S.J. & Wirgin, I. (1991). Cytochrome P4501A mRNA expression in feral Hudson River tomcod. *Environmental Research*, **55**, 64–78.

Melancon, M.J., Yeo, S. & Lech, J.J. (1987). Induction of hepatic microsomal monooxygenase activity in fish by exposure to river water. *Environmental Toxicology and Chemistry*, **6**, 127–35.

Mix, M.C. (1986). Cancerous diseases in aquatic animals and their association with environmental pollutants: a critical literature review. *Marine Environmental Research*, **20**, 1–141.

Nebert, D.W. & Gonzalez, F.J. (1987). P450 genes. Structure, evolution and regulation. *Annual Review of Biochemistry*, **56**, 945–93.

Parke, D.V., Ioannides, C. & Lewis, D.F.V. (1990). The role of the cytochromes P450 in the detoxication and activation of drugs and other chemicals. *Canadian Journal of Physiology*, **69**, 537–49.

Payne, J.F., Fancey, L.L., Rahmitula, A.D. & Porter, E.L. (1987). Review and perspective on the use of mixed function oxygenase enzymes in biological monitoring. *Comparative Biochemistry and Physiology*, **86C**, 233–45.

Poland, A. & Knutson, J.C. (1982). 2,3,7,8-tetrachlorodibenzo-*p*-dioxin and related halogenated aromatic hydrocarbons: examination of the mechanism of toxicity. *Annual Review of Pharmacolgy and Toxicology*, **22**, 517–54.

Roberts, L. (1991). Dioxin risks revisited. *Science*, **251**, 624–6.

Safe, S. (1990). Polychlorinated biphenyls (PCBs), dibenzo-*p*-dioxins (PCDDs), dibenzofurans (PCDFs) and related compounds: environmental and mechanistic considerations which support the development of toxic equivalency factors (TEFs). *Critical Reviews in Toxicology*, **21**, 51–8.

Stegeman, J.J. & Hahn, M.E. (1994). Biochemistry and molecular biology of monooxygenases: current perspectives on forms, functions and regulation of cytochrome P450 in aquatic species. Chapter 3, pp. 87–206. In *Aquatic Toxicology: Molecular, Biochemical and Cellular*

Perspectives. Malins, D. & Ostrander, G. (eds.). Lewis Publishers, Boca Raton.

Swanson, H.I. & Bradfield, C.A. (1993). The AH-receptor: genetics, structure and function. *Pharmacogenetics*, **3**, 213–30.

van der Weiden, M.E.J., Bleumink, R., Seinen, W. & van den Berg, M. (1994*a*). Concurrence of P450 1A induction and toxic effects in the mirror carp (*Cyprinus carpio*), after administration of a low dose of 2,3,7,8-tetrachlorodibenzo-*p*-dioxin. *Aquatic Toxicology*, **29**, 147–62.

van der Weiden, M.E.J., Celander, M., Seinen, W., van den Berg, M., Goksoyr, A. & Forlin, L. (1993). Induction of cytochrome P450 1A in fish treated with 2,3,7,8-tetrachlorodibenzo-*p*-dioxin or chemically contaminated sediment. *Environmental Toxicology and Chemistry*, **12**, 989–99.

van den Weiden, M.E.J., de Vries, L.P., Fase, K., Celander, M., Seinen, W. & van den Berg, M. (1994*b*). Relative potencies of polychlorinated dibenzo-*p*-dioxins (PCDDS), dibenzofurans (PCDFS) and biphenyls (PCBS) for cytochrome P450 1A induction in the mirror carp (*Cypinus carpio*) *Aquatic Toxicology*, **29**, 163–82.

Walker, C.H. (1990). Review: persistent pollutants in fish-eating seabirds – bioaccumulation, metabolism and effects. *Aquatic Toxicology*, **17**, 293–324.

Whitlock, J.P. (1990). Genetic and molecular aspects of 2,3,7,8-tetrachlorodibenzo-*p*-dioxin action. *Annual Review of Pharmacology and Toxicology*, **30**, 251–77.

Index